Breeding for Enhanced Nutrition and Bio-Active Compounds in Food Legumes

Debjyoti Sen Gupta • Sanjeev Gupta
Jitendra Kumar
Editors

Breeding for Enhanced Nutrition and Bio-Active Compounds in Food Legumes

Springer

Editors
Debjyoti Sen Gupta
ICAR-Indian Institute of Pulses Research
Kanpur, Uttar Pradesh, India

Sanjeev Gupta
ICAR-Indian Institute of Pulses Research
Kanpur, Uttar Pradesh, India

Jitendra Kumar
ICAR-Indian Institute of Pulses Research
Kanpur, Uttar Pradesh, India

ISBN 978-3-030-59214-1   ISBN 978-3-030-59215-8 (eBook)
https://doi.org/10.1007/978-3-030-59215-8

This Springer imprint is published by the registered company Springer Nature Switzerland AG
The registered company address is: Gewerbestrasse 11, 6330 Cham, Switzerland

Preface

Food legumes are increasingly becoming popular in the developing and developed countries for their high protein and fiber content, low glycemic index, and high levels of cancer-fighting antioxidants. Limited efforts have gone into breeding for higher protein and other quality traits. However, recent screenings of germplasm have shown great variability for protein content and many health-promoting bioactive compounds, which paves the way to strengthen research programs on increasing nutritional quality as well as health-promoting factors. The focus should be placed on increasing the concentration of micronutrients along with increasing their bioavailability. This can be achieved by enhancing the promoters that stimulate the absorption of minerals and by reducing the concentrations of anti-nutrients that interfere with absorption. Therefore, high micronutrient content must be included as a core trait of breeding programs by concatenation of micronutrient-dense parental lines and identification of superior lines with increased bioavailability of important micronutrients like iron, zinc, and selenium.

During the past years, significant advances have been made on different aspects of nutritional and bioactive compounds in food legume crops that can be used in the breeding of nutrient-rich improved varieties. However, this voluminous information has been scattered over different journals and books, and to date, no single publication is available that has focused comprehensive insight into this literature with respect to breeding for enhancing the nutrition and bioactive compounds in food legumes. Therefore, the objective of editing this book was to address recent advances in research on nutritional and quality improvement of food legumes and to make available this information for individual food legume crops in specific order at one platform.

Breeding for Enhanced Nutrition and Bio-active Compounds in Food Legumes comprises 11 chapters authored by scientists/researchers who are actively involved in improving legumes for nutritional and quality traits. These chapters have covered identification of priority traits for genetic biofortification, genetic variability of available germplasm, biochemistry of the identified traits, analytical methods, and genomic approaches for improving quality traits, which are described in detail for each food legume crop in respective chapters. Advances in exploring nutraceuticals and milling or baking applications of food legume crops are also covered in this book. The contribution of authors for this book is enormous in presenting up-to-date information on the subject. We are extremely thankful to all our experienced authors who gave their valuable time for writing these chapters with great responsibility and care. In addition to this, there are many people around the globe who were always available during the entire developmental period of this book, influencing positively to make this project feasible: Dr. Shiv Kumar, ICARDA, Rabat, Morocco; Dr. J. Souframanien, BARC, India; and Dr. Cliffard Hall, NDSU, Fargo, USA.

We editors are thankful to our parent organization, Indian Council of Agricultural Research (ICAR), New Delhi, for supporting our scientific pursuit in the form of a book *Breeding for Enhanced Nutrition and Bio-active Compounds in Food Legumes*. We are highly thankful to Dr. T. Mohapatra, Director General, ICAR, and Secretary, DARE, Ministry of Agriculture and Farmers' Welfare, Government of India, and Dr. T.R. Sharma, Deputy Director General (Crop Science), ICAR, for their constant support and guidance in this endeavor. Dr. N. P. Singh, Director, ICAR-IIPR; Dr. Shiv Sewak, former Head, Division of Crop Improvement, ICAR-IIPR, Kanpur; and Dr. Farindra Singh, Head, Division of Crop Improvement, ICAR-IIPR, deserve special thanks for supporting and encouraging us to undertake this task.

We thank our families for being patient and supportive in this long journey, without their moral support, it would not be possible. We thank the entire team at Springer, especially Mr. Kenneth Teng, Publishing Editor, Life Science; Mr. Shabib Shaikh, Project Coordinator, Books, Springer Nature; and Mr. Menas Donal Kiran, Production Editor, who have always been cooperative to make this publication a reality. They have been very generous in accommodating even last-minute changes and deserve our genuine appreciations. We hope that this book will absolutely serve its purpose and provide a latest and comprehensive treatise to the readers in furthering their academic and research pursuits.

Kanpur, Uttar Pradesh, India
July 31, 2020

Debjyoti Sen Gupta
Jitendra Kumar
Sanjeev Gupta

Contents

Breeding for Enhanced Nutrition Status in Food Legumes: Retrospects and Prospects

Sanjeev Gupta, Debjyoti Sen Gupta, and Jitendra Kumar

Abstract Food legumes are rich source of dietary proteins, calories, folates, vitamins, minerals and fibre contents. With very low fat and considerable proportion of slowly digestible starch, food legumes are highly beneficial for human health. In recent years, research on the active presence of many bioactive compounds like glycosides, tannins, saponins and alkaloids in legume seeds increased the importance of this group of crops from biomedical point of view. However, some of these if consumed over a threshold quantity for prolonged period may behave toxic, unpalatable or indigestible, and these molecules are collectively known as 'anti-nutritional factors' (ANFs). Enhancing nutrition status and reducing ANFs among food legumes can be achieved through genetic manipulation utilising natural intra- or interspecific variation within a breeding population or any mutant population or by engineering respective biosynthetic pathways. It is important to mention that ANFs have certain role to play in plant metabolism and physiology. Therefore, breeding for high nutrient and low or null ANFs may require close attention to the overall plant physiology or agronomic performance or any other biological implication.

Keywords Biofortification · Iron · Zinc · Selenium · Anti-nutrients · HarvestPlus · Micronutrients · Food legumes

Introduction

The use of food legumes with high protein content, low glycaemic index, high fibre content and presence of many other bioactive compounds is increasingly becoming popular with higher incidences of noninfectious chronic diseases like diabetes, heart problems and cancers worldwide. However, little efforts have gone into breeding for developing more nutritious cultivars in food legumes, although considerable

S. Gupta (✉) · D. Sen Gupta (✉) · J. Kumar (✉)
ICAR-Indian Institute of Pulses Research, Kanpur, UP, India

D. S. Gupta et al. (eds.), *Breeding for Enhanced Nutrition and Bio-Active Compounds in Food Legumes*, https://doi.org/10.1007/978-3-030-59215-8_1

variability was reported for many nutrients like protein, micronutrients and vitamins. Therefore, pulses in combination with cereals may supply adequate calorie as well as essential nutrients. Besides proteins, food legumes supply 15 essential minerals required by humans (Wang et al. 2009). High quantity of iron and zinc concentration in food legumes makes them potential candidate for micronutrient biofortification. Besides Asia, there is recent trend in Europe and America for food legume-based gluten-free products, ready-to-eat baked goods, mixes, soups, sauces and several other products. Therefore, there is a need to orient breeding programmes to develop varieties having richness of more nutrients along with superior agronomic performance. Nutrient bioavailability is another important factor while breeding for enhanced nutrients. Bioavailability is the proportion of the nutrient that is metabolised by human body (Srinivasan 2001). Nutrient availability may be altered by changing the promoter-to-antinutrient ratio. By increasing promoters or by reducing antinutrients, bioavailability of a nutrient may be improved. The HarvestPlus Biofortification Project (https://www.harvestplus.org) has been initiated in multiple crops including lentil and common bean, and target traits are iron and zinc concentration. Food legumes are rich in protein ranging from 25 to 30% or even higher in many instances. Food legumes are deficient in sulphur-containing amino acids (Wang et al. 2003). However, there is scope to improve protein quality by modulating proportion of sulphur-containing amino acids. Conventional breeding approaches may be employed to explore gene banks with global germplasm holdings to find out donors for such traits, or mutation breeding or transgenic approaches may be utilised.

Nutrition Enhancement

Food legumes are cheaper sources of high-quality protein compared to animal sources. The protein content is almost double that of cereals, and amino acid profile is rich in lysine and tryptophan. Cereal proteins are deficient in these amino acids (Nielsen et al. 1993; Yamauchi and Minamikawa 1998). Therefore, food legumes are regarded as perfect complementary protein sources served with cereals (Sharma 1988; Shewry and Pandya 1999). The major protein fraction in legume seeds is comprised of globulins (60–90%), which are storage proteins rich in arginine, glutamic acid, aspartic acid and their amides. Among food legumes, lentils are the richest in protein which generally varies between 16% and 32% (Kumar et al. 2016). In chickpea, the protein content varies from 17% to 22% (before dehulling) and 25% to 29% (after dehulling) (Jukanti et al. 2012). Pigeonpea accounts for the protein content of 20–25%. Saxena and Sawargaonkar (2016) reported that newly bred pigeonpea lines have protein between 28% and 30%. Mungbean and blackgram generally have considerable amount of protein and contain 20–22% protein in their grain. Many pea genotypes having more than 30% seed protein have been identified (Bing 2015; Shen et al. 2016; Demirbasß 2018). Pea protein is known for its better digestibility, less allergenic responses or negative health implications (Owusu-Ansah and McCurdy 1991; Allred et al. 2004; Roy et al. 2010; Boye et al. 2010;

Lam et al. 2018). Seed protein content and quality are complex traits and therefore controlled by polygenes or QTLs. More research is required in this direction for identification of genic factors controlling protein quantity and quality in food legumes.

Food legumes are quite rich in dietary fibres. Such content is around 18% in peas, lentils and chickpeas and 28% in beans (Devos 1988). High consumption of refined foodstuff and less fibre consumption lead to many chronic diseases or disorders like constipation, diverticulosis, haemorrhoid, diabetes, obesity, intestinal cancer and cardiovascular problems (Trowell et al. 1985; Pekşen and Artık 2005). The studies indicated that soluble dietary fibres had significant positive impacts by reducing low-density lipoprotein or harmful cholesterol levels (Glore et al. 1994). Clinical studies also revealed that soluble dietary fibres control type II diabetes by reducing postprandial blood sugar, insulin and blood serum lipid levels (Tabatabai and Li 2000; Pekşen and Artık 2005). Overall, soluble and insoluble dietary fibres have positive impacts on obesity (Anderson and Bryant 1986; Marlett et al. 2002). Consumption of insoluble dietary fibres may reduce the risks of colon cancer (Hughes 1991; Marlett et al. 2002; Pekşen and Artık 2005). Therefore, dietary fibre contents are priority traits in breeding for enhanced nutrition in food legumes.

Legumes usually have low fat content (except for soybean and peanut), and they do not contain cholesterol. The absence of cholesterol makes consumption of food legumes better for cardiovascular health. The crude fat content was high in chickpea (5.2%), followed by cowpea (4.8%), lentil (3.2%) and green pea (1.5%) (Iqbal et al. 2006). Fat content while testing 15 common bean accessions was reported to be between 0.33% and 1.33% (Celmeli et al. 2018). Legume fats are usually polyunsaturated with high linoleic acid contents. Moreover, fats in legumes are quite less influenced by processing practices (Devos 1988; Pekşen and Artık 2005).

Legumes are also quite rich in micro- and macronutrients, especially in potassium, phosphorus, calcium and iron. "Hidden hunger" (Muthayya et al. 2013) arising from micronutrient and vitamin deficiency in over 2 billion people is ranked one of the top ten risk factors contributing to disease burden globally (Yang et al. 2007). Children, premenopausal and lactating women and people with poor diet are most susceptible to malnutrition caused by the deficiency of Fe, Zn and vitamin A. Zn deficiency retards normal growth and learning capacity in children and increases the risk of infections, cancer and DNA damage (Institute of Medicine 2001; Prasad 2004; Wang and Busbey 2005; Maret and Sandstead 2006; Shahzad et al. 2014). In the recent years, genetic potential for increasing the concentrations of Fe and Zn in grains of various food crops such as maize, rice, wheat, common bean and field pea has been reviewed in detail (Frossard et al. 2000; Gomez-Galera et al. 2010; Mayer et al. 2008; Welch and Graham 2004; Amarakoon et al. 2012). The efforts have been made towards the screening of existing released varieties for iron and zinc content under the HarvestPlus Challenge Program. The HarvestPlus Program that is a part of CGIAR Research Program on Agriculture for Nutrition and Health has set the targets for iron and zinc for only two food legumes, for example, increasing iron to 94 ppm from baseline of 50 ppm in common bean and to 70 ppm from baseline of 40 ppm in lentil. Target for zinc has not been set as zinc has been considered as associated trait to Fe content in these legumes (Bouis and Saltzman 2017). However, for national programmes, similar targets need to set for enhanced nutrition status in other legumes as the target levels are set based on the average quantity consumed on a daily basis. Recently, lentil and chickpea cultivars having both iron and zinc in high amounts have been identified (Joshi-Saha and Reddy 2014). Also, these varieties are being disseminated to farmers on a fast-track mode through national programmes. For example, in Bangladesh, the government has taken a massive dissemination program to promote promising lentil varieties having high Fe and Zn, Barimasur 5 and Barimasur 6, in traditional and nontraditional areas. Similarly, in Nepal, lentil varieties such as Sishir, Khajurah 2, Khajurah 1 and Shital are spreading fast in the Terai region. In India, Pusa Vaibhav of lentil rich in Fe is becoming popular in north-western part of the country (HarvestPlus 2014). A high-yielding variety of lentil IPL 220 having high concentration of Fe (73–114 ppm) and Zn (51–65 ppm) has been released as biofortified variety in India for cultivation in north-eastern regions of country. Common bean varieties showed Fe concentrations ranging between 30 and 120 ppm (Graham et al. 1999; Beebe et al. 2000; Guzmán-Maldonado et al. 2000; Moraghan et al. 2002; Islam et al. 2002) and zinc concentrations varying between 20 and 60 ppm (Moraghan and Grafton 1999; Welch et al. 2000; House et al. 2002; Hacisalihoglu et al. 2004). Harmankaya et al. (2010) reported noteworthy variations for protein and minerals among the studied genotypes. The protein content in pea varied from 21.13 to 27.05, potassium from 562.8 to 937.8 mg/100 g, phosphorus from 163.4 to 374.2 mg/100 g, calcium from 45.91 to 157.40 mg/100 g, magnesium from 47.31 to 102.81 mg/100 g, sulphur from 75.69 to 194.4 mg/100 g, iron from 2.19 to 5.84 mg/100 g and zinc from 2.10 to 5.71 mg/100 g. A single serving of 100 g field pea can provide 26–78% of adult RDA of Fe, Zn and Mg. Peas are also a good source of selenium, and high Se may be advantageous for areas of the world where Se deficiency is prominent (Reichert and MacKenzie 1982). Genetics of micronutrients should be worked out before breeding for high micronutrient content in any food legume (Cichy et al. 2005). To date, inheritance of Fe and Zn content in few food legumes is only known. With these reports, the mineral uptake from soil by roots and transport to seeds is now known in detail. Most studies have indicated multigenic inheritance of micronutrient traits (Blair and Izquierdo 2012), while a few initial reports suggested that the inheritance of zinc concentration in common beans might be by a few genes. The genotype × environment interaction has been reported to have a notable effect on both Fe and Zn contents (Cichy and Raboy 2009; Tryphone and Msolla 2010; Kumar et al. 2018). The bioavailability is crucial issue in biofortified varieties. The impact of recently developed biofortified varieties in providing the mineral micronutrient also needs to be assessed in clinical trials. Such studies have been conducted for iron-biofortified beans and have shown promising results (Bouis and Saltzman 2017; De Moura et al. 2014). In a recent study, Fe absorbed from biofortified broad beans was almost 50% more than that of common beans (Petry et al. 2016). Although biofortification research is instrumental in developing varieties with high iron and zinc content, one in nine people in the world still suffers from hunger (FAO 2018). Thus, global efforts are to be accelerated to develop and scale up micronutrient-rich staple food crops including food legumes.

Role of Anti-nutritional Factors Regulating Bioavailability of Nutrients

The increased bioavailability of nutrient is an important aspect of breeding for quality traits in food legumes. Bioavailability of a nutrient depends on the chemical interaction of nutrient with other molecules, collectively known as antinutrients. Anti-nutritional factors (ANFs), such as phytic acid, trypsin inhibitors, lectins, tannins, saponins, oligosaccharides, non-protein amino acids (NPAAs), alkaloids, cyanogenic glycosides, pyrimidine glycosides and isoflavones, are present in food legumes (Pusztai and Bardocz 1996; Enneking and Wink 2000). Many of the ANFs are toxic, unpalatable or indigestible; hence, its reduction is required to improve grain quality of food legumes. Lectins (haemagglutinins) are specific sugars on the surface of cells in the intestinal wall which can regulate the breakdown and absorption of nutrients. Pyrimidine glycosides like vicine and convicine are seen in *Vicia faba* and cause favism (Khazaei et al. 2019). Non-protein amino acid β-N-oxalyl-L-α, β-diaminopropionic acid (β-ODAP) in *Lathyrus* spp. is reported to cause lathyrism, also known as neurolathyrism (Adiga et al. 1962, 1963; Kuo et al. 1988; Nunn et al. 1987; Padmanaban 1980; Rao et al. 1964; Ross et al. 1989; Roy et al. 1963; Spencer et al. 1986; Wang et al. 2000; Zhao et al. 1999; Kumar et al. 2011). For ODAP content, studies have shown a wide range of variation within the existing germplasm, ranging from 0.02% to 2.59% (Pandey et al. 1997; Hanbury et al. 1999).

Protease inhibitors in leguminous seeds inhibit the function of digestive enzymes such as trypsin and chymotrypsin (Birk 1983). Protease inhibitors have a molecular weight of ~8000 and denaturation and proteolysis proof S–S bonds (Birk 1983). Legume seeds have protease inhibitors of both Kunitz and Bowman-Birk families (Lajolo et al. 2004; Srinivasan et al. 2005). The protease inhibitors of both families can inhibit trypsin and chymotrypsin. Since protease inhibitors are heat-labile, processing or cooking can inhibit their activity. Protease inhibitors also have defensive role against different pests (Birk 1983). It is suggested that diets rich in protease inhibitors may have an anti-carcinogenic effect (Clemente et al. 2004).

The most common oligosaccharides in plants are α-galactosides, of which raffinose group oligosaccharides are the most common (Kadlec et al. 2000). Raffinose family of oligosaccharides (RFOs) is a group of soluble, nonreducing sugars with α-1-6-glycosidic bonds, accumulated in seeds during their development. Raffinose group oligosaccharides, raffinose (trisaccharide), stachyose (a tetrasaccharide), verbascose (a pentasaccharide) and ajugose (a hexasaccharide), are present in legumes (Kotiguda et al. 2006; Muzquiz and Wood 2007; Han and Baik 2006; Muzquiz et al. 2012). These oligosaccharides are present significantly and in varying amounts in different legumes (Rao and Belavady 1978; Martinez-Villaluenga et al. 2008). It has been reported that raffinose oligosaccharides are effective in drought and frost tolerance in plants (Arora 1983; Castonguay et al. 1995). Raffinose is predominantly present in monocots

(Peterbauer and Richter 2001). Human body cannot hydrolyse raffinose family oligosaccharides as it lacks α-glycosidase that hydrolyses RFOs. The undigested RFOs in the gastrointestinal tract undergo fermentation by microflora in the large intestine to produce gases like carbon dioxide, hydrogen and methane which cause flatulence and discomfort (Naczk et al. 1997). However, RFOs have important role in seed physiology and play a role in seed germination (Koster and Leopold 1988; Bentsink et al. 2000). In addition, RFOs also play an important role in stress tolerance (Egert et al. 2013; Gangl and Tenhaken 2016). In chickpea, raffinose content progressively increases during seed development and gradually declines as seed germinates (Salvi et al. 2016). Recently, in *Arabidopsis*, total RFOs, RFO/sucrose ratio, were found to be positively correlated, while absolute individual RFO amounts were not correlated to seed vigour (Li et al. 2017). Therefore, a proper balance in RFO contents in plants needs to be maintained for regulation of plant metabolism and growth and makes legumes easy to digest by reducing the RFOs.

Phytic acid (*myo*-inositol-1,2,3,4,5,6-hexa*kis*phosphate, $InsP_6$) is the most abundant form of phosphorus occurring in seeds and other plant tissues. Due to its chemical structure (highly negatively charged at physiological pH), PA easily binds important mineral cations (positively charged) such as iron, zinc, potassium, calcium and magnesium, thereby making them unavailable. It is reported that PA can reduce bioavailability of Zn, Ca, Mg and Fe by 5–15% (Bohn et al. 2004; Phillippy 2006; Das et al. 2012). PA content varies from one food legume to another. Phytic acid content in seeds of mungbean ranged from 6.17 to 12 mgg^{-1} (Chitra et al. 1995; Raboy 1997; Sompong et al. 2010; Tajoddin et al. 2011; Sompong et al. 2012; Dahiya et al. 2013; Dhole and Reddy 2015), while in blackgram it ranged from 0.06 to 13.7 mgg^{-1} (Duhan et al. 1989; Chitra et al. 1995; Suneja et al. 2011; Singh et al. 2017a, b). Recently, there is an increasing interest in the development of crops with low PA (*lpa*) content to enhance the bioavailability of minerals and other nutrients. Earlier breeding efforts have identified several *lpa* mutants resulting in reduction of seed PA phosphorus from 50% to >95% in crops, such as barley, wheat, maize, soybean and common bean (Raboy and Gerbasi 1996; Larson et al. 1998; Guttieri et al. 2004; Yuan et al. 2007; Campion et al. 2009). Recent studies indicate the involvement of PA in biotic and abiotic stress tolerance. They are known to be involved in plant defence mechanism against biotic and abiotic stresses; therefore, instead of breeding for low or null PA, genetic biofortification with moderate PA can be a better breeding target. Moreover, in a clinical trial, it was found that low phytic acid beans were difficult to cook and cause adverse gastrointestinal symptoms (Petry et al. 2016). Biofortified beans and low phytic acid beans provided more bioavailable iron than control beans. Gastrointestinal side effects of low phytic acid beans were likely caused by L residues of phytohemagglutinin. From this experiment, it was unclear to what extent the associated digestive problems reduced iron bioavailability (Petry et al. 2016). For low or no ANFs in future varieties, recent advances in biotechnology and genomics-assisted breeding approaches are pertinent to deploy in food legumes.

Bioactive Compounds

Legume seeds have a large array of bioactive compounds with known different biochemical structures and functions. Many antinutrients also have bioactive properties. Knowledge about these bioactive molecules is gaining momentum as the research progresses in this field. The roles of these bioactive molecules have come to the limelight with different in vivo or in vitro studies. Bioactive molecules in food legumes balance blood sugar and thus reduce the risk of cardiovascular diseases, diabetes and obesity-related problems. Health benefits with food legume consumption are associated with high fibre contents, low glycaemic index and presence of other molecules such as phytosterols, saponins, oligosaccharides, isoflavones and other factors in trace quantity (Lang et al. 1999; Rizkalla et al. 2002; Scarafoni et al. 2005). In some studies, many forms of cancers were found to be regulated with the sufficient consumption food legumes in diet (Jain et al. 1999). Few research studies indicated the possibilities of their use as probiotic intestinal, metabolic and hormonal regulators (Muzquiz et al. 2012). Lectins are another group of bioactive compounds which are used to treat obesity and partially inhibit tumour formation (Pusztai et al. 2004). Research efforts in common bean revealed that common beans possess enormous quantities of polyphenols and other metabolites, have antioxidant activities, have major role in health-promoting effects and protect against various diseases including diabetes, cardiovascular diseases, cancer and microbial infections. Alpha-amylases (α-1,4-glucanohydrolases) are produced by the synthesis of glucose and other sugars, which are energy sources in humans and animals, and play an important role in carbohydrate metabolism. Among the α-amylase inhibitors (αAI) found in plants, attention is paid to the α-amylase inhibitors found in legumes, especially in beans (Lajolo and Genovese 2002; Muzquiz et al. 2012). α-Amylase inhibitors reduce the amylase activity and starch digestion in the intestine (Singh et al. 1982) when administered orally, and it has been found to be beneficial in the fight against obesity or diabetes as it prevents the increase blood glucose after taking meal. α-Amylase inhibitors are also harmful to many harmful pests in plants. Bruchid resistance may have role of α-amylase inhibitors. Transgenic plants obtained by transferring these genes to some plants have been reported to be more resistant to insect damage (Gatehouse 2011).

Epilogue

The role of food legumes has been recognised in human food nutrition since ancient times. Recently, as more and more research-based evidences are emerging on bioactive and nutraceutical properties of the food legumes, popularity of this group of crops is increasing. Efforts to increase nutrients and reduce antinutrients should be integrated into conventional breeding programs to further improve the quality of food legumes. High micronutrient (Fe and Zn) content must be included as a core trait of breeding programmes by phenotyping all advanced breeding lines for high iron and zinc concentrations. The influence of environment on nutritional traits is predominant; therefore, growing conditions including soil nutrient status should be taken into consideration. Therefore, to detect genetic variation for these traits, there is also a need to ensure a uniform inorganic components or raw material supply to the plants. For example, genetic increase in protein in legumes requires an adequate supply of overall nitrogen from soil or growing media to the plants (Bhatia 1983). The use of high-throughput phenotyping technologies and molecular markers linked to biofortification traits can accelerate genetic gain for quality improvement in food legumes. Molecular markers have been identified for biofortification traits in few pulse crops. QTLs and/or SNP markers associated with Fe and/or Zn concentrations have been identified in peas (Ma et al. 2017; Gali et al. 2018), chickpeas (Upadhyaya et al. 2016), common beans (Blair et al. 2011) and lentils (Khazaei et al. 2017); those can be tested for their usefulness in marker-assisted selection. The detailed genetics for biofortification traits including Fe, Zn, selenium, carotenoids and folates in different pulse crops were reviewed by Jha and Warkentin (2020). More efforts should be made by the governmental and nongovernmental agencies to create interest among the growers to cultivate biofortified varieties. Marketing strategies should be developed so that biofortified products reach consumers readily. A number of initiatives made quantified impacts on reducing global malnutrition like Nutrition International, Iron Deficiency Project Advisory Service (IDPAS), New Partnership for Africa's Development (NEPAD), UNICEF-Micronutrients, Global Alliance for Improved Nutrition, Helen Keller International, CGIAR Research Program on Agriculture for Nutrition and Health (A4NH) – HarvestPlus and Global Alliance for Improved Nutrition (GAIN). However, still malnutrition is a global issue despite progress over the last few decades. Provided the recent growing interest among different international as well as national agricultural research systems remains instrumental, the development of next-generation biofortified crops will surely be a success.

Naturally, food legumes are rich in nutritional traits, and a wide range genetic variability exists among the germplasm of different food legume crops, in the past years efforts have been made towards the breeding of nutritionally rich cultivars in these crops. Breeding of these cultivars has been discussed in different chapters of this book. However, intensive breeding efforts are required for developing the more biofortified varieties by public sector institutions. In addition to this, there is a need to refine policies for ensuring a significant increase in the adoption and acceptance of biofortified food legume varieties. For this, supply of seeds of biofortified varieties will be required to strengthen for popularisation of biofortified varieties. During the seed production of these varieties, it will be necessary to ensure the genetic purity for maintaining the nutritional quality trait intact.

During the past years, significant advancements have made in the development of genomics tool and techniques. This led to make food legume crop rich with genomic resources. These genomic resources along with the advanced tools and techniques like genome editing and speed breeding can help to develop the biofortified varieties of food legumes more rapidly through genomics-assisted breeding.

For those traits, significant variability is not available in the germplasm; transgenic approach can be used as well-established genetic transformation protocols are available in these crops. Therefore, a better coordination and collaboration among breeders, biotechnologists, biochemists, seed technologists, agronomists and post-harvest technologists would further accelerate the development of biofortified varieties of food legumes in a more effective way. Such efforts should be done in a properly managed coordinated mode to expedite delivery of biofortified cultivars. The possibility of utilising improved genome editing tools to target specific genes of biofortification traits does exist. With the availability of whole genome sequences of the food legumes (Varshney et al. 2012, 2013; Kang et al. 2014; Schmutz et al. 2014; Souframanien et al. 2020; Emmrich et al. 2020), targeting specific genes becomes easier than before; therefore, biotechnological interventions are more likely. In conclusion, biofortified varieties developed this way will surely aid in eliminating malnutrition for millions of people worldwide who do not have enough access to commercially fortified foods, diversified diets and food supplements.

References

Adiga PR, Padmanaban G, Rao SLN, Sarma PS (1962) The isolation of a toxic principle from *Lathyrus sativus* seeds. J Sci Ind Res (India) 21C:284–286

Adiga PR, Rao SLN, Sarma PS (1963) Some structural features and neurotoxic action of a compound from *Lathyrus sativus* seeds. Curr Sci 32:133–135

Allred CD, Allred KF, Ju YH, Goeppinger TS, Doerge DR, Helferih WG (2004) Soy processing influences growth of estrogen-dependent breast cancer tumors. Carcinogenesis 25(9):1649–1657. https://doi.org/10.1093/carcin/bgh178

Amarakoon D, McPhee K, Thavarajah P (2012) Iron-, zinc-, and magnesium-rich field peas (*Pisum sativum* L.) with naturally low phytic acid: a potential food-based solution to global micronutrient malnutrition. J Food Compos Anal 27:8–13

Anderson JW, Bryant CA (1986) Dietary fiber: diabetes and obesity. Am J Gastroenterol 81:898–906

Arora SK (1983) Chemistry and biochemistry of legumes. Edward Arnold, London

Beebe S, Gonzalez AV, Rengifo J (2000) Research on trace minerals in the common bean. Food Nutr Bull 21:387–391

Bentsink L, Alonso-Blanco C, Vreugdenhil D et al (2000) Genetic analysis of seed-soluble oligosaccharides in relation to seed storability of Arabidopsis. Plant Physiol 124:1595–1604

Bhatia CR (1983) Possibilities of seed protein improvement in tropical and sub-tropical legumes. In: Gottschalk W, Müller HP (eds) Seed proteins. Advances in agricultural biotechnology, vol 2. Springer, Dordrecht

Bing DJ (2015) Breeding field pea cultivars with improved protein content. In: Eucarpia international symposium on protein crops. Pontevedra, Spain, April 5–7, abstracts, pp 31–32

Birk Y (1983) Protease inhibitors from legume seeds: chemistry and implications in nutrition, pest control and medicine. In: Chambers CM, Chambers PL, Gitter S (eds) Toxicology in the use, misuse, and abuse of food, drugs, and chemicals. Archives of toxicology (supplement), vol 6. Springer, Berlin/Heidelberg

Blair MW, Izquierdo P (2012) Use of the advanced backcross- QTL method to transfer seed mineral accumulation nutrition traits from wild to Andean cultivated common beans. Theor Appl Genet 125:1015–1031

Blair MW, Astudillo C, Rengifo J, Beebe SE, Graham R (2011) QTL analyses for seed iron and zinc concentrations in an intra-gene pool population of Andean common beans (Phaseolus vulgaris L.). Theor Appl Genet 122(3):511–521

Bohn T, Davidsson L, Walczyk T, Hurrell RF (2004) Phytic acid added to white-wheat bread inhibits its fractional apparent magnesium absorption in humans. Am J Clin Nutr 79:418–423

Bouis HE, Saltzman A (2017) Improving nutrition through biofortification: a review of evidence from HarvestPlus, 2003 through 2016. Glob Food Sec 12:49–58

Boye J, Zare F, Pletch A (2010) Pulse proteins: processing, characterization, functional properties and applications in food and feed. Food Res Int 43(2):414–431

Campion B, Sparvoli F, Doria E, Tagliabue G, Galasso-Fileppi IM, Bollini R, Nielsen E (2009) Isolation and characterization of an *lpa* (low phytic acid) mutant in common bean (*Phaseolus vulgaris* L.). Theor Appl Genet 118:1211–1221

Castonguay Y, Nadeau P, Lechasseur P, Chouinard L (1995) Differential accumulation of carbohydrates in alfalfa cultivars of contrasting winter hardiness. Crop Sci 35(2):509–516

Celmeli T, Sari H, Canci H, Sari D, Adak A, Eker T, Toker C (2018) The nutritional content of common bean (*Phaseolus vulgaris* L.) landraces in comparison to modern varieties. Agronomy 8(9):166

Chitra U, Vimala V, Singh U et al (1995) Variability in phytic acid content and protein digestibility of grain legumes. Plant Foods Hum Nutr 47:163–172

Cichy K, Raboy V (2009) Evaluation and development of low phytate crops. Modification of seed composition to promote health and nutrition. Krishnan H (ed). Agron Monogr 51:177–200

Cichy KA, Forster S, Grafton KF, Hosfield GL (2005) Inheritance of seed zinc accumulation in navy bean. Crop Sci 45(3):864–870

Clemente A, Mackenzie DA, Johnson IT, Domoney C (2004) Investigation of legume seed protease inhibitors as potential anti-carcinogenic proteins. In: 5th, European conference on grain legumes; legumes for the benefit of agriculture, nutrition and the environment, Dijon, France

Dahiya PK, Linnemann AR, Nout MJR, Boekel MAJS, Grewal RB (2013) Nutrient composition of selected newly bred and established mungbean varieties. LWT Food Sci Technol 54:249–256

Das A, Raychaudhuri U, Chakraborty R (2012) Cereal based functional food of Indian subcontinent: a review. J Food Sci Technol 49(6):665–672

De Moura FF, Palmer AC, Finkelstein JL et al (2014) Are biofortified staple food crops improving vitamin a and iron status in women and children? New evidence from efficacy trials. Adv Nutr 5:568–570

Demirbasß A (2018) Micro and macronutrients diversity in Turkish pea (Pisum sativum) germplasm. Int J Agric Biol 20(4):701–710

Devos P (1988) Nutritional value of lentils and chickpeas and changes during processing. In: Lentil for everyone symposium, 174–196. Ankara, Turkey: Toprak Mahsulleri Ofisi, 29–30 September 1988 Marmaris, Turkey

Dhole VJ, Reddy KS (2015) Genetic variation for phytic acid content in mungbean (*Vigna radiata* L. Wilczek). Crop J 3:157–162

Duhan A, Chauhan BM, Kapoor AC (1989) Phytic acid content of chickpea (*Cicer arietinum*) and black gram (*Vigna mungo*): varietal differences and effect of domestic processing and cooking methods. J Sci Food Agric 49:449–455

Egert A, Keller F, Peters S (2013) Abiotic stress-induced accumulation of raffinose in *Arabidopsis* leaves is mediated by a single raffinose synthase (RS5, At5g40390). BMC Plant Biol 13:218

Emmrich PM, Sarkar A, Njaci I, Kaithakottil GG, Ellis N, Moore C, Edwards A, Heavens D, Waite D, Cheema J, Trick M (2020) A draft genome of grass pea (*Lathyrus sativus*), a resilient diploid legume. bioRxiv

Enneking D, Wink M (2000) Towards elimination of anti-nutritional factors in grain legumes. In: Knight R (ed) Linking research and marketing opportunities for pulses in the 21st century. Kluwer Academic Publishers, Dordrecht, pp 671–683

FAO (2018) World food and agriculture – statistical pocketbook 2018. FAO, Rome, p 254

Frossard E, Bucher M, Mächler F, Mozafar A, Hurrell R (2000) Potential for increasing the content and bioavailability of Fe, Zn and Ca in plants for human nutrition. J Sci Food Agric 80(7):861–879

Gali KK, Liu Y, Sindhu A et al (2018) Construction of high-density linkage maps for mapping quantitative trait loci for multiple traits in field pea (*Pisum sativum* L.) (2016). BMC Plant Biol 18:172

Gangl R, Tenhaken R (2016) Raffinose family oligosaccharides act as galactose stores in seeds and are required for rapid germination of Arabidopsis in the dark. Front Plant Sci 7:1115. https://doi.org/10.3389/fpls.2016.01115

Gatehouse JA (2011) Prospects for using proteinase inhibitors to protect transgenic plants against attack by herbivorous insects. Curr Protein Pept Sci 12(5):409–416

Glore SR, Van Treeck DV, Knehaus AW, Gild M (1994) Soluble fiber and serum lipids: a literature review. J Am Diet Assoc 94:425–436

Gomez-Galera S, Rojas E, Sudhakar D, Zhu CF, Pelacho AM, Capell T et al (2010) Critical evaluation of strategies for mineral fortification of staple food crops. Transgenic Res 19:165–180. https://doi.org/10.1007/s11248-009-9311-y

Graham RD, Senadhira D, Beebe SE, Iglesias C, Ortiz-Monasterio I (1999) Breeding for micronutrient density inedible portions of staple food crops: conventional approaches. Field Crop Res 60:57–80

Guttieri M, Bowen D, Dorsch JA, Raboy V, Souza E (2004) Identification and characterization of low phytic acid wheat. Crop Sci 44:418–424

Guzman-Maldonado SH, Acosta-Gallegos J, Paredes-Lopez O (2000) Protein and mineral content of a novel collection of wild and weedy common bean (*Phaseolus vulgaris* L). J Sci Food Agric 80(13):1874–1881

Hacisalihoglu G, Osturk L, Cakmak I, Welch RM, Kochian L (2004) Genotypic variation in common bean in response to zinc deficiency in calcareous soil. Plant Soil 259:71–83. https://doi.org/10.1023/B:PLSO.0000020941.90028.2c

Han IH, Baik BK (2006) Oligosaccharide content and composition of legumes and their reduction by soaking, cooking, ultrasound, and high hydrostatic pressure. Cereal Chem 83(4):428–433

Hanbury CD, Siddique KHM, Galwey NW, Cocks PS (1999) Genotype-environment interaction for seed yield and ODAP concentration of *Lathyrus sativus* L. and L. cicera L. in Mediterranean-type environments. Euphytica 110:45–60

Harmankaya M, Özcan MM, Karadaş S, Ceyhan E (2010) Protein and mineral contents of pea (Pisum sativum L.) genotypes grown in Central Anatolian region of Turkey. South West J Hortic Biol Environ 1(2):159–165

HarvestPlus (2014) Biofortification progress briefs. Iron and Zinc lentils, p 17

House WA, Welch RM, Beebe S, Cheng Z (2002) Potential for increasing the amounts of bioavailable zinc in dry beans through plant breeding. J Sci Food Agric 82:1452–1457. https://doi.org/10.1002/jsfa.1146

Hughes JS (1991) Potential contribution of dry bean dietary fiber to health. Food Technol 45:122–126

Institute of Medicine, Food and Nutrition Board (2001) Dietary reference intakes for vitamin a, vitamin k, arsenic, boron, chromium, copper, iodine, iron, manganese, molybdenum, nickel, silicon, vanadium, and zinc. National Academy Press, Washington, DC

Iqbal A, Khalil IA, Ateeq N, Khan MS (2006) Nutritional quality of important food legumes. Food Chem 97(2):331–335

Islam FM, Basford KE, Jara C, Redden RJ, Beebe S (2002) Seed compositional and disease resistance differences among gene pools in cultivated common bean. Genet Resour Crop Evol 49(3):285–293

Jain PC, Kushwaha PS, Dhaked US, Khan H, Trivedi SK (1999) Response of chickpea (*Cicer arietinum* L.) to phosphorus and biofertilizer. Indian J Agron 22:241–244

Jha AB, Warkentin TD (2020) Biofortification of pulse crops: status and future perspectives. Plants 9(1):73. https://doi.org/10.3390/plants9010073

Joshi-Saha A, Reddy KS (2014) Assessment of mineral micronutrient and phytic acid content in chickpea. In: Dixit GP, Singh J, Sing NP (eds) Pulses: challenges and opportunities under changing climatic scenario (Proceedings of the national conference on pulses: challenges and opportunities under changing climatic scenario, Sep 29 to Oct 01, 2014, JNKVV, Jabalpur), Indian Society of Pulses Research and Development, ICAR-Indian Institute of Pulses Research, Kanpur-208024, UP, India, pp 322–327

Jukanti AK, Gaur PM, Gowda CLL et al (2012) Nutritional quality and health benefits of chickpea (*Cicer arietinum* L.): a review. Br J Nutr 108:S11–S26

Kadlec P, Kaasová J, Bubník Z, Pour V (2000) Effect of germination and microwave treatment on chemical composition of pea. In: Proceedings of international conference on microwave chemistry. Antibes, Francie, pp 241–243

Kang YJ, Kim SK, Kim MY, Lestari P, Kim KH, Ha BK, Jun TH, Hwang WJ, Lee T, Lee J, Shim S (2014) Genome sequence of mungbean and insights into evolution within *Vigna* species. Nat Commun 5:5443

Khazaei H, Podder R, Caron CT et al (2017) Marker-trait association analysis of iron and zinc concentration in lentil (*Lens culinaris* Medik.) seeds. Plant Genome 10:2

Khazaei H, Purves RW, Hughes J, Link W, O'Sullivan DM, Schulman AH et al (2019) Eliminating vicine and convicine, the main anti-nutritional factors restricting faba bean usage. Trends Food Sci Technol *91*:549–556

Koster KL, Leopold AC (1988) Sugars and desiccation tolerance in seeds. Plant Physiol 88:829–832

Kotiguda G, Peterbauer T, Mulimani VH (2006) Isolation and structural analysis of ajugose from Vigna mungo L. Carbohydr Res 341(12):2156–2160

Kumar S, Bejiga G, Ahmed S, Nakkoul SH, Sarker A (2011) Genetic improvement of grass pea for low neurotoxin (ODAP) content. Food Chem Toxicol 49:589–600

Kumar J, Thavarajah D, Kant R, Sarker A, Singh NP (2016) Genetic variability for mineral nutrients in lentil. J Food Legum 29(2):94–101

Kumar J, Kumar S, Sarker A, Singh NP (2018) Analysis of genetic variability and genotype× environment interactions for iron and zinc content among diverse genotypes of lentil. J Food Sci Technol 55:3592–3605

Kuo TM, Van Middlesworth JF, Wolf WJ (1988) Content of raffinose oligosaccharides and sucrose in various plant seeds. J Agric Food Chem 36:32–36

Lajolo FM, Genovese MI (2002) Nutritional significance of lectins and enzyme inhibitors from legumes. J Agric Food Chem 50(22):6592–6598

Lajolo FM, Genovese MI, Pryme IF, Dale TM (2004) Beneficial (antiproliferative) effects of different substances. Publ Eur Assoc Anim Prod 110:123–136

Lam ACY, Karaca AC, Tyler RT, Nickerson MT (2018) Pea protein isolates: structure, extraction and functionality. Food Rev Int 34(2):126–147. https://doi.org/10.1080/87559129.2016.1242135

Lang V, Bornet FR, Vaugelade P, van Ypersele de Strihou M, Luo J, Pacher N, Rossi F, La Droitte P, Duée PH, Slama G (1999) Euglycemic hyperinsulinemic clamp to assess posthepatic glucose appearance after carbohydrate loading. 2. Evaluation of corn and mung bean starches in healthy men. Am J Clin Nutr 69(6):1183–1188

Larson SR, Young KA, Cook A, Blake TK, Raboy V (1998) Linkage mapping of two mutations that reduce phytic acid content of barley grain. Theor Appl Genet 97:141–146

Li T, Zhang Y, Wang D, Liu Y, Dirk LM, Goodman J, Zhao T (2017) Regulation of seed vigor by manipulation of raffinose family oligosaccharides in maize and Arabidopsis thaliana. Mol Plant 10(12):1540–1555

Ma Y, Coyne CJ, Grusak MA et al (2017) Genome-wide SNP identification, linkage map construction and QTL mapping for seed mineral concentrations and contents in pea (*Pisum sativum* L.). BMC Plant Biol 17:43

Maret W, Sandstead HH (2006) Zinc requirements and the risks and benefits of zinc supplementation. J Trace Elem Med Biol 20:3–18

Marlett JA, McBurney MI, Slavin JL (2002) Health implications of dietary fiber. J Am Diet Assoc 102:993–1000. https://doi.org/10.1016/S0002-8223(02)90228-2

Martinez-Villaluenga C, Frias J, Vidal-Valverde C (2008) Alpha-galactosides: antinutritional factors or functional ingredients? Crit Rev Food Sci Nutr 48:301–316

Mayer JE, Pfeiffer WH, Bayer P (2008) Biofortified crop to alleviate micronutrient malnutrition. Curr Opin Plant Biol 11:166–170

Moraghan JT, Grafton K (1999) Seed zinc concentration and the zinc-efficiency trait in navy bean. Soil Sci Soc Am J63:918–922

Moraghan JT, Padilla J, Etchevers JD, Grafton K, Acosta-Gallegos JA (2002) Iron accumulation in seed of common bean. Plant Soil 246:175–183. https://doi.org/10.1023/A:1020616026728

Muthayya S, Rah JH, Sugimoto JD et al (2013) The global hidden hunger indices and maps: an advocacy tool for action. PLoS One 8:e67860

Muzquiz M, Wood JA (2007) Antinutritional factors. Chickpea Breed Manag:143–166

Muzquiz M, Varela A, Burbano C, Cuadrado C, Guillamón E, Pedrosa MM (2012) Bioactive compounds in legumes: pronutritive and antinutritive actions. Implications for nutrition and health. Phytochem Rev 11:227–244

Naczk M, Amarowicz R, Shahidi F (1997) α-Galactosides of sucrose in foods: composition, flatulence-causing effects, and removal. In: Shahidi F (ed) Antinutrients and phytochemicals in food, vol 8. American Chemical Society, Washington, DC, pp 127–151

Nielsen SS, Brandt WE, Singh BB (1993) Genetic variability for nutritional composition and cooking time of improved cowpea lines. Crop Sci 33:469–472

Nunn PB, Seelig M, Zagoren JC, Spencer PS (1987) Stereospecific acute neurotoxicity of uncommon plant amino acids linked to human motor-system diseases. Brain Res 410:375–379

Owusu-Ansah YJ, McCurdy SM (1991) Pea proteins: a review of chemistry, technology of production, and utilization. Food Rev Int 7(1):103–134

Padmanaban G (1980) Lathyrogens. In: Leiner IE (ed) Toxic constituents of plant foodstuffs. Academic, New York, pp 244–250

Pandey R, Chitale MW, Sharma RN, Geda AK (1997) Evaluation and characterization of germplasm of grass pea (*Lathyrus sativus*). J Med Aromat Plant Sci 19:14–16

Pekşen E, Artık C (2005) Antibesinsel maddeler ve yemeklik tane baklagillerin besleyici değerleri. Anadolu Tarım Bilimleri Dergisi 20(2):110–120

Peterbauer T, Richter A (2001) Biochemistry and physiology of raffinose family oligosaccharides and galactosyl cyclitols in seeds. Seed Sci Res 11:185–197

Petry N, Rohner F, Gahutu JB et al (2016) In Rwandese women with low iron status, iron absorption from low-phytic acid beans and biofortified beans is comparable, but low-phytic acid beans cause adverse gastrointestinal symptoms. J Nutr. https://doi.org/10.3945/jn.115.223693

Phillippy BQ (2006) Transport of calcium across Caco-2 cells in the presence of inositol hexakisphosphate. Nutr Res 26:146–149. https://doi.org/10.1016/j.nutres.2006.02.008

Prasad AS (2004) Zinc deficiency: its characterization and treatment. Met Ions Biol Syst 41:103–137

Pusztai A, Bardocz S (1996) Biological effects of plant lectins on the gastrointestinal tract. Trends Glycosci Glycotechnol 8(41):149–165

Pusztai A, Bardocz S, Martín-Cabrejas MA (2004) The mode of action of ANFs on the gastrointestinal tract and its microflora. Publ Eur Assoc Anim Prod 110:87–100

Raboy V (1997) Accumulation and storage of phosphate and minerals. In: Larkins BA, Vasil IK (eds) Cellular and molecular biology of plant seed development. Kluwer Academic Publishers, Dordrecht, pp 441–477

Raboy V, Gerbasi P (1996) Genetics of myo-inositol phosphate synthesis and accumulation. In: Myo-inositol phosphates, phosphoinositides, and signal transduction. Springer, Boston, pp 257–285

Rao SLN, Adiga PR, Sarma PS (1964) The isolation and characterization of β-N-Oxalyl-L-α, β-Diaminopropionic acid: a neurotoxin from the seeds of *Lathyrus sativus*. Biochemistry 3:432–436

Rao PU, Belavady B (1978) Oligosaccharides in pulses: varietal differences and effects of cooking and germination. J Agric Food Chem 26:316–319
Reichert RD, MacKenzie SL (1982) Composition of peas (*Pisum sativum*) varying widely in protein content. J Agric Food Chem 30(2):312–317
Rizkalla SW, Bellisle F, Slama G (2002) Health benefits of low glycaemic index foods, such as pulses, in diabetic patients and healthy individuals. Br J Nutr 88(S3):255–262
Ross SM, Roy DN, Spencer PS (1989) β-N-oxalyl amino-L-alanine action on glutamate receptors. J Neurochem 53:710–715
Roy DN, Nagarajan V, Gopalan C (1963) Production of neurolathyrism in chicks by the injection of *Lathyrus sativus* concentrates. Curr Sci 32:116–118
Roy F, Boye JI, Simpson BK (2010) Bioactive proteins and peptides in pulse crops: pea, chickpea and lentil. Food Res Int 43(2):432–442
Salvi P, Saxena SK, Petla BP (2016) Differentially expressed galactinol synthase(s) in chickpea are implicated in seed vigor and longevity by limiting the age induced ROS accumulation. Sci Rep 6:35088. https://doi.org/10.1038/srep35088
Saxena KB, Sawargaonkar SL (2016) Genetic enhancement of seed proteins in pigeonpea – methodologies, accomplishments, and opportunities. Int J Sci Res 4:3–7. ISSN No.2277-8179
Scarafoni A, Kumar J, Magni C, Sironi E, Duranti M (2005) Biologically active molecules and nutraceutical properties of legume seeds. IFLRC IV:18–22
Schmutz J, McClean PE, Mamidi S, Wu GA, Cannon SB, Grimwood J, Jenkins J, Shu S, Song Q, Chavarro C, Torres-Torres M (2014) A reference genome for common bean and genome-wide analysis of dual domestications. Nat Genet 46(7):707–713
Shahzad Z, Rouached H, Rakha A (2014) Combating mineral malnutrition through iron and zinc biofortification of cereals. Compr Rev Food Sci Food Saf 13:329–346
Sharma B (1988) İnsanbeslenmesindemercimekvenohut: Bugünküdurumuvebeklentiler (Lentils and chickpeas in human nutrition conditions: present state and prospects). HerkesİçinMercimekSempozyumu (Lentils for everyone symposium) (29–30 Eylül 1988), Marmaris/Muğla, pp 157–171
Shen S, Hou HW, Ding CB, Bing DJ, Lu ZX (2016) Protein content correlates
Shewry PR, Pandya MJ (1999) The 2S Albumin storage proteins. In: Shewry PR, Casey R (eds) . Kluwer Academic Publishers, Seed proteins, pp 563–586
Singh U, Kherdekar MS, Jambunathan R (1982) Studies on desi and kabuli chickpea (*Cicer arietinum* L.) cultivars. The levels of amylase inhibitors, levels of oligosaccharides and in vitro starch digestibility. J Food Sci 47(2):510–512
Singh B, Singh JP, Shevkani K, Singh N, Kaur A (2017a) Bioactive constituents in pulses and their health benefits. J Food Sci Technol 54:858–870
Singh J, Kanaujia R, Srivastava AK, Dixit GP, Singh NP (2017b) Genetic variability for iron and zinc as well as antinutrients affecting bioavailability in black gram (Vigna mungo (L.) Hepper). J Food Sci Technol *54*(4):1035–1042
Sompong U, Kaewprasit C, Nakasathienl S, Srinives P (2010) Inheritance of seed phytate in mungbean (*Vigna radiata* L. Wilczek). Euphytica 171:389–396
Sompong U, Somta P, Raboy V et al (2012) Mapping of quantitative trait loci for phytic acid and phosphorus contents in seed and seedling of mungbean (*Vigna radiata* (L.) Wilczek). Breed Sci 62:87–92
Souframanien J, Raizada A, Punniyamoorthy D, Penna S (2020) Draft genome sequence of the pulse crop blackgram [*Vigna mungo* (L.) Hepper] reveals potential R-genes. bioRxiv
Spencer PS, Roy DN, Ludolph A, Hugon J, Schaumburg HH (1986) Lathyrism: evidence for role of the neuroexcitatory amino acid BOAA. Lancet 2:1066–1067
Srinivasan VS (2001) Bioavailability of nutrients: a practical approach to in vitro demonstration of the availability of nutrients in multivitamin-mineral combination products. J Nutr *131*(4):1349S–1350S

Srinivasan A, Giri AP, Harsulkar AM et al (2005) A Kunitz trypsin inhibitor from chickpea (*Cicer arietinum* L.) that exerts anti-metabolic effect on pod borer (*Helicoverpa armigera*) larvae. Plant Mol Biol 57:359

Suneja Y, Kaur S, Gupta AK, Kaur N (2011) Levels of nutritional constituents and antinutritional factors in black gram (*Vigna mungo* L. Hepper). Food Res Int 44:621–628

Tabatabai A, Li S (2000) Dietary fiber and type 2 diabetes. Clin Excell Nurse Pract 4(5):272–276

Tajoddin MD, Shinde M, Lalitha J (2011) In vivo reduction the phytic acid content of mungbean (*Phaseolus aureus* L.) cultivars during germination. Am-Euras J Agric Environ Sci 10:127–132

Trowell H, Burkitt D, Heaton K (1985) Dietary fibre, fibre-depleted foods and disease. Academic

Tryphone GM, Msolla N (2010) Diversity of common bean (*Phaseolus vulgaris* L.) genotypes in iron and zinc contents under greenhouse conditions. Afr J Agric Res 5:738–747

Upadhyaya HD, Bajaj D, Das S et al (2016) Genetic dissection of seed-iron and zinc concentrations in chickpea. Sci Rep 6:24050

Varshney RK, Chen W, Li Y, Bharti AK, Saxena RK, Schlueter JA, Donoghue MT, Azam S, Fan G, Whaley AM, Farmer AD (2012) Draft genome sequence of pigeonpea (*Cajanus cajan*), an orphan legume crop of resource-poor farmers. Nat Biotechnol 30(1):83–89

Varshney RK, Song C, Saxena RK, Azam S, Yu S, Sharpe AG, Cannon S, Baek J, Rosen BD, Tar'an B, Millan T (2013) Draft genome sequence of chickpea (*Cicer arietinum*) provides a resource for trait improvement. Nat Biotechnol 31(3):240–246

Wang LC, Busbey S (2005) Images in clinical medicine. Acquired acrodermatitis enteropathica. N Engl J Med 352:1121

Wang F, Chen X, Chen Q, Qin X, Li Z (2000) Determination of neurotoxin β-N-oxalyl-L-α, β-diaminopropionic acid and nonprotein amino acids in *Lathyrus sativus* by precolumn derivatization with 1-fluoro-2, 4-dinitrobenzene. J Chromatogr A883:113–118

Wang TL, Domoney C, Hedley CL, Casey R, Grusak MA (2003) Can we improve the nutritional quality of legume seeds? Plant Physiol 131:886–891

Wang N, Hatcher DW, Toews R, Gawalko EJ (2009) Influence of cooking and dehulling on nutritional composition of several varieties of lentils (*Lens culinaris*). Lebensmittel-Wissenschaft+Technologie 42:842–848. https://doi.org/10.1016/j.lwt.2008.10.007

Welch RM, Graham RD (2004) Breeding for micronutrients in staple food crops from a human nutrition perspective. J Exp Bot 55:353–364. https://doi.org/10.1093/jxb/erh064

Welch RM, House WA, Beebe S, Cheng Z (2000) Genetic selection for enhanced bioavailable levels of iron in bean (*Phaseolus vulgaris* L.). J Agric Food Chem 48:3576–3580

Yamauchi D, Minamikawa T (1998) Improvement of the nutritional quality of legume seed storage proteins by molecular breeding. J Plant Res 111:1–6

Yang XE, Chen WR, Feng Y (2007) Improving human micronutrient nutrition through biofortification in the soil plant system: China as a case study. Environ Geochem Health 29:413–428

Yuan S, Zhao FJ, Ren HJ, Zhu XL, Fu SL, Shu XJ (2007) Generation and characterization of two novel low phytate mutations in soybean (*Glycine max* L. Merr.). Theor Appl Genet 115:945–957

Zhao L, Chen XG, Hu ZD, Li QF, Chen Q, Li ZX (1999) Analysis of β-N-oxalyl-L-α, β-diaminopropionic acid and homoarginine in *Lathyrus sativus* by capillary zone electrophoresis. J Chromatogr 857:295–302

Quality Improvement in Chickpea

Archana Joshi-Saha, Golu Misra, and Kandali S. Reddy

Abstract Grain legumes- also known as "pulses" are nutritionally rich and serve as the sole source of dietary proteins for a large population that is primarily vegetarian by choice or nonavailability of affordable animal proteins. In many countries, pulses constitute an essential part of daily diet and are consumed in a variety of forms. Chickpea (gram) is one of the widely grown pulse crop ranking second in area and third in production worldwide. Apart from proteins it is also a good source of carbohydrates and dietary fibers. Due to its wider cultivation and consumption, chickpea is also a good target for genetic biofortification- i.e. nutritional enhancement through genetic intervention, particularly through breeding. The present chapter discusses the target traits for biofortification, assessment of chickpea germplasm for variability in the target traits, identification of quantitative trait loci and markers associated with those traits. In addition, to increase the nutrient content, there is a need to reduce the antinutritional compounds present in chickpea seed. However, some of these compounds play an essential role in plant growth and development as well as biotic and abiotic stress tolerance. This conflict between reducing the antinutrients and compromising plant yields is also addressed in the present chapter.

Keywords *Cicer arietinum* · Iron · Zinc · Protein · Raffinose family oligosaccharides · Phytic acid · Genetic biofortification

A. Joshi-Saha (✉) · K. S. Reddy
Nuclear Agriculture and Biotechnology Division, Bhabha Atomic Research Centre, Mumbai, India

Homi Bhabha National Institute, Mumbai, India
e-mail: archanaj@barc.gov.in

G. Misra
Nuclear Agriculture and Biotechnology Division, Bhabha Atomic Research Centre, Mumbai, India

D. S. Gupta et al. (eds.), *Breeding for Enhanced Nutrition and Bio-Active Compounds in Food Legumes*, https://doi.org/10.1007/978-3-030-59215-8_2

Introduction

A rapid increase in world population has led to food scarcity in different parts of the world, particularly in developing countries. Therefore, providing sufficient quantity of food that can provide adequate nutrition (as per the National Food Security Act, 2013) is a major challenge today. Added to this, is the problem of inadequate nutrition particularly vitamin and mineral micronutrients malnutrition collectively known as "hidden hunger" that globally affects more than 2 billion people, particularly women and children (Muthayya et al. 2013).

"Quality," as defined by oxford English dictionary is "*the standard of something as measured against other things of a similar kind; the degree of excellence of something*"; or "*a distinctive attribute or characteristic possessed by someone or something,*" in context with grain legumes includes parameters related to nutrition in terms of proteins, carbohydrates, and mineral micronutrients and their bioavailability. In addition, quality also concerns with respect to antinutrients that limit the bioavailability of nutrients or render the food toxic/or less digestible or unpalatable. They include raffinose family of oligosaccharides (RFO's), trypsin/protease inhibitors, and phytic acid to name a few.

To improve the nutritional value of the food, there are many alternatives including supplementation of diet with the nutrient, and fortification of the food grain with the required nutrient. The fortification of the grain can be achieved either through agronomic practices by improving soil fertility (*agronomic biofortification*) or through genetic interventions that include conventional breeding methods and/or transgenic approaches (*genetic biofortification*). Among the various strategies, genetic biofortification has been considered as the most suitable, cost-effective, and sustainable method to improve the nutritional status of the crops (Bouis et al. 2011; Saltzman et al. 2013).

So far, the efforts of biofortification have been made in cereals, as they are the staple crops consumed as substantial portion of food. However, grain legumes, also known as "pulses," are also important food and feed crops and are integral part of diet in many developing countries serving as a rich and affordable source of vegetable proteins. Due to their symbiotic association with Rhizobium, they fix nitrogen and are the backbone of sustainable agriculture. In the developing countries, traditionally a variety of pulses are consumed on a daily basis (Saltzman et al. 2013). There have been a few efforts for genetic biofortification of grain legumes particularly common beans and lentils (Bouis and Saltzman 2017; Kumar et al. 2016). Chickpea is an important grain legume cultivated in over 50 countries and is one of the earliest domesticated "founder" crops (Kerem et al. 2007). Very recently, quality improvement programs in chickpea have also been initiated and the crop is being evaluated for various quality traits (Tan et al. 2017).

Prior to the boom in genomic sequencing technology, pulses were often considered as "orphan" crops. However, with the availability to cheaper sequencing platforms, a number of pulse crops including cultivated and wild chickpea (Gupta et al. 2017; Jain et al. 2013; Varshney et al. 2013) have been sequenced and a large number of genomic tools are being made available for their utilization in genetic improvement programs (Pandey et al. 2016). The following sections will deal with target traits for quality improvement in chickpea, the variability available in chickpea germplasm for the target traits, biochemistry of the target traits, and molecular tools that are presently being developed considering the breeding programs for quality improvement.

Priority Traits for Quality Improvement in Chickpea

Chickpea is an important grain legume, grown worldwide in over 50 countries. During 2013, the global chickpea area was 13.57 M ha with the production of 13.12 M t. India ranks first in chickpea production accounting for almost 67% of the world production (8.8 M t from an area of 9.6 M ha in year 2013) (E pulse databook, http://www.iipr.res.in/e-pulse-data-book-country-wise.html). India is also the largest consumer of pulses, where they are used as an integral part of daily diet and consumed along with cereals to provide for a cheaper and rich source of proteins for a large population that is vegetarian by choice or economic constraints. Chickpea (*Cicer arietinum* L.) belongs to the galagoid clade (cool season legumes) of pulses (Cronk et al. 2006) and are of two main biotypes: *kabuli* (*macrosperma*) and *desi* (*microsperma*) with *kabuli* types possibly evolving from *desi* (Moreno and Cubero 1978) or from wild *Cicer reticulatum* (Toker 2009). Chickpea is not only a rich source of proteins; it is also rich in carbohydrates, dietary fibers, vitamins, and minerals (Jukanti et al. 2012). Due to its wide cultivation and consumption, chickpea is gaining importance as a target for quality improvement particularly for improving mineral micronutrient content and proteins. These recent efforts, factors governing these traits, and limitations to them are discussed in the following sections.

Target Trait: Protein Content

Legumes that are integral part of the regular diet in many Asian and African countries are rich source of proteins and are often termed as “poor man’s meat.” However, assessment of protein quality based on growth of laboratory animals, mainly rat, indicated poor protein efficiency ratios. This was attributed to a very high requirement of methionine in rats that is not fulfilled by the inherently low sulfur-containing amino acids in legumes (Messina 1999). A number of other methods are now being adopted for assessing protein quality in legumes, which include amino acid composition, their bioavailability, and protein digestibility (Boye et al. 2012) and indicate variability for protein quality based on species, cultivar, and processing methods; however, there is more emphasis on breeding for increased protein content rather than protein quality (Vaz Patto et al. 2015). In addition, legume storage proteins are deficient in sulfur-containing amino acids including essential amino acid methionine

(Yamauchi and Minamikawa 1998). To ensure improvement in protein quantity and quality without yield compromise, adequate and balanced supply of nitrogen and photosynthates is needed. In chickpea, it is estimated that for each percent point increase in protein; nitrogen demand is increased by around 3.5% (Bhatia 1983). Similarly, metabolic cost of methionine production in terms of consumed mol of ATP per molecule produced is found to be highest as compared to other amino acids in *E. coli* (Kaleta et al. 2013), and in legumes construction cost of methionine is about 64% higher than that of glutamic acid, which is the most abundant amino acid (Bhatia 1983). In legumes, nitrogen supply is also dependent on Rhizobium–legume symbiosis that needs energy in addition to that needed by developing seeds. Therefore, carbon assimilation becomes an important factor in governing yield and protein quality in grain legumes.

Biochemistry

Classically the seed storage proteins have been grouped in four major groups based on their solubility and extraction: albumins (water soluble), globulins (soluble in dilute saline), prolamins (soluble in alcohol/water mixture), and glutelins (soluble in dilute acid or alkali) (Osborne 1924), of which prolamins are predominantly found in cereals (Shewry et al. 1995). Globulins are present as major fractions in legumes (Duranti 2006). Globulins are the most widely distributed conserved, storage proteins that are synthesized during seed development and are stored in protein bodies and are hydrolyzed during seed germination to provide nitrogen and carbon to the growing seedling (Wang et al. 2003). Based on the sedimentation coefficients globulins are further subclassified as 7/8S (trimers and variously known as vicilin, phaseolin, etc., depending on the plant species) and 11/12S (oligomers, usually hexamers, variously known as legumin, glycinin, etc., depending on the plant species) (Casey 1999). Albumins are water-soluble proteins that include several "housekeeping" proteins (Wang et al. 2003). Based on their sedimentation coefficient, they were classified as 2S albumins that are widely distributed in dicots (Shewry et al. 1995). Since the 2S albumin of Brazilian nut was found to have quite high (about 17 mol%) methionine (Shewry and Pandya 1999), a chimeric gene encoding it was over expressed in potato and tobacco to increase the methionine content in them (Altenbach et al. 1989; Tu et al. 1998). Such genetic engineering strategies have also found application in legumes to increase their methionine content (Müntz et al. 1998). However, in the transgenic seeds, the abundance of some endogenous sulfur-rich seed proteins decreased possibly due to competition for a limiting supply of sulfur amino acids in developing seeds (Tabe et al. 2012). Other strategies like increasing sulfur supply or cysteine content were not very successful in increase the free methionine in developing seeds, or for accumulation of sulfur-rich seed storage proteins in mature seeds (Tabe et al. 2012). There is a scope to increase the free methionine content through mutations; however, such mutation can be pleiotropic and have to be screened extensively (Shen et al. 2002).

Genetic Variability

The amounts of proteins present in seeds vary from about 10% (in cereals) to almost 40% in some legumes and oilseeds). In chickpea the protein content varies from 17% to 22% (before dehulling) and 25.3 to 28.9% (after dehulling) of total dry seed mass (Jukanti et al. 2012) and the reference therein). In a recent study, in 336 chickpea accessions, protein content varied between 15–22% with no significant difference between *kabuli* and *desi* accessions (Upadhyaya et al. 2016b). Fractions of proteins in chickpea seeds have been evaluated with globulins being in highest percentage; however, there is some discrepancy over the percentage of albumins, glutelins, and prolamins with higher percentage of glutelins or prolamine reported in some cases (Table 1, Chang et al. 2012). The protein profiling using SDS PAGE and reverse phase HPLC identified globulin proteins 11S legumins and 7S vicilins as the major and 2S albumin as minor protein fractions in chickpea seeds (Chang et al. 2012; Singh and Matta 2003). Protein content of 47 chickpea genotypes grown in four locations indicated that the locations have significant influence on protein content, yet the cultivars X location interactions were nonsignificant with good correlations among locations suggesting that breeding for improved seed protein content in chickpea could be effectively carried out at a single location (Singh et al. 1983). A large variability for seed protein content in wild *Cicer* species was also observed; although, protein content was found to be negatively correlated to harvest index (Table 1, Ocampo et al. 1998). Amino acid profiling indicates that the globulin fraction contains less amount of sulfur-containing amino acids as compared to glutelins,

Table 1 Distribution of seed proteins in chickpea

Sr. No.	Reference	Total protein	Globulins	Albumins	Prolamins	Glutelins
1.	Dhawan et al. (1991) (N = 6)	20.9–25.27%	53.44–60.29%	8.39–12.31%	19.38–24.40%	3.12–6.89%
2.	Ocampo et al. (1998) (N = 228, 8 wild species and 20 cultivars)	Wild: 16.8–26.8% (mean: 21.7%)	n.d.	n.d.	n.d.	n.d.
3.	da Silva et al. (2001)	n..d.	41.79%	16.18%	0.48%	9.99%
4.	Sharma et al. (2013) (N = 9) cultivars	18–31%[a]				
5.	Torutaeva et al. (2014) (N = 23)	14.5–26.9%	n.d.	n.d.	n.d.	n.d.
6.	Jadhav et al. (2015) (N = 187)	13.25–26.77% (mean = 20.16%)	n.d.	n.d.	n.d.	n.d.
7	Upadhyaya et al. (2016b) (N = 336)	15.6–22.4% (mean = 17.6%)[b]	n.d.	n.d.	n.d.	n.d.

N Number of genotypes
[a]Protein content of *kabuli* (N = 4) significantly higher than that of *desi* (N = 5) biotypes
[b]no significant difference in protein content between *desi* (N = 206) and *kabuli* (N = 120) biotypes

and it was suggested that identification of chickpea cultivars with higher glutelin to globulin ratios would help to improve protein quality (Singh 1985).

Marker-Assisted Approaches

Seed protein content and quality of protein is a complex trait; therefore, identification of gene networks regulating the content and quality of proteins will be useful in developing marker-assisted approaches for breeding for high protein content or enhanced quality in terms of increase in sulfur-containing amino acids, particularly methionine. Recently efforts are being made to analyze the protein content in diverse accession of chickpea and to study the marker trait associations for protein content. Using a set of 187 genotypes, SSR markers associated with seed proteins content and potential candidate genes have been identified that will be useful in developing markers for marker-assisted breeding (Jadhav et al. 2015). Upadhyaya et al. (2016b) identified SNP allelic variants in six potential genes (encode ATP-dependent RNA helicase DEAD-box, cystathionine-beta synthase, ABC transporter transmembrane domain, CMP, and dCMP deaminases, and G10 and zinc finger protein) regulating seed protein content trait in chickpea using a combinatorial strategy involving genome-wide association study, selective genotyping in mapping population, and differential gene expression profiling. These candidate genes can be further explored in breeding programs.

Target Trait: Mineral Micronutrients

Mineral micronutrients serve as essential components of human nutrition especially iron (Fe) and zinc (Zn). Iron and zinc are two important micronutrients that act as cofactors for several proteins including hemoglobin, enzymes, and various transcription factors and are required for human growth and development (Abbaspour et al. 2014; McCall et al. 2000). Over 2 billion people worldwide suffer from vitamin and mineral malnutrition collectively termed as "hidden hunger" (Muthayya et al. 2013). The deficiency of Fe leads to iron deficiency anemia (IDA) affecting physical and mental development as well as learning capacity (Abbaspour et al. 2014). Zinc deficiencies, although not well documented, are suspected to be equally severe leading to retarded growth, skeletal abnormalities, delayed wound healing, increased abortion risk, and diarrhea (Rahman et al. 2016; Wessells and Brown 2012). So far, the efforts of biofortification have been mainly focused on cereals like rice, wheat, and maize, which are staple crops worldwide (Shahzad et al. 2014). There is a need to incorporate nutrient-dense traits, particularly high Fe and/or Zn, in various pulse crops, including chickpea.

Biochemistry

The mineral micronutrients have to be taken up from the soil and transported from root, shoot, and leaf tissues to their final destination, in seeds. There are several steps that need to be regulated by the plant to ensure proper distribution of mineral micronutrients. The steps include uptake, transport, remobilization, and storage and are influenced by speciation of metal ion in soil and in circulation in various plant tissues.

Uptake: The uptake of divalent cations like Fe^{+2} and Zn^{+2} from soil is energetically favored and does not require active transport because of the presence of a highly negative membrane potential that can maintain about ten thousand-fold concentration difference of a divalent cation across the plasma membrane (Olsen and Palmgren 2014). Iron in the rhizosphere can be present in ferrous (Fe^{+2}) and/or ferric (Fe^{+3}) forms depending on the soil type, pH, etc., and the agricultural practices. Compounds containing Fe^{+3} form have low solubility and hence less bioavailable. Therefore, in order to uptake this form of iron, two distinct strategies for iron uptake in plants have been proposed viz. Strategy I (also known as reducing strategy, involving acidification, reduction of Fe^{+3} to Fe^{+2}, and uptake of Fe^{+2} through IRT1) and Strategy II (also known as chelating strategy, involving secretion of chelators and uptake of chelated Fe^{+3} though specific transporters) (Fig. 1). The reducing strategy was thought to predominate in all plants except for the Poaceae family while, chelating strategy being followed in Poaceae members including barley, rice and maize (Connorton et al. 2017). However, under iron deficiency and alkaline conditions secretions of metabolites like chelators particularly coumarin-derived phenolics in Arabidopsis and flavins in Medicago facilitating Fe mobilization through IRT1/FRO have recently been reported (Fourcroy et al. 2016; Rodríguez-Celma et al. 2013; Schmid et al. 2014; Siso-Terraza et al. 2016). Additionally, in rice, under iron deficiency conditions expression of two functional Fe^{2+} transporters, OsIRT1 and OsIRT2, in roots was observed, and it was suggested that this could be an adaptive mechanism of rice under flooding conditions where Fe^{+2} is more abundant that Fe^{+3} (Ishimaru et al. 2006; Ricachenevsky and Sperotto 2014; Walker and Connolly 2008). Zinc is taken from soil solution primarily as Zn^{+2} or as Zn-phytosiderophore complexes (Broadley et al. 2007). Zinc competes with Fe for transport (Rosen et al. 1977). In Arabidopsis, IRT-1 is a multi-cation transporter that can transport Fe^{+2}, Zn^{+2}, Mn^{+2}, $Co^{+}2$, and Cd^{+2} (Korshunova et al. 1999). Other members the ZIP family (ZRT-IRT- like Protein) of which IRT-1(Iron Regulated Transporter) is also a member are also implicated in Zn transport; however, their role in uptake from soil to root cells is not clear (Milner et al. 2013).

Radial translocation in root: Once taken up by root epidermal cells, both Fe and Zn have to be translocated radially till xylem loading. However both the metals can be toxic, and moreover Fe is also redox active, and therefore, to ensure proper homeostasis under the cellular conditions of near-neutral pH, these metals bind to numerous organic molecules including proteins, glutathione, phytochelatins, and nicotinamines (White and Broadley 2011). Such Fe- Zn- complexes can

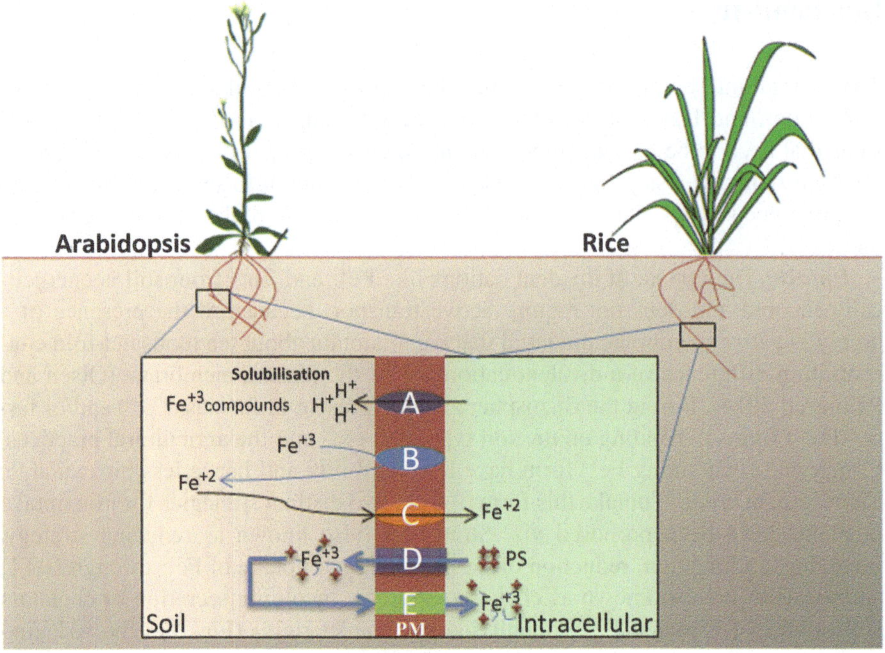

Fig. 1 Overview of strategies for Fe uptake in Arabidopsis and rice across plasma membrane (PM). Classically Strategy I is used by all plants except those of Poaceae family and comprises of three components i. Increasing solubility of Fe^{+3} though exudation of protons (A: AHA2 proton pump in Arabidopsis, Santi and Schmidt 2009) ii. Reduction of Fe^{+3} to Fe^{+2} by a membrane-bound reductase (B: FRO2, Ferric reduction oxidase 2 in Arabidopsis, Robinson et al. 1999) iii. Transport of Fe^{+2} across the membrane through membrane transporter (C: iron regulated transporter 1, IRT1, Eide et al. 1996). Strategy II is predominantly used by Poaceae members, which includes i. synthesis of small molecular weight compounds of the mugineic acid family called phytosiderophores (PS) that are released into the rhizosphere though specific transporters (D: TOM1/OsZIFL4, transporter of mugineic acid family of phytosiderophores in rice and barley (Nozoye et al. 2011, Ricachenevsky et al. 2011) ii. importing Fe^{+3} –PS chelates though membrane transporters (E: oligopeptide transporter YS1 in maize (Curie et al. 2001), YSL15 in rice (Inoue et al. 2009)

symplastically diffuse from cell to cell connected through plasmodesmata till they are loaded in the xylem vessels. These complexes can also travel apoplastically till they reach casparian strip barrier that surrounds endodermis in dicots (Naseer et al. 2012) and additionally in exodermis above cortex and below endodermis (Olsen and Palmgren 2014). Here, the Fe or Zn complexes have to enter the cells and get transported symplastically. Interestingly, the endodermal suberization is regulated via ethylene signalling with respect to nutrient status, with Fe- and Zn deficient conditions delaying suberization (Barberon et al. 2016). The homeostasis for these metals in root symplast is maintained though their sequestration and release in the vacuoles. Many tonoplast-located transporters are implicated in this process (Peng and Gong 2014). The import in the vacuole requires an active transport while the export from vacuole is energetically favored due to substantial membrane potential gradients (Olsen and Palmgren 2014).

Xylem loading: Once Fe- or Zn complexes have passed through the endodermis, they have to be loaded into the xylem for transport to the shoot. This is carried out by pericycle cells that are present inside the endodermis. The complexes have to be actively transported from the symplast to apoplastic spaces of the dead cells of xylem vessels. In xylem sap, iron was found bound largely to citrate (Fe^{+3}-citrate) (around 65%) and slightly to deoxymugineic acid (DMA) (around 5%) and some as free ions (Yoneyama et al. 2015). Therefore, the iron that is taken up mostly as Fe^{+2} must also be oxidized to Fe^{+3} (Connorton et al. 2017).

Plasma membrane localized Fe-efflux transporters IRON REGULATED1 (IREG1/FPN1) and IREG2/FPN2 are two orthologs of Ferroportin (iron efflux transporter in animals) that are expressed in Arabidopsis stele and are implicated in xylem loading (Morissey et al. 2009). In addition, YSL2 is implicated in translocation of Fe-NA complexes in the vasculature in Arabidopsis (DiDonato et al. 2004) while other members of YSL family translocate Fe-Phytosiderophore (PS) complexes in cereals like barley, maize, and rice (Pinto and Ferreira 2015). In xylem sap zinc was found as free cation or partially bound to unidentified chelators (Yoneyama et al. 2015). HMA2 and HMA4 are primary active zinc pumps involved in the efflux of Zn from root symplast to apoplast for xylem loading (Hussain et al. 2004; Sinclair et al. 2007). Additionally, plant cadmium resistance 2 (PCR2), a Zn-efflux transporter, mediates detoxification of Zn in the root epidermis at high Zn concentrations and radial transport of Zn from the epidermis to the xylem parenchyma under low Zn concentrations (Song et al. 2010).

Xylem unloading: Leaf is an important sink tissue where Fe and Zn are required for many enzymatic activities. Transpiration pull can carry micronutrients to the leaf xylem parenchyma from where the Fe and/or Zn complexes enter leaf symplast (as sink). The transport proteins involved in zinc transport for this step are not yet identified; however, members of the IRT and ZIP family are implicated (Pinto and Ferreira 2015). Similarly, mechanisms for acquisition of Fe from xylem to leaf symplast are not clear, but the components of Fe uptake by strategy I (as described above) are implicated that includes genes coding for IRT1 and FRO (Kim and Guerinot 2007). The Fe which is mostly in the form of Fe^{+3} in the xylem is reduced to Fe^{+2} and reduction of Fe^{+3} is a prerequisite for Fe uptake into leaf cells (Nikolic and Römheld 1999).

Phloem loading: In the leaf symplast, the ions are either assimilated by participating in physiological reactions in the cytosol as well as cellular organelles like chloroplast and mitochondria, or if in excess have to be stored in vacuoles (Pinto and Ferreira 2015; Sinclair and Krämer 2012). Thus, for remobilization from leaf tissues, Zn and/or Fe should first be transported out of chloroplastic, mitochondrial, or vacuolar membranes. The transporters involved in transporting Fe out of chloroplastic and mitochondrial membranes are not yet identified, although, AtFRO7 localized in chloroplast envelope and AtFRO3 and AtFRO8 localized in mitochondrial membrane are implicated in maintaining Fe homeostasis in these organelles (Jeong and Connolly 2009), AtNRAMP3 and AtNRAMP4are Fe efflux transporters present in vacuolar membranes and are essential for seed germination (Lanquar et al. 2005). Similarly for zinc, AtHMA1 localizes to the chloroplastic membrane and are involved in export of zinc (Kim et al. 2009).

In order to translocate Fe and/or Zn to seed, they have to be loaded into the phloem from leaf tissue. This requires transportation of these minerals from leaf mesophyll cells to leaf apoplastic spaces followed by loading into phloem tissue (Olsen and Palmgren 2014). How Fe and/or Zn are exported out of cytoplasm and loaded into phloem is not known as no plasma membrane bound Fe and/or Zn efflux transporter have yet been characterised. However, IREG2/FPN2, expressed in Arabidopsis stele has been implicated in vasculature loading of Fe (Morissey et al. 2009). Yellow Stripe-Like (YSL) family proteins are implicated in transport of Zn-NA complexes in phloem (Olsen and Palmgren 2014).

The pH of phloem is slightly alkaline and in phloem of Arabidopsis, tomato, and rice, Zn is translocated bound to nicotinamine (Zn-NA). Fe is bound to nicotinamine in Arabidopsis phloem, while in rice phloem Fe-2′-deoxymugineic acid complex has been identified as major species (Ling et al. 1999; Nishiyama et al. 2012; White and Broadley 2011).

Transport into seed: The minerals in seed can accumulate via root uptake and supply to developing seed or stored minerals in leaves get remobilized during senescence. A mineral partitioning study in peas indicated that both continued uptake and translocation (via xylem) and remobilization of stored minerals (via phloem) occurs (Grillet et al. 2014; Sankaran and Grusak 2014). Seed comprises of two types of tissues: maternal (seed coat) and filial (embryo and endosperm). Legumes including chickpea have exalbuminous seed that contain very little or no endosperm in the mature seed, as it is fully consumed during seed development (Wood et al. 2011; Yan et al. 2014). While the seed coat and pod wall both have a vascularised structure of mother plant, developing embryo has no direct symplastic connections to these mother tissue; thus water and nutrient transport to the embryo has to occur via seed apoplast (Shackel and Turner 2000). In chickpea, in addition to this symplastic isolation of embryo, there is an additional barrier to water exchange between pod wall and seed coat implying that phloem and/or apoplastic pathway may be a major route for uptake in seeds (Shackel and Turner 2000). The mechanisms of transport of Fe and Zn from symplast to seed apoplast surrounding the seed embryo are not well characterized, yet an active transport system is required (Olsen and Palmgren 2014). Recently, two Zn transporting P1B-ATPases actively exporting Zn from the seed coat (tissue from mother plant) to the filial tissues were identified. Mutant plants that lack both Zn pumps accumulated Zn in the seed coat and consequently have vastly reduced amounts of Zn inside the seed (Olsen et al. 2016). In dicot seeds, iron is stored within plastids bound to ferritin that accounted for upto 90% of iron in peas and other legumes (Marentes and Grusak 1998). However in Arabidopsis and cereals, ferritin does not constitute the major iron pool (Ravet et al. 2009). In Arabidopsis, Fe is primarily sequestered in vacuoles of endodermal cells surrounding the vasculature that account for about 50% of Fe in seed (Grillet et al. 2014; Kim et al. 2006; Schnell Ramos et al. 2013). In contrast to the ferritin-bound Fe in legume that is bioavailable (Hallberg 2001), the vacuolar globoids localized Fe and Zn may be sequestered as phytate-complexes and thus reducing their bioavailability (Lanquar et al. 2005; Otegui et al. 2002). A few genes have been implicated in Fe translocation to seeds. METAL TOLERANCE PROTEIN8 (MTP8), which is expressed in

Arabidopsis embryo was recently found to be involved in Fe and manganese (Mn) accumulation in a cell-specific manner (Chu et al. 2017; Eroglu et al. 2017).

Genetic Variability

Biofortification of chickpeas for improvement in Fe and Zn content is a relatively new area of research; therefore, so far no targets have been set with respect to grain Fe and Zn content in this crop. To set the targets, one of the requirement is to assess the content in the available germplasm, particularly in the popular cultivars of a particular region. There have been limited studies for the evaluation of chickpea germplasm for mineral micronutrient content particularly in the popular cultivars of a region (Table 2). The locally available chickpeas were found to contain 2.0 and 2.7 mg/100 g of Zn in whole or decorticated seeds, respectively; while the Fe content

Table 2 Iron and zinc content in chickpea genotypes

Sr. No.	Reference	Number of samples (n), and location	Fe (mg/kg or ppm)	Zn (mg/kg or ppm)
1	Avancini et al. (1992)[a]	n.a.	n.a.	38.6–44.2
2	Sika et al. (1995)	n = n.a. Morocco	64.5	36.3
3	Ibáñez et al. (1998)	n = 37, Sweden	45.1 ± 3.4 (D: n-16), 44.6 ± 6.3 (K: n = 21)	35.7 ± 3.0 (D; n = 16), 35.0 ± 4.1(K; n = 21)
4	Rao and Deosthale (1981)	n = n.a.	46	61.1
5	Wang and Daun (2004)	n = n.a.	59 (46–70, D) 55 (43–76, K)	36 (28–51, D) 44 (36–56, K)
6	Hemalatha et al. (2007)	n = n.a., India	49.5 (whole) 50.5 (decorticated)	20.3 (whole) 26.8 (decorticated)
7	Grusak (2006)[b]	n = 239	42–133	45-123
8	Bueckert et al. (2011)	n = 10, Canada,	95–104 (D, n = 6) 100–108 (K, n = 4)	33 (D, n = 6) 37–41 (K, n = 4)
9	Diapari et al. (2014)	n = 94, Canada	42.8 ± 3.2–55.8 ± 0.4[c]	27.1 ± 3.5–44.5 ± 0.2[c]
10	Kahraman et al. (2015)	n = 10, Turkey, n.a.	58.9–88.36	25.46–30.47
11	Upadhyaya et al. (2016a)	n = 92, in 2 locations India	63.3 ± 13.3 (40.2–91)	46.2 ± 9.1 (26.8–61.8)

n.a. information not available
[a] as cited in Ibáñez et al. 1998
[b] as cited in White and Broadley 2009
[c] range of means over 2 different locations and 2 years

was about 5 mg/100 g in both the cases (Hemalatha et al. 2007). A set of 94 accessions (consisting of 23 *desi* and 71 *kabuli* types) grown in Canada in two different years showed an overall range of 3.81–8.64 mg/100 and 2.52–6.23 mg/100 g iron and zinc content, respectively (Diapari et al. 2014). Recently, a set of 92 accessions (consisting of 39 *desi*and 53 *kabuli* biotypes) grown in India were analyzed and were found to contain 4.02–9.1 mg/100 g and 2.68–6.18 mg/100 g of iron and zinc, respectively (Upadhyaya et al. 2016a). The iron and zinc content of chickpea cultivars grown in USA ranged from 4.6–6.7 mg/100 g and 3.7–7.4 mg/100 g respectively (Thavarajah and Thavarajah 2012). The mineral micronutrient content is highly influenced by environment, as indicated by significant G X E interactions (Diapari et al. 2014; Upadhyaya et al. 2016a). In a set of 19 genotypes, iron (79–120 mg/kg) and zinc (56–137 mg/kg) content in the leaves was found to be quite high (Ibrikci et al. 2003), yet the final concentrations in grain were lower (Table 2), possibly due to low mobility of these minerals in phloem as well as translocation to seeds. Moreover, the total mineral content of the soil as well as the phyto-available minerals and their acquisition by roots often limit their uptake by the plants (White and Broadley 2011).

Marker-Assisted Approaches

Genetic dissection of iron and/or zinc content in chickpea has been started recently, and there are only two major reports on identification of markers associated and/or linked with the trait. Using a panel of 94 chickpea accessions, marker trait association studied for 398 SNPs and 9 SNPs was found to be significantly associated with iron and/or zinc concentrations in chickpea seeds (Diapari et al. 2014). Another recent association mapping based on SNP genotyping of 92 accessions identified 16 genomic loci/genes associated (29% combined PVE) with seed-Fe and Zn concentrations, of which 11 trait-associated SNPs were validated using SNP-based high-resolution QTL maps harboring robust QTLs in eight major genomic regions of *kabuli* genome governing seed-Fe and Zn concentrations (Upadhyaya et al. 2016a). Among the SNPs significantly associated in both the studies, based on gene annotations, SNPs in two genes coding for elongations factor and IQ-domain 1-like were common; although, the physical position of these SNPs did not show any correspondence. Nevertheless, the validated SNPs have potential in marker-assisted breeding for high Fe and or zinc content in chickpea.

Target Trait: Phytic Acid Content

Phytic acid (PA) is one of the most potent chelating agents present in food crops including cereals and pulses. As compared to vegetative tissue it is accumulated about a 1000 times more in seeds during their development and in legume seeds more than 95% of phytic acid is accumulated in cotyledons (Sparvoli and Cominelli 2015). Chemically, PA is inositol hexa*kis* phosphate (IP_6 or $InsP_6$) that serves as the primary reservoir of phosphates in seeds, and is mainly stored in globoids as phytate (Otegui et al. 2002). Phosphorous, inositol, and minerals are made available to the seedlings during their germination by the action of phytases on phytates (Mullaney and Ullah 2003). Monogastric animals including humans lack phytase, because of which phytate remains undigested and most of the minerals, chelated by phytates due to its highly negative charge, remain unabsorbed (Schlemmer et al. 2009). Phytate that escapes digestion by monogastric animals is excreted in the environment, causing concerns on eutrophication and management of phosphorus for sustainable and environment-friendly agricultural production (Raboy 2009). Phytic acid is considered as an "antinutritional" compound as it limits the bioavailability of mineral micronutrients, and attempts have been made in a number of crops to reduce its levels in the seed (Cichy and Raboy 2009). However, levels of phytic acid have been found to be associated with biotic and abiotic stress tolerance in Arabidopsis (Murphy et al. 2008), mungbean (Dhole and Reddy 2016), and chickpea (Joshi-Saha and Reddy 2015). In vitro studies using Arabidopsis guard cells and studies using low phytic acid genotypes/mutants indicate that perturbations in phytic acid biosynthesis and accumulation affects stress signaling pathways as well as shows negative impact on yield and plant performance (Doria et al. 2009; Lemtiri-Chlieh et al. 2000, 2003; Meis et al. 2003; Naidoo et al. 2012b; Oltmans et al. 2005; Raboy 2009; Raboy et al. 2015). Instead of targeting to reduce phytic acid, that may compromise the plant responses to stress tolerance and performance, genetic biofortification to increase the mineral micronutrients in a moderate phytic acid background can be a better alternate to overcome the problem of mineral chelation by phytic acid. In a recent feeding trial, biofortified beans (with 1.5-fold iron than *low-phytic acid lpa* bean) with normal phytic acid content provided similar levels of bioavailable iron as that of *lpa* beans (having only 10% phytic acid as that of biofortified bean) (Petry et al. 2016).

Genetic Variability

A wide variability for phytic acid has been reported in chickpea (Table 3). However, as compared to other pulses, low to moderate levels of phytic acid have been reported in chickpeas (Chitra et al. 1995; Sathe 1996).

Biochemistry

There are several recent reviews dealing with biosynthesis of PA, its regulation and signaling (Gillaspy 2011; Joshi-Saha and Reddy 2016; Sparvoli and Cominelli 2015). Briefly, there are two broad pathways for PA biosynthesis: i. lipid dependent

Table 3 Diversity in chickpea phytic acid content

Sr. No.	Reference (N[a])	PA range (mg/g)
1.	Duhan et al. (1989) (N = 8)	7.48–8
2.	Chitra et al. (1995)	
	Desi: N = 13	7.7–12.3
	Kabuli: N = 3	5.41–11.4
3.	Bueckert et al. (2011) (N = 10)	3.8–9
4.	Thavarajah and Thavarajah (2012) (N = 10)	5.8–13.6
5.	Kaur et al. (2016) (N = 16)	14.77–28.97
6.	Dwivedi et al. (2017) (N = 52)	12.26–27.9
7.	Misra et al. (2017) (N = 83)	Year 2013: 8.81–21.97 mg/g Year 2014: 10.47–20.5 mg/g

[a]*N* number of genotypes

(presumably ubiquitous in all the eukaryotes, involves formation of phosphatidylinositol (PtdIns) – a lipid with inositol headgroup- its sequential phosphorylation and generation of inositol triphosphate (Ins (1,4,5)P3) and subsequent phosphorylations to generate PA) and ii. lipid independent (present in slime molds and plants with no involvement of lipid derivatives, but occurs via sequential phosphorylation of inositol ring to $InsP_6$ (Brearley and Hanke 1996, Stephens and Irvine 1990)). Irrespective of the pathway, the biosynthesis of PA starts with the conversion of glucose-6-phosphate to *myo*-inositol in two steps by the sequential action of D-*myo*-inositol 3-phosphate synthase (MIPS) and inositol monophosphatase (IMP). MIPS is a highly conserved enzyme that catalyzes the rate-limiting step of the de novo pathway that provides inositol ring in all organisms (Geiger and Jin 2006). IMP belongs to the superfamily of metal-dependent phosphatase and catalyzes the dephosphorylation of *myo*-inositol 1-phosphate (Atack et al. 1995). They have broad substrate specificity, at least in soybean and chickpea, with one of the substrate being PA itself (Islas-Flores and Villanueva 2007; Saxena et al. 2013). The significance of this finding is yet not clear; however, increased expression of IMP and increased PA content in seeds has recently been found to be associated with drought tolerance in chickpea (Joshi-Saha and Reddy 2015).

Marker-Assisted Approaches

Many efforts are being made in cereals, especially maize, to identify markers for marker-assisted selection of low phytic acid content (Naidoo et al. 2012a; Sureshkumar et al. 2014). Among grain legumes, efforts for genetic analysis and identification of markers linked to phytic acid content have been reported in common bean (Blair et al. 2012), mungbean (Sompong et al. 2012), and pea (Shunmugam et al. 2015). There are very limited reports on genetic analysis of phytic acid content in chickpea. A SSR marker present in the promoter region of IMP gene was found to be associated with phytic acid content in chickpea (Dwivedi et al. 2017; Joshi-Saha and Reddy 2015). In addition, polygenic inheritance and identification of transgressive segregants for phytic acid content was reported in chickpea recently (Misra et al. 2017).

Target Trait: Raffinose Family of Oligosaccharides

Raffinose family of oligosaccharides (RFOs) is a group of soluble, non reducing sugars with α 1–6 glycosidic bonds, accumulated in seeds during their development. They are galactosyl derivatives of sucrose, formed due to the action of galactosyl-transferases that add galactose moieties from galactinol to sucrose (Peterbauer and Richter 2001). They include raffinose (the first member, predominantly present in monocots) followed by stachyose(tetrasaccharide) and verbascose (pentasaccharide), the latter two being predominantly present in dicots (Peterbauer and Richter 2001), They are categorized as antinutrient because monogastric animals lack α-glycosidase that hydrolyses RFOs. The undigested RFOs in the gastrointestinal tract are acted upon by microflora of the large intestine to produce gases like Carbon dioxide, hydrogen and methane to cause flatulence and discomfort (Naczk et al. 1997). They are important as storage carbohydrates in seed physiology and play a role in early stages of seed germination as a form of energy source and for desiccation tolerance and longevity of seeds (Koster and Leopold 1988; Bentsink et al. 2000). In addition, they also play an important role in stress tolerance (Egert et al. 2013, Gangl and Tenhaken 2016). In chickpea, galactinol synthase activity as well as galactinol and raffinose content, with potential role in maintaining seed vigor and longevity, progressively increase during seed development and gradually declines as seed germinates (Salvi et al. 2016). Recently in Arabidopsis total RFOs, RFO/sucrose ratio, were found to positively correlated while, absolute individual RFO amounts were not correlated to seed vigor. In the same study, distinct requirements for RFOs in modulating seed vigor in a monocot and a dicot were identified (Li et al. 2017).

Genetic Variability

RFO content has extensively been studied in a wide collection of chickpeas germplasm accessions (Table 4). Stachyose was found to be the major RFO followed by raffinose and verbascose (Gangola et al. 2012). The RFO content under controlled conditions was found to be lower as compared to field grown *desi* and *kabuli* chickpea supporting their role in abiotic stress tolerance (Table 4, Gangola et al. 2013).

In addition to RFOs, chickpea also contains another class of α-galactosides called galactosyl cyclitols of which ciceritol-a trisaccharide- is an important member (Quemener and Brillouet 1983). The hydrolysis of ciceritol by α-galactosidase

Table 4 Variability for RFOs in chickpea

Sr. No.	Reference (N*)	Sucrose	Total RFOs	Raffinose	Stacchyose	Verbascose	Ciceritol
1.	Mulimani et al. (1997) (N=11)	0.76–0.88 g/100g	–	0.4–1.18 g/100g	0.49–1.58 g/100g (Stachyose+ verbascose)		–
2.	Wang and Daun (2004)	3.10–4.41 g/100g (K) 1.56–2.85 g/100g (D)	–	0.48–0.73 g/100g (K) 0.46–0.77 g/100g(D)	1.76–2.72 g/100g (K) 1.25–1.98 g/100g (D)	–	–
3.	Han and Baik (2006) (N=1, cv. Dwelly)	–	–	50.2 mg/g	27 mg/g	nd	67.7 mg/g
4.	Aguilera et al. (2009) (N=1)	15.2 g/kg	48.5 g/kg	3.2 g/kg	17.7 g/kg	–	26.7g/kg
5.	Xiaoli et al. (2008) (N=19)	1.8–4.9% dry matter (w/w)	[a]6.35–8.68% dry matter (w/w)	0.46–0.91% dry matter (w/w)	0.64–3.09% dry matter (w/w)	0.27–0.70% dry matter (w/w)	3.04–5.06% dry matter (w/w)
6.	Gangola et al. (2012) (N=152)	–	1.58–4.67 mmole/100g	–	–	–	–

7.	Gangola et al. (2013) (N=171)	0.60–3.59 g/100 g	1.58–5.31 mmoles/100g (D)	D GH: 0.27–0.95 g/100g F2009: 0.09–1.1g/100g F2010: 0.4–1.19g/100g	D GH: 0.43–1.86 g/100g F2009: 0.18–2.36g/100g F2010: 0.78–1.99g/100g	D GH:0.01–0.11 g/100g F2009: 0.02–0.11/100g F2010: 0.01–0.13/100g	–
			2.11–5.83 mmoles/100g (K)	K GH: 0.27–0.95 g/100g F2009: 0.69–1.17g/100g F2010: 0.58–1.08g/100g	K GH: 0.40–1.65 g/100g F2009: 1.31–2.38g/100g F2010: 1.06–2.17g/100g	K GH: 0.01–1.11 g/100g F2009: 0.05–0.13g/100g F2010: 0.04–0.12g/100g	
8.	Konsam et al. (2014) (N=50)	–	–	2.94–26.04[b]	7.44–53.96[b]	–	–
9.	Ramadoss et al. (2015) (N=213)	3.57–54.12mg/g	[c]15.53–177.34 mg/g	0.16–15.13mg/g	2.77–59.43mg/g	–	4.36–90.65mg/g

**N* number of genotypes, *K Kabuli*, *D Desi*, *nd* not detected, *a* total α-galactosides, *b* units not defined, *c* total sugars action is limited due to the occurrence of a pinitol moiety (Quemener and Brillouet 1983). Moreover, a recent study has demonstrated that ciceritol can be a potential prebiotic (Zhang et al. 2017).

Biochemistry

The metabolism of RFOs have extensively been reviewed elsewhere (Peterbauer and Richter 2001). Here we overview the major pathways and recent information pertaining to legumes, particularly chickpea. The biosynthesis of RFOs occurs with the transfer of galactosyl moiety from a donor primarily galactinol and RFOs themselves in certain plants (Bachmann et al. 1994; Gilbert et al. 1997) to an acceptor (sucrose, raffinose or its higher homologs). Synthesis of galactinol by the action of galactinol synthase (GolS) is the first committed step of the RFO biosynthetic pathway, where GolS transfers the galactosyl residue from UDP-D-galactose (obtained from primary metabolism) to *myo*-inositol leading to the formation of galactinol (Sprenger and Keller 2000). The reason for UDP-D-galactose not directly acting as galactosyl donor to sucrose is possibly to separate the storage mechanism of metabolites from primary carbohydrate metabolism (Peterbauer and Richter 2001). The first member of the series raffinose (trisaccharide) is synthesized by the action of raffinose synthase on sucrose (as galactosyl acceptor) and galactinol (as galactosyl donor), while raffinose is converted to stachyose (tetrasaccharide) by the action of stachyose synthase. Sucrose synthase or a similar enzyme carries out the further addition of galactosyl residue to stachyose leading to the formation of verbascose (pentasaccharide) (Sengupta et al. 2015). Comparison of high- and low-RFO-containing genotypes of chickpea revealed an increased accumulation of *myo*-inositol and sucrose during early seed development and 2–three-fold higher activities of all the RFO biosynthetic enzymes in genotypes containing high RFOs (Gangola et al. 2016). Inositol metabolism is known to affect the RFO pathway and levels of inositol, rather than GolS activity shows a correlation with RFO and galactinol accumulation (Karner et al. 2004). Similar observations were made in chickpea where accumulation of RFOs occurs before accumulation of phytic acid (Zhawar et al. 2011). Galactinol can also glycosylate cyclitols like ononitol and pinitol to glycosyl ononitol and galactopinitol A, respectively, that can further act as galactosyl donors for biosynthesis of stachyose (Richter et al. 2000).

Marker-Assisted Approaches

The biochemical pathway for RFO synthesis is well known, yet there are very limited studies on developing molecular markers for low RFO content. Alleles of raffinose synthase gene (RS2) for reduced raffinose and stachyose content in soybean have been identified and allele-specific markers have been developed (Yang et al.

2014). Chickpea genotypes with varying RFO content have been grouped in different clusters-based genetic analyses using SSR markers (Konsam et al. 2014).

Future Outlook

Chickpea is a rich source of proteins, carbohydrates as well as dietary fibers. A multidisciplinary and multifaceted approach is needed to popularize this grain legume as a functional food among the consumers. On one hand, studies have to be focused on the distribution of various nutrients, their bioavailability and enhancement of their content through conventional or biotechnological tools. At the same time, nutritionists/food technologists have to develop various products to suit the consumer preferences with minimum loss of nutrition. Efforts are underway to improve the nutritional quality particularly mineral micronutrient content in chickpea. Breeding for enhancing nutrient quality of food crops, genetic biofortification, has been suggested as a relatively cost-effective, sustainable, and long-term means for providing micronutrients to the individuals increasing the daily adequacy throughout life span (Bouis et al. 2011; Saltzman et al. 2013). Following aspects should be considered for genetic biofortification:

1. *Setting the target level of the micronutrient for breeding:* The harvest plus program that is a part of CGIAR Research Program on Agriculture for Nutrition and Health has set the targets for iron and zinc in some of their mandate crops, for example, increasing iron to 94 ppm from baseline of 50 ppm in bean, and to 77 ppm from baseline of 47 ppm in pearl millet. Target for zinc have been set only in rice (28 ppm from baseline of 16 ppm) and wheat (37 ppm from baseline of 25 ppm) (Bouis and Saltzman 2017). Similar targets have not yet been set for chickpea. Although the target levels have to be set based on the average quantity consumed on a daily basis. However, considering bean as reference, some of the chickpea germplasm accessions contain this level of iron and zinc (Table 2), recently chickpea cultivars having both iron and zinc in high amounts have been identified (Joshi-Saha and Reddy 2014).
2. *Influence of environment and mapping of soil nutrient status in the target area*: All the quality traits are influenced by environment. Therefore, to genetically enhance the quality parameter, there is also a need to ensure an adequate supply of inorganic components or raw materials to the plants. For example, genetic increase in protein nitrogen in legumes requires an adequate supply of overall nitrogen to the plants (Bhatia 1983). Since legumes also fix nitrogen symbiotically, therefore, by increasing the nitrogen fixation through rhizobium–legume symbiosis or by efficient mobilization of leaf nitrogen to developing seed this can be improved. Notable here, are the recent efforts made in increasing the efficiency of plant N usage by over-expressing AMINO ACID PERMEASE1 (AAP1) in the phloem and embryos of pea plants. Such plants showed improved source-to-sink allocation of amino acids and led to increased seed yield and seed

storage protein levels when the plants were grown with highly abundant N nutrition (Zhang et al. 2015). Such plants also performed better in terms of seed yield in various N regimes (low, moderate and high) indicating that manipulation of source-to-sink N transport is a promising approach for increasing plant productivity while optimizing N usage (Perchlik and Tegeder 2017). Similarly, in case of genetic biofortification of minerals, there is a need to ascertain the soil mineral levels in the target environment. Analysis of soil and plant samples has indicated that 49% and 12% of soils in India are potentially deficient in Zn and Fe, respectively (Singh 2008). Therefore, suitable agronomic practices have to be recommended to the farmers such that adequate Zn and Fe supply is ensured to optimize the potential of genetically biofortified cultivars being developed for the target environments.

3. *Assessment of biofortified crops*: The impact of recently developed biofortified crops in providing the mineral micronutrient also need to be assessed in population trials. Such studies have been conducted for iron biofortified beans and cereals like iron biofortified pearl millet, zinc biofortified wheat etc. and have shown promising results (Bouis and Saltzman 2017; De Moura et al. 2014). In a recent study, Fe absorbed from biofortified beans (BB) was almost 50% more than that of control beans (CB) (Petry et al. 2016); however, previous study in the same study population attributed either no increase (Petry et al. 2012) or only moderate (around 19%) increase (Petry et al. 2014) in absorbed Fe from BB as compared to CB. This was attributed to differences in the PA contents of the cultivars tested, with higher PA contents reducing Fe bioavailability (Petry et al. 2016). However, low phytic acid beans (lpa) did not show any significant difference in absorbed Fe as compared to BB despite a significantly lower PA as compared to BB, making this study inconclusive about the role of lpa beans (Petry et al. 2016). Also, there is a need to develop new robust trials and more sensitive assay methods for evaluation of biofortified crops (Bouis and Saltzman 2017).

In addition, reduction in antinutrients, particularly phytic acid, has been considered as a tool to enhance the bioavailability of mineral micronutrients. However, recent studies indicate the involvement of PA in biotic and abiotic stress tolerance, therefore, instead of breeding for low PA, genetic biofortification with moderate PA can be a better breeding target. Also, low phytic acid bean were difficult to cook and cause adverse gastrointestinal symptoms (Petry et al. 2016). More over many of the food processing treatments have found to be effective in reducing the antinutrients, particularly PA and RFOs (Alajaji and El-Adawy 2006; Girigowda et al. 2005; Han and Baik 2006).

Moreover, despite massive efforts for their characterization, the large germplasm collections of chickpea are underutilized (Upadhyaya et al. 2011). In addition to the use of germplasm, wild species can also be involved in prebreeding for quality improvement. However, the major reason for nonpreference of such germplasms by breeders is apprehensions regarding their adaptability and association of linkage drag (Upadhyaya 2015). In such context, mutation breeding can prove to be an important supplementary breeding activity to isolate mutants with enhanced quality (high methionine/high Fe and/or Zn, or low antinutritional compounds). Such quality mutants have been isolated in cereals, for example, high lysine-containing maize (Arruda et al. 1978), barley (Eggum 1978) and rice (Kumamaru et al. 1997). Although there are several reports on the pleiotropic effects of such mutations affecting the quality and/or grain yield, there is a possibility to improve these aspects through selection (Oram and Doll 1981; Sarika et al. 2018; Zhang et al. 2016). These mutants isolated in a background of agronomically important cultivars can be directly useful in breeding programs.

References

Abbaspour N, Hurrell R, Kelishadi R (2014) Review on iron and its importance for human health. J Res Med Sci 19:164–174

Aguilera Y, Martín-Cabrejas MA, Benítez V et al (2009) Changes in carbohydrate fraction during dehydration process of common legumes. J Food Compos Anal 22:678–683

Alajaji SA, El-Adawy TA (2006) Nutritional composition of chickpea (*Cicer arietinum* L.) as affected by microwave cooking and other traditional cooking methods. J Food Compos Anal 19:806–812

Altenbach SB, Pearson KW, Meeker G et al (1989) Enhancement of methionine content of seed proteins by the expression of a chimeric gene encoding a methionine-rich protein in transgenic plants. Plant Mol Biol 11:717–739

Arruda P, da Silva WJ, Teixeira JPF (1978) Protein and free amino acids in a high lysine maize double mutant. Phytochemistry 17:1217–1218

Atack JR, Broughton HB, Pollack S (1995) Structure and mechanism of inositol monophosphatase. FEBS Lett 361:1–7

Avancini SRP, Sales AM, Aguirre JM et al (1992) Chemical composition and nutritional value of chickpea cultivars grown in the state of Sao Paulo. Coletanea do Instituto de Tecnologia de Alimentos 22:145–153

Bachmann M, Matile P, Keller F (1994) Metabolism of the Raffinose family oligosaccharides in leaves of *Ajuga reptans* L. (Cold Acclimation, translocation, and sink to source transition: discovery of chain elongation enzyme). Plant Physiol 105:1335–1345

Barberon M, Vermeer JEM, De Bellis D et al (2016) Adaptation of root function by nutrient-induced plasticity of endodermal differentiation. Cell 164:447–459

Bentsink L, Alonso-Blanco C, Vreugdenhil D et al (2000) Genetic analysis of seed-soluble oligosaccharides in relation to seed storability of Arabidopsis. Plant Physiol 124:1595–1604

Bhatia CR (1983) Possibilities of seed protein improvement in tropical and sub-tropical legumes. In: Gottschalk W, Müller HP (eds) Seed Proteins, Advances in agricultural biotechnology, vol 2. Springer, Dordrecht

Blair MW, Herrera AL, Sandoval TA et al (2012) Inheritance of seed phytate and phosphorus levels in common bean (*Phaseolus vulgaris* L.) and association with newly-mapped candidate genes. Mol Breed 30:1265–1277

Bouis HE, Hotz C, Mc Clafferty B et al (2011) Biofortification: a new tool to reduce micronutrient malnutrition. Food Nutr Bull 32:31S–40S

Bouis HE, Saltzman A (2017) Improving nutrition through biofortification: a review of evidence from HarvestPlus, 2003 through 2016. Glob Food Sec 12:49–58

Boye J, Wijesinha-Bettoni R, Burlingame B (2012) Protein quality evaluation twenty years after the introduction of the protein digestibility corrected amino acid score method. Br J Nutr 108:S183–S211

Brearley CA, Hanke DE (1996) Metabolic evidence for the order of addition of individual phosphate esters in the *myo*-inositol moiety of inositol hexaphosphate in the duckweed *Spirodela polyrhiza* L. Biochem J 314:227–233

Broadley MR, White PJ, Hammond JP et al (2007) Zinc in plants. New Phytol:677–702

Bueckert RA, Thavarajah D, Thavarajah P et al (2011) Phytic acid and mineral micronutrients in field-grown chickpea (*Cicer arietinum* L.) cultivars from western Canada. Eur Food Res Technol 233:203–212

Casey R (1999) Distribution and some properties of seed globulins. In: Shewry PR, Casey R (eds) Seed proteins. Springer, Dordrecht, pp 159–169

Chang Y-W, Alli I, Molina AT et al (2012) Isolation and characterization of chickpea (*Cicer arietinum* L.) seed protein fractions. Food Bioprocess Technol 5:618–625

Chitra U, Vimala V, Singh U et al (1995) Variability in phytic acid content and protein digestibility of grain legumes. Plant Foods Hum Nutr 47:163–172

Chu HH, Car S, Socha AL et al (2017) The Arabidopsis MTP8 transporter determines the localization of manganese and iron in seeds. Sci Rep 7:11024

Cichy K, Raboy V (2009) Evaluation and development of low phytate crops. In: Modification of seed composition to promote health and nutrition, Agronomy monograph 51

Connorton JM, Balk J, Rodriguez-Celma J (2017) Iron homeostasis in plants- a brief overview. Metallomics 9:813–823

Cronk Q, Ojeda I, Pennington RT (2006) Legume comparative genomics: progress in phylogenetics and phylogenomics. Curr Opin Plant Biol 9:99–103

Curie C, Panaviene Z, Loulergue C et al (2001) Maize yellow stripe1 encodes a membrane protein directly involved in Fe(III) uptake. Nature 409:346–349

da Silva MA, Neves VA, Lourenço EJ (2001) Protein fractions and major globulin from chickpea (*Cicer arietinum* L.). Alimentose Nutriçåo 12:131–149

De Moura FF, Palmer AC, Finkelstein JL et al (2014) Are biofortified staple food crops improving vitamin a and Iron status in women and children? New evidence from efficacy trials. Adv Nutr 5:568–570

Dhawan K, Malhotra S, Dahiya BS et al (1991) Seed protein fractions and aminoacid composition in gram (*Cicer arietinum*). Plant Foods Hum Nutr 41:225–232

Dhole VJ, Reddy KS (2016) Association of phytic acid content with biotic stress tolerance in mungbean (*Vigna radiata* L. Wilezek). Phytoparasitica. https://doi.org/10.1007/s12600-016-0514-5

Diapari M, Sindhu A, Bett K et al (2014) Genetic diversity and association mapping of iron and zinc concentrations in chickpea (*Cicer arietinum* L.). Genome 57:459–468

DiDonato RJ, Roberts A, Sanderson T et al (2004) Arabidopsis *Yellow Stripe-Like2* (*YSL2*): a metal-regulated gene encoding a plasma membrane transporter of nicotinamine-metal complexes. Plant J 39:403–414

Doria E, Galleschi L, Calucci L et al (2009) Phytic acid prevents oxidative stress in seeds: evidence from a maize (*Zea mays* L.) low phytic acid mutant. J Exp Bot 60:967–978

Duhan A, Chauhan BM, Punia B et al (1989) Phytic Acid content of chickpea (*Cicer arietinum*) and black gram (*Vigna mungo*): varietal differences and effect of domestic processing and cooking methods. J Sci Food Agric 49:449–455

Duranti M (2006) Grain legume proteins and nutraceutical properties. Fitoterapia 77:67–82

Dwivedi V, Parida SK, Chattopadhyay D (2017) A repeat length variation in myo-inositol monophosphatase gene contributes to seed size trait in chickpea. Sci Rep 7:4764

Egert A, Keller F, Peters S (2013) Abiotic stress-induced accumulation of raffinose in Arabidopsis leaves is mediated by a single raffinose synthase (RS5, At5g40390). BMC Plant Biol 13:218

Eggum BO (1978) Protein quality of induced high lysine mutants in barley. Adv Exp Med Biol 105:317–341

Eide D, Broderius M, Fett J et al (1996) A novel iron-regulated metal transporter from plants identified by functional expression in yeast. Proc Natl Acad Sci U S A 93:5624–5628

Eroglu S, Giehl RFH, Meier B et al (2017) Metal Tolerance Protein 8 Mediates Manganese homeostasis and iron reallocation during seed development and germination. 174: 1633–1647

Fourcroy P, Tissot N, Gaymard F et al (2016) Facilitated Fe nutrition by phenolic compounds excreted by the Arabidopsis ABCG37/PDR9 transporter requires the IRT1/FRO2 high-affinity root Fe^{+2} transport systems. Mol Plant 9:485–488

Gangl R, Tenhaken R (2016) Raffinose family oligosaccharides act as galactose stores in seeds and are required for rapid germination of Arabidopsis in the dark. Front Plant Sci 7:1115. https://doi.org/10.3389/fpls.2016.01115

Gangola MP, Khedikar YP, Gaur PM et al (2012) Variation in important seed constituents among various chickpea genotypes. Qual Assur Safe Crops Foods 4:139. https://doi.org/10.1111/j.1757-837X.2012.00143.x

Gangola MP, Khedikar YP, Gaur PM et al (2013) Genotype and growing environment interaction shows a positive correlation between substrates of raffinose family oligosaccharides (RFO) biosynthesis and their accumulation in chickpea (*Cicer arietinum* L.) seeds. J Agric Food Chem 61:4943–4952

Gangola MP, Jaiswal S, Kannan U et al (2016) Galactinol synthase enzyme activity influences raffinose family oligosaccharides (RFO) accumulation in developing chickpea (*Cicer arietinum* L.) seeds. Phytochemistry 125:88–98

Geiger JH, Jin X (2006) The structure and mechanism of myo-Inositol-1-phosphate synthase. In: Majumder AL, Biswas BB (eds) Biology of inositols and phosphoinositides, Subcellular biochemistry, vol 39. Springer, Boston

Gilbert GA, Wilson C, Madore MA (1997) Root-zone salinity alters raffinose oligosaccharide metabolism and transport in *Coleus*. Plant Physiol 115:1267–1276

Gillaspy GE (2011) The cellular language of *myo*-inositol signalling. New Phytol 192:823–839

Girigowda K, Prashanth SJ, Mulimani VH (2005) Oligosaccharins of black gram (*Vigna mungo* L.) as affected by processing methods. Plant Foods Hum Nutr 60:173–180

Grillet L, Mari S, Schmidt W (2014) Iron in seeds – loading pathways and subcellular localization. Front Plant Sci 4:535

Gupta S, Nawaz K, Praveen S et al (2017) Draft genome sequence of *Cicer reticulatum* L., the wild progenitor of chickpea provides a resource for agronomic trait improvement. DNA Res 24:1–10

Hallberg L (2001) Perspectives on nutritional iron deficiency. Annu Rev Nutr 21:1–21

Han IH, Baik BK (2006) Oligosaccharide content and composition of legumes and their reduction by soaking, cooking, ultrasound, and high hydrostatic pressure. Cereal Chem 83:428–433

Hemalatha S, Platel K, Srinivasan K (2007) Zinc and iron contents and their bioaccessibility in cereals and pulses consumed in India. Food Chem 102:1328–1336

Hussain D, Haydon MJ, Wang Y et al (2004) P- type ATpase heavy metal transporters with role in essential zinc homeostasis in Arabidopsis. Plant Cell 16:1327–1339

Ibáñez MV, Rincón F, Amaro M et al (1998) Intrinsic variability of mineral composition of chickpea (*Cicer arietinum* L.). Food Chem 63:55–60

Ibrikci H, Knewtson SJB, Grusak MA (2003) Chickpea leaves as a vegetable green for humans: evaluation of mineral composition. J Sci Food Agric 83:945–950

Inoue H, Kobayashi T, Nozoye T et al (2009) Rice OsYSL15 is an iron-regulated iron(III)-deoxymugineic acid transporter expressed in the roots and is essential for iron uptake in early growth of the seedlings. J Biol Chem 284:3470–3479

Ishimaru Y, Suzuki M, Tsukamoto T et al (2006) Rice plants take up iron as an Fe^{3+}- phytosiderophore and as Fe^{2+}. Plant J 45:335–346

Islas-Flores I, Villanueva MA (2007) Inositol-1 (or 4)-monophosphatase from *Glycine max* embryo axes is a phosphatase with broad substrate specificity that include phytate dephosphorylation. Biochim Biophys Acta 1770:543–550

Jadhav AA, Rayate SJ, Mhase LB et al (2015) Marker-trait association study for protein content in chickpea (*Cicer arietinum* L.). J Genet 94:279–286

Jain M, Misra G, Patel RK et al (2013) A draft genome sequence of the pulse crop chickpea (*Cicer arietinum* L.). Plant J 74:715–729

Jeong J, Connolly EL (2009) Iron uptake mechanisms in plants: functions of the FRO family of ferric reductases. Plant Sci 176:709–714

Joshi-Saha A, Reddy KS (2014) Assessment of mineral micronutrient and phytic acid content in chickpea. In: Dixit GP, Singh J and Sing NP (eds) Pulses: challenges and opportunities under changing climatic scenario (Proceedings of the national conference on pulses: challenges and opportunities under changing climatic scenario, Sep 29 to Oct 01, 2014, JNKVV, Jabalpur), Indian Society of Pulses Research and Development, ICAR-Indian Institute of Pulses Research, Kanpur-208024, UP, India, pp 322–327

Joshi-Saha A, Reddy KS (2015) Repeat length variation in the 5′UTR of *myo*-inositol monophosphatase gene is related to phytic acid content and contributes to drought tolerance in chickpea (*Cicer arietinum* L.). J Exp Bot. 66: 5683–5690 https://doi.org/10.1093/jxb/erv156

Joshi-Saha A, Reddy KS (2016) Phytic acid boon or bane?: conflict between animal/human nutrition and plant health. J Food Legumes 29:163–173

Jukanti AK, Gaur PM, Gowda CLL et al (2012) Nutritional quality and health benefits of chickpea(*Cicer arietinum* L.): a review. Br J Nutr 108:S11–S26

Kahraman A, Ceyhan E, Harmankaya M (2015) Nutritional variation and drought tolerance in chickpeas (*Cicer arietinum* L.). J Elem 20:331–341

Kaleta C, Schäuble S, Rinas U et al (2013) Metabolic costs of amino acid and protein production in *Escherichia coli*. Biotechnol J 8:1105–1114

Karner U, Peterbauer T, Raboy V et al (2004) *myo*-Inositol and sucrose concentrations affect the accumulation of raffinose family oligosaccharides in seeds. J Exp Bot 55:1981–1987

Kaur S, Kaur S, Gupta AK et al (2016) Physiochemical and nutritional attributes of raw and soaked seeds of chickpea (*Cicer arietinum* L.) genotypes. Legume Res 39:359–369

Kerem Z, Lev-Yadun S, Gopher A et al (2007) Chickpea domestication in the Neolithic Levant through the nutritional perspective. J Archaeol Sci 34:1289–1293

Kim SA, Guerinot ML (2007) Mining iron: iron uptake and transport in plants. FEBS Lett 581:2273–2280

Kim SA, Punshon T, Lanzirotti A et al (2006) Localization of iron in Arabidopsis seed requires the vacuolar membrane transporter VIT1. Science 314:1295–1298

Kim Y-Y, Choi H, Segami S et al (2009) AtHMA1 contributes to the detoxification of excess Zn(II) in Arabidopsis. Plant J 58:737–753

Konsam S, Chellapilla B, Ram G et al (2014) Molecular diversity of chickpea (*Cicer arietinum* L.) genotypes differing in their raffinose family oligosaccharides viz., raffinose and stachyose content as revealed through SSR markers. Aust J Crop Sci 8:1175–1185

Korshunova Y, Eide D, Clark W et al (1999) The IRT1 protein from *Arabidopsis thaliana* is a metal transporter with a broad substrate range. Plant Mol Biol 40:37–44

Koster KL, Leopold AC (1988) Sugars and desiccation tolerance in seeds. Plant Physiol 88:829–832

Kumamaru T, Sato H, Satoh H (1997) High-lysine mutants of rice, *Oryza sativa* L. Plant Breed 116:245–249

Kumar J, Kant R, Quatadah SM et al (2016) Biofortification through breeding interventions in lentil. In: Singh U, Praharaj C, Singh S, Singh N (eds) Biofortification of food crops. Springer, New Delhi

Lanquar V, Lelièvre F, Bolte S et al (2005) Mobilization of vacuolar iron by AtNRAMP3 and AtNRAMP4 is essential for seed germination on low iron. EMBO J 24:4041–4051

Lemtiri-Chlieh F, MacRobbie EA, Brearley CA (2000) Inositol hexakisphosphate is a physiological signal regulating the K^+–inward rectifying conductance in guard cells. Proc Natl Acad Sci U S A 97:8687–8692

Lemtiri-Chlieh F, MacRobbie EA, Webb AA et al (2003) Inositol hexakisphosphate mobilizes an endomembrane store of calcium in guard cells. Proc Natl Acad Sci U S A 100:10091–10095

Li T, Zhang Y, Wang D (2017) Regulation of seed vigor by manipulation of raffinose family oligosaccharides in maize and *Arabidopsis thaliana*. Mol Plant 10:1540–1555

Ling HQ, Koch G, Baumlein H et al (1999) Map-based cloning of *chloronerva*, a gene involved in iron uptake of higher plants encoding nicotianamine synthase. Proc Natl Acad Sci U S A 96:7098–7103

Marentes E, Grusak MA (1998) Iron transport and storage within the seed coat and embryo of developing seeds of pea (*Pisum sativum* L.). Seed Sci Res 8:367–375

McCall KA, Huang C-C, Fierke CA (2000) Function and mechanism of zinc metalloenzymes. J Nutr 130:1437S–1446S

Meis SJ, Fehr WR, Schnebly SR (2003) Seed source effect on field emergence of soybean lines with reduced phytate and raffinose saccharides. Crop Sci 43:1336–1339

Messina MJ (1999) Legumes and soybeans: overview of their nutritional profiles and health effects. Am J Clin Nutr 70:439–450

Milner M, Seamon J, Craft E et al (2013) Transport properties of members of the ZIP family in plants and their role in Zn and Mn homeostasis. J Exp Bot 64:369–381

Misra G, Joshi Saha A, Reddy KS (2017) Distribution and genetic analysis of phytic acid content in chickpea (*Cicer arietinum* L.). In: National Symposium on pulses for nutritional security and agricultural sustainability. Dec 2–4. ICAR-Indian Institute of Pulses Research, Kanpur, India, p 211

Moreno M, Cubero JI (1978) Variation in *Cicer arietinum* L. Euphytica 27:465–485

Morissey J, Baxter IR, Lee J et al (2009) The ferroportin metal efflux protein function in iron and cobalt homeostasis in Arabidposis. Plant Cell 21:3326–3338

Mulimani VH, Thippeswamy S, Ramalingam (1997) Changes in carbohydrate content during chickpea cooking: a new approach of teaching of carbohydrate chemistry in biochemical course. Biochem Educ 25:44–45

Mullaney EJ, Ullah AHJ (2003) The term phytase comprises several different classes of enzymes. Biochem Biophys Res Commun 312:179–184

Müntz K, Christov V, Saalbach G et al (1998) Genetic engineering for high methionine grain legumes. Nahrung 42:125–127

Murphy AM, Otto B, Brearley CA et al (2008) A role for inositol hexakisphosphate in the maintenance of basal resistance to plant pathogens. Plant J 56:638–652

Muthayya S, Rah JH, Sugimoto JD et al (2013) The global hidden hunger indices and maps: an advocacy tool for action. PLoS One 8:e67860

Naczk M, Amarowicz R, Shahidi F (1997) α-Galactosides of sucrose in foods: composition, flatulence-causing effects, and removal. In: Shahidi F (ed) Antinutrients and phytochemicals in food, vol 8. American Chemical Society, Washington, DC, pp 127–151

Naidoo R, Tongoona P, Derera J et al (2012a) Combining ability of low phytic acid (*lpa1-1*) and quality protein maize (QPM) lines for seed germination and vigour under stress and non-stress conditions. Euphytica 185:529

Naidoo R, Watson GMF, Derera J et al (2012b) Marker-assisted selection for low phytic acid (*lpa1-1*) with single nucleotide polymorphism marker and amplified fragment length polymorphisms for background selection in a maize backcross breeding programme. Mol Breed 30:1207–1217

Naseer S, Lee Y, Lapierre C et al (2012) Casparian strip diffusion barrier in Arabidopsis is made of a lignin polymer without suberin. Proc Natl Acad Sci U S A 109:10101–10106

Nikolic M, Römheld V (1999) Mechanism of Fe uptake by the leaf symplast: is Fe inactivation in leaf a cause of Fe deficiency chlorosis? Plant Soil 215:229

Nishiyama R, Kato M, Nagata S, Yanagisawa S, Yoneyama T (2012) Identification of Zn-nicotianamine and Fe-2′-Deoxymugineic acid in the phloem sap from rice plants (*Oryza sativa* L.). Plant Cell Physiol 53:381–390

Nozoye T, Nagasaka S, Kobayashi T et al (2011) Phytosiderophore efflux transporters are crucial for iron acquisition in graminaceous plants. J Biol Chem 286:5446–5454

Ocampo B, Robertson LD, Singh KB (1998) Variation in seed protein content in the annual wild *Cicer* species. J Sci Food Agric 78:220–224

Olsen LI, Palmgren MG (2014) Many rivers to cross: the journey of zinc from soil to seed. Front Plant Sci 5:1–6

Olsen LI, Hansen TH, Larue C et al (2016) Mother-plant-mediated pumping of zinc into the developing seed. Nat Plants 2:16036
Oltmans SE, Fehr WR, Welke GA et al (2005) Agronomic and seed traits of soybean lines with low–phytate phosphorus. Crop Sci 45:593–598
Oram RN, Doll H (1981) Yield improvement in high lysine barley. Aust J Agric Res 32:425–434
Osborne TB (1924) The vegetable proteins. Longmans, Green, London
Otegui MS, Capp R, Staehelin LA (2002) Developing seeds of Arabidopsis store different minerals in two types of vacuoles and in the endoplasmic reticulum. Plant Cell 14:1311–1327
Pandey MK, Roorkiwal M, Singh VK et al (2016) Emerging genomic tools for legume breeding: current status and future prospects. Front Plant Sci 7:455
Peng J-S, Gong J-M (2014) Vacuolar sequestration capacity and long-distance metal transport in plants. Front Plant Sci 5:19
Perchlik M, Tegeder M (2017) Improving plant nitrogen use efficiency through alteration of amino Acid transport processes. Plant Physiol 175:235–247
Peterbauer T, Richter A (2001) Biochemistry and physiology of raffinose family oligosaccharides and galactosyl cyclitols in seeds. Seed Sci Res 11:185–197
Petry N, Egli I, Gahutu JB et al (2012) Stable iron isotope studies in Rwandese women indicate that the common bean has limited potential as a vehicle form iron biofortification. J Nutr 142:492–497
Petry N, Egli I, Gahutu JB et al (2014) Phytic acid concentration influences iron bioavailability from biofortified beans in Rwandese women with low iron status. J Nutr 144:1681–1687
Petry N, Rohner F, Gahutu JB et al (2016) In Rwandese women with low iron status, iron absorption from low-phytic acid beans and biofortified beans is comparable, but low-phytic acid beans cause adverse gastrointestinal symptoms. J Nutr. https://doi.org/10.3945/jn.115.223693
Pinto E, Ferreira IMPLVO (2015) Cation transporters/channels in plnats: tools for nutrient biofortification. J Plant Physiol 179:64–82
Quemener B, Brillouet JM (1983) Ciceritol, a pinitol digalactoside from seeds of chickpea, lentil and white lupin. Phytochemistry 22:1745–1751
Raboy V (2009) Approaches and challenges to engineering seed phytate and total phosphorus. Plant Sci 177:281–296
Raboy V, Peterson K, Jackson C (2015) A substantial fraction of barley (*Hordeum vulgare* L.) *low phytic acid* mutations have little or no effect on yield across diverse production environments. Plan Theory 4:225–239
Ramadoss BR, Balraj R, Agasimani S et al (2015) Determination of oligosaccharide fraction in a worldwide germplasm collection of chickpea (*Cicer arietinum* L.) using high performance liquid chromatography. Aust J Crop Sci 9:605–613
Rao DSS, Deosthale YG (1981) Mineral composition of four Indian food legumes. J Food Sci 46:1962–1963
Ravet K, Touraine B, Boucherez J et al (2009) Ferritins control interaction between iron homeostasis and oxidative stress in Arabidopsis. Plant J 57:400–412
Rahman S, Ahmed T, Rehman AS et al (2016) Status of zinc nutrition in Bangladesh: the underlying associations. J Nutr Sci e25 5:1–9
Ricachenevsky FK, Sperotto RA, Menguer PK et al (2011) ZINC-INDUCED FACILITATOR-LIKE family in plants: lineage-specific expansion in monocotyledons and conserved genomic and expression features among rice (*Oryza sativa*) paralogs. BMC Plant Biol 11:20. https://doi.org/10.1186/1471-2229-11-20
Ricachenevsky FK, Sperotto RA (2014) There and back again, or always there? The evolution of rice combined strategy for Fe uptake. Front Plant Sci 5:1–5
Richter A, Hoch G, Puschenreiter M et al (2000) The role of stachyose synthase in the oligosaccharide metabolism of legume seeds. In: Black M, Bradford KJ, Vázquez-Ramos J (eds) Seed biology: advances and applications. CAB International, Wallingford, pp 75–84
Robinson NJ, Procter CM, Connolly EL et al (1999) A ferric-chelate reductase for iron uptake from soils. Nature 397:694–697

Rodríguez-Celma J, Lin WD, Fu GM et al (2013) Mutually exclusive alterations in secondary metabolism are critical for the uptake of insoluble iron compounds by Arabidopsis and *Medicago truncatula*. Plant Physiol 162:1473–1485

Rosen JA, Pike CS, Golden ML (1977) Zinc, Iron, and chlorophyll metabolism in zinc-toxic corn. Plant Physiol 59:1085–1087

Saltzman A, Birol E, Bouis HE et al (2013) Biofortification: progress toward a more nourishing future. Glob Food Sec 2:9–17

Salvi P, Saxena SK, Petla BP (2016) Differentially expressed galactinol synthase(s) in chickpea are implicated in seed vigor and longevity by limiting the age induced ROS accumulation. Sci Rep 6:35088. https://doi.org/10.1038/srep35088

Sankaran RP, Grusak MA (2014) Whole shoot mineral partitioning and accumulation in pea (*Pisum sativum*). Front Plant Sci 5:149. https://doi.org/10.3389/fpls.2014.00149

Santi S, Schmidt W (2009) Dissecting iron deficiency-induced proton extrusion in Arabidopsis roots. New Phytol 183:1072–1084

Sarika K, Hossain F, Muthusamy V et al (2018) Opaque16, a high lysine and tryptophan mutant, does not influence the key physico-biochemical characteristics in maize kernel. PLoS One 13:e0190945

Sathe SK (1996) The nutritional value of selected Asiatic pulses: chickpea, black gram, mung bean and pigeon pea. In: Nwokolo E, Smartt J (eds) Food and feed from legumes and oilseeds. Springer, Boston, MA

Saxena SC, Salvi P, Kaur H et al (2013) Differentially expressed *myo*-inositol monophosphatase gene (*CaIMP*) in chickpea (*Cicer arietinum* L.) encodes a lithium-sensitive phosphatise enzyme with broad substrate specificity and improves seed germination and seedling growth under abiotic stresses. J Exp Bot 64:5623–5629

Schlemmer U, Frølich W, Prieto RF et al (2009) Phytate in foods and significance for humans: food sources, intake, processing, bioavailability, protective role and analysis. Mol Nutr Food Res 53:S330–S375

Schmid NB, Giehl RF, Döll S et al (2014) Feruloyl-CoA 6-Hydroxylase1-dependent coumarins mediate iron acquisition from alkaline substrates in Arabidopsis. Plant Physiol 164:160–172

Schnell Ramos M, Khodja H, Mary V (2013) Using μPIXE for quantitative mapping of metal concentration in *Arabidopsis thaliana* seeds. Front Plant Sci 4:168

Sengupta S, Mukherjee S, Basak P et al (2015) Significance of galactinol and raffinose family oligosaccharide synthesis in plants. Front Plant Sci 6:656. https://doi.org/10.3389/fpls.2015.00656

Shackel KA, Turner NC (2000) Seed coat turgor in chickpea is independent of changes in plant and pod water potential. J Exp Bot 51:895–900

Shahzad Z, Rouached H, Rakha A (2014) Combating mineral malnutrition through iron and zinc biofortification of cereals. Compr Rev Food Sci Food Saf 13:329–346

Sharma S, Yadav N, Singh A et al (2013) Nutritional and antinutritional profile of newly developed chickpea (*Cicer arietinum* L) varieties. Int Food Res J 20:805–810

Shen B, Li C, Tarczynski MC (2002) High free-methionine and decreased lignin content result from a mutation in the Arabidopsis S-adenosyl-L-methionine synthetase 3 gene. Plant J 29:371–380

Shewry PR, Napier JA, Tatham AS (1995) Seed storage proteins: structures and biosynthesis. Plant Cell 7:945–956

Shewry PR, Pandya MJ (1999) The 2S Albumin storage proteins. In: Shewry PR, Casey R (eds) Seed proteins. Kluwer Academic Publishers, Dordrecht, pp 563–586

Shunmugam ASK, Liu X, Stonehouse R et al (2015) Mapping seed phytic acid concentration and iron bioavailability in a pea recombinant inbred line population. Crop Sci 55:1–9

Sika M, Terrab A, Swan PB et al (1995) Composition of selected Moroccan cereals and legumes: comparison with the FAO table for use in Africa. J Food Comp Anal 8:62–70

Sinclair SA, Krämer U (2012) The zinc homeostasis network of land plants. Biochim Biophys Acta 1823:1553–1567

Sinclair SA, Sherson SM, Jarvis R et al (2007) The use of zinc fluorophore, Zinpyr-1, in the study of zinc homeostasis in Arabidopsis root. New Phytol 174:39–45

Singh MV (2008) Micronutrient deficiencies in crops and soils in India. In: Alloway BJ (ed) Micronutrient deficiencies in global crop production. Springer, Dordrecht

Singh R, Matta NK (2003) The four protein fractions in chickpea seeds (*Cicer arietinum* L.). Physiol Mol Biol Plants 9:233–237

Singh U (1985) Nutritional quality of chickpea (*Cicer arietinum* L.): current status and future research needs. Qual Plant Plant Foods Hum Nutr 35:339–351

Singh U, Kumar J, Gowda CLL (1983) The protein content of chickpea (*Cicer arietinum* L.) grown at different locations. Plant Food Hum Nutr 32:179–184

Siso-Terraza P, Luis-Villarroya A, Fourcroy P et al (2016) Accumulation and secretion of coumarin oliganans and other caumarins in *Arabidopsis thaliana* roots in response to iron deficiency at high pH. Front Plant Sci 7:1711

Sompong U, Somta P, Raboy V et al (2012) Mapping of quantitative trait loci for phytic acid and phosphorus contents in seed and seedling of mungbean (*Vigna radiata* (L.) Wilczek). Breed Sci 62:87–92

Song W-Y, Choi KS, Kim DY et al (2010) Arabidopsis PCR2 is a zinc exporter involved in both zinc extrusion and long-distance zinc transport. Plant Cell 22:2237–2252

Sparvoli F, Cominelli E (2015) Seed biofortification and phytic acid reduction: a conflict of interest for the plants? Plan Theory 4:728–755

Sprenger N, Keller F (2000) Allocation of raffinose family oligosaccharides to transport and storage pools in *Ajuga reptans*: the roles of two distinct galactinol synthases. Plant J 21:249–258

Stephens LR, Irvine RF (1990) Stepwise phosphorylation of *myo*-inositol activity in *Dictyostelium*. Biochem J 272:201–210

Sureshkumar S, Tamilkumar P, Senthil N et al (2014) Marker assisted selection of low phytic acid trait in maize (*Zea mays* L.). Hereditas 151:20–27

Tabe L, Popelka C, Chiaiese P et al (2012) Manipulating the sulfur composition of seeds. In: De Kok L et al (eds) Sulfur metabolism in plants. Proceedings of the international plant sulfur workshop, vol 1. Springer, Dordrecht

Tan GZH, Das Bhowmik SS, Hoang TML et al (2017) Finger on the pulse: pumping Iron into chickpea. Front Plant Sci 8:1755

Thavarajah D, Thavarajah P (2012) Evaluation of chickpea (*Cicer arietinum* L.) micronutrient composition: biofortification opportunities to combat global micronutrient malnutrition. Food Res Int 49:99–104

Toker C (2009) A note on the evolution of kabuli chickpeas as shown by induced mutations in *Cicer reticulatum* Ladizinsky. Genet Resour Crop Evol 56:7

Torutaeva E, Asanaliev A, Prieto-Linde M-L et al (2014) Evaluation of microsatellite-based genetic diversity, protein and mineral content in chickpea accessions grown in Kyrgyzstan. Hereditas 151:81–90

Tu HM, Godfry LW, Sun SS (1998) Expression of the Brazelian nut methionine-rich protein and mutants with increased methionine in transgenic potato. Plant Mol Biol 37:829–838

Upadhyaya HD (2015) Establishing core collections for enhanced use of germplasm in crop improvement. Ekin J Crop Breed Genet 1:1–12

Upadhyaya HD, Bajaj D, Das S et al (2016a) Genetic dissection of seed-iron and zinc concentrations in chickpea. Sci Rep 6:24050

Upadhyaya HD, Bajaj D, Narnoliya L et al (2016b) Genome-wide scans for delineation of candidate genes regulating seed-protein content in chickpea. Front Plant Sci 7:302

Upadhyaya HD, Thudi M, Dronavalli N et al (2011) Genomic tools and germplasm diversity for chickpea improvement. Plant Genet Resour-C 9:45–58

Varshney RK, Song C, Saxena RK et al (2013) Draft genome sequence of chickpea (*Cicer arietinum*) provides a resource for trait improvement. Nat Biotechnol 31:240–246

Vaz Patto MC, Amarowicz R, Aryee ANA et al (2015) Crit Rev Plant Sci 34:105–143

Walker EL, Connolly EL (2008) Time to pump iron: iron-deficiency-signaling mechanisms of higher plants. Curr Opin Plant Biol 11:530–535

Wang N, Daun JK (2004) The chemical composition and nutritive value of Canadian pulses. Canadian Grain Commission Report:19–29

Wang TL, Domoney C, Hedley CL et al (2003) Can we improve the nutritional quality of legume seeds? Plant Phyiol 131:886–891

Wessells KR, Brown KH (2012) Estimating the global prevalence of zinc deficiency: results based on zinc availability in national food supplies and the prevalence of stunting. PLoS One 7:e50568

White PJ, Broadley MR (2009) Biofortification of crops with seven mineral elements often lacking in human diets – iron, zinc, copper, calcium, magnesium, selenium and iodine. New Phytol 182: 49–84

White PJ, Broadley MR (2011) Physiological limits to zinc biofortification of edible crops. Front Plant Sci 2:1–11

Wood JA, Knights EJ, Choct M (2011) Morphology of chickpea seeds (*Cicer arietinum* L.): comparison of *desi* and *kabuli* types. Int J Plant Sci 172:632–643

Xiaoli X, Liyi Y, Shuang H et al (2008) Determination of oligosaccharide contents in 19 cultivars of chickpea (*Cicer arietinum* L) seeds by high performance liquid chromatography. Food Chem 111:215–219

Yamauchi D, Minamikawa T (1998) Improvement of the nutritional quality of legume seed storage proteins by molecular breeding. J Plant Res 111:1–6

Yan D, Duermeyer L, Leoveanu C et al (2014) The functions of the endosperm during seed germination. Plant Cell Physiol 55:1521–1533

Yang K, Ko J-M, Ha TJ (2014) Development of molecular markers for low raffinose and stachyose in Korean soybean cultivars. Plant Breed Biotech 2:151–157

Yoneyama T, Ishikawa S, Fujimaki S (2015) Route and regulation of zinc, cadmium, and iron transport in rice plants (*Oryza sativa* L.) during vegetative growth and grain filling: metal transporters, metal speciation, grain cd reduction and Zn and Fe biofortification. Int J Mol Sci 16:19111–19129

Zhang L, Garneau MG, Majumdar R (2015) Improvement of pea biomass and seed productivity by simultaneous increase of phloem and embryo loading with amino acids. Plant J 81:134–146

Zhang Y, Su D, He J et al (2017) Effects of ciceritol from chickpeas on human colonic microflora and the production of short chain fatty acids by in vitro fermentation. LWT – Food Sci Technol 79:294–299

Zhang Z, Zheng X, Yang J et al (2016) Maize endosperm-specific transcription factors O2 and PBF network the regulation of protein and starch synthesis. Proc Natl Acad Sci U S A 113:10842–10847

Zhawar VK, Kaur N, Gupta AK (2011) Phytic acid and raffinose series oligosaccharides metabolism in developing chickpea seeds. Physiol Mol Biol Plants 17:355–362

Potential of Field Pea as a Nutritionally Rich Food Legume Crop

A. K. Parihar, G. P. Dixit, U. Singh, Anil K. Singh, Nitin Kumar, and Sanjeev Gupta

Abstract Malnutrition is becoming a serious problem due to dearth of proteins, carbohydrates, vitamins, and macro and micronutrients in the daily diet of human beings mainly in developing countries. The micronutrient malnutrition in human body known as "hidden hunger" impelled loads of health-related problems including low birth weight, anaemia, learning disabilities, increased morbidity and death rates, poor work efficiency, and soaring healthcare expenses. Overall more than 2 billion people from developing countries suffer by micronutrient starvation, while worldwide more than 3 billion people are facing micronutrient deficiencies. In recent years, sincere efforts have been made to overcome the problems of malnutrition using different approaches like dietary supplementation, food fortification and biofortification. Biofortification of food crop with essential micronutrients is one of the best strategies to stride against micronutrient deficiencies through conventional plant breeding and modern genomics and agronomical approaches. Among pulses, field pea is one of the crops targeted for biofortification and has long been recognized as a valuable and nutritious food crop for the human diet. Field pea is a very important, economic, and nutritive crop and is often regarded as "poor man's meat" due to high protein, vitamin, minerals, and prebiotic carbohydrate content, and it has enormous genetic variability for these traits in existing germplasm stock. More specifically, field pea is naturally rich in iron, zinc, and Se; consequently, could be used to address most of the common micronutrient deficiencies in the world. Therefore, field pea crop has been recognized a candidate crop for micronutrient biofortification and a potential complete food solution to the global micronutrient malnutrition. Therefore, in the present chapter, efforts have been made to present the current progress made in field pea for nutritional enrichment using different approaches.

Keywords *Pisum sativum* · Iron · Zinc · Protein · Raffinose family oligosaccharides · Phytic acid · Genetic biofortification

---

A. K. Parihar (✉) · G. P. Dixit · A. K. Singh · N. Kumar · S. Gupta
ICAR-Indian Institute of Pulses Research, Kanpur, Uttar Pradesh, India

U. Singh
Agriculture University, Jodhpur, Rajasthan, India

D. S. Gupta et al. (eds.), *Breeding for Enhanced Nutrition and Bio-Active Compounds in Food Legumes*, https://doi.org/10.1007/978-3-030-59215-8_3

Introduction

Undernourishment is a severe problem that persists worldwide owing to scarcity of proteins, carbohydrates, vitamins, micronutrients, and presence of anti-nutritional components in the daily diet of human beings predominantly in developing countries (Kumar et al. 2016). Human health relies primarily upon 49 micronutrients including minerals and vitamins to ensure perfect metabolic activities (Welch and Graham 2004). The micronutrient malnutrition in human body, commonly called as "hidden hunger", impelled loads of health-related problems. Deficiency of micronutrients is gradually becoming a serious problem, because currently people eat more cereals-based carbohydrate-loaded diet worldwide, which is deficient in micronutrients and vitamins (Stewart et al. 2010; Cakmak et al. 2010; Bouis et al. 2011). Overall more than 2 billion people of developing countries are being influenced by micronutrient starvation, while worldwide more than 3 billion people are facing micronutrient deficiencies (Welch 2003; Cakmak et al. 2010; Depar et al. 2011; Kumar et al. 2016). Due to the deficiency of micronutrients several health problems occur, including low birth weight, anaemia, learning disabilities, increased morbidity and mortality rates, poor work efficiency, and soaring healthcare costs (Welch and Graham 1999; Batra and Seth 2002). Deficiencies of iron (Fe), zinc (Zn), selenium (Se), and iodine (I) are commonly found worldwide particularly in the rural residents of developing countries. It has been reported that around 60, 33, and 15 percent population of world is suffering from Fe, Zn, and Se deficiency, respectively (Arthur 2003; Yang et al. 2007; Hotz and Brown 2004; FAOSTAT 2007; WHO 2009). In addition, deficiencies of vitamin A, Zn, Fe, and/or I collectively cause around 20 percent deaths of below five years age children (Prentice et al. 2008). Importantly, pre-school kids and pregnant women are more vulnerable to the zinc and iron deficiency (Welch and Graham 1999; White and Broadley 2009; WHO 2012). Iron deficiency causes anaemia, poor work efficiency, fatigue, and retard mental growth, and may lead to the risk of women casualty during childbirth (Umbreit 2005; Shivay et al. 2016). It mainly exists in the vegetarian population that depends on plant-based diets, since these people have less bioavailable iron than non-vegetarians. In the case of non-vegetarians, their diets comprise pork, liver, and fish, which contain about 30–40% of iron, and in beef, lamb, and chicken 40–60% of iron is heme iron, which has 15–35% absorption value (Monsen et al. 1978). On the other hand, plant foods contain non-heme iron and its absorption is less than 10% (Zimmermann and Hurrell 2007). Similarly, Zn insufficiency responsible for growth retard, lowers immunity, and also promotes risk of diarrhoeal disease and respiratory infections. Notably, it has been reported that more than 48% and 70% of children in India under the age of 5 are suffering from zinc and iron deficiency, respectively (Shivay et al. 2016). According to the World Health Organization (WHO) about 25% population of the world suffers from anaemia, while 17.3% population has health risk owing to insufficient Zn intake (WHO 2008; Wessells and Brown 2012). Zinc deficiency is prevalent in rural populations where rice is the major source of calories in the diet (Hefferon 2019). Zn deficiency causes approximately 433,000 annual deaths of below five years age children worldwide (WHO

2009). Another important micronutrient is Se whose deficiency is prevalent in countries where soils are deprived of Se (100 to 2000 μgkg^{-1}) (Fordyce 2013; Lyons et al. 2005; Spallholz et al. 2008). It is considered as an important micronutrient since it subsidizes the cytotoxic effect of arsenic due to mutual detoxification (Biswas et al. 1999; Holmberg Jr and Ferm 1969). Due to Se deficiency more than one billion people suffer from two serious diseases i.e. Keshan disease (cardiomyopathy) and Kashin–Beck disease (osteoarthropathy) (Reilly 1996). The daily recommended intake of Zn and Se is 15 mg and 55 mg per day, respectively (Elmadfa 2009). However, on the basis of clinical testings a regular oral dose of 200 mg Se per day has been suggested to lessen the risk of certain cancers, cardiomyopathy, and free-radical-induced diseases and to guard against HIV (Fairweather-Tait et al. 2011). In addition, the deficiency of folate or vitamin B9 occurs worldwide among millions of people across the countries and responsible for several other health problems (Gupta et al. 2013). Beta-carotene and folate deficiencies are also affecting about three million children around the world by developing vitamin A deficiency, and annually more than half million children lose their eyesight (Reifen 2002). Vitamin A plays an important role in vision, bone growth, reproduction, cell division, and cell differentiation in mammals (Stephens et al. 1996). Beta-carotene is the carotenoid widely distributed in plants and most efficiently transformed to vitamin A (Reifen 2002). Other carotenoids like lutein and zeaxanthin do not have provitamin activity but display biological activity in relation to human health. Lutein and its stereoisomer zeaxanthin are the only carotenoids present in the macula region of the retina where they are effective against senile macular degeneration (Krinsky et al. 1990; Meydani et al. 1994; Olmedilla et al. 2001). Lutein may also play important role to defend skin from ultraviolet (UV) radiation-induced damage and decrease the risk of cardiovascular disease and cataracts (Alves-Rodrigues and Shao 2004; Moeller et al. 2000). Carotenoids may help in protecting humans from skin disorders and different types of cancer (Bramley 2000; Snodderly 1995).

Most of the micronutrient deficiency or hidden hunger (Pfeiffer and McClafferty 2007) is prevalent in the Asia and Africa regions, wherein daily diet rice (*Oryza sativa* L.) or wheat (*Triticum aestivum* L.) consumption is more and both are deprived sources of essential micronutrients as compared to pulses. Besides that, cereal grains are inherently low in Zn content and its bioavailability, mainly in Zn-deficient soils compared to pulses (Cakmak et al. 2010; Prasad et al. 2014; Singh et al. 2016; Shivay et al. 2016). Food legumes offer essential nutrients and usually contain higher concentrations of mineral nutrients than cereals and root crops (Blair et al. 2011). Among the pulses, field pea (*Pisum sativum* L.) is being considered as a rich and cheaper source of protein and has a good macro- and micronutrient profile. Field pea has higher potential than cereals to increase both Se and Zn contents in grain (Poblaciones et al. 2013, 2014a; Gomez-Coronado et al. 2016; Poblaciones and Rengel 2016). The high nutrient density of pea makes it a valuable food commodity, capable of meeting the nutritionally rich dietary needs of the estimated 800–900 million undernourished individuals worldwide (FAO 2011; Dahl et al. 2012). Field pea is a very important, economic and nutritive crop and is often regarded as "poor man's meat" due to its high levels of protein, vitamins and minerals, and prebiotic carbohydrate content availability at affordable rate for poorer consumers (Amarakoon et al. 2012). More specifically, field pea is naturally rich in iron and zinc, and thus could address two of the most common micronutrient deficiencies in the world (Amarakoon et al. 2012; Demirbas 2018). Field pea is one of the crops targeted for biofortification and has long been recognized as a valuable, nutritious food for the human diet (Amarakoon et al. 2012). This is being cultivated around the world in about 94 countries (Smýkal et al. 2012). The total production and area of dry peas worldwide is at present approximated to be 16.20 mt and 8.14 mha, respectively (FAOSTAT 2019). In the recent past, sincere efforts have been made to overcome the problems of malnutrition using different approaches like dietary supplementation, food fortification, and biofortification. Biofortification is the process of nutritional enrichment of the staple food crops through agronomical approaches and plant breeding to provide essential micronutrients and vitamins to poor population through daily diet (White and Broadley 2009; Bouis et al. 2011). Therefore, in the present chapter, efforts have been made to present the current progress made in field pea for nutritional enrichment using different approaches.

Nutritional Composition of Field Pea

Pea is an annual cool season and high-valued food legume cultivated extensively worldwide. It is a good source of protein, carbohydrates, minerals, and vitamins, and plays an instrumental role in human nutrition (Harmankaya et al. 2010; Singh et al. 2013; Parihar et al. 2016; Shivay et al. 2016). A number of previous studies have described its nutritional composition, which is presented in Table 1. It is a cheap source of digestible protein in the diets of millions of vegetarian populations who cannot afford non-vegetarian product as a source of protein for balanced nutrition. Most of the protein in round-seeded peas is storage or globulins type, and the amino acid profile of these proteins determines their nutritional value (Bourgeois et al. 2011; Boye et al. 2011). In addition to protein, it is a good source of arginine, valine, and methionine relative to human requirements (Tömösközi et al. 2001). Carbohydrates are the major component of pea found in variable amounts of the dry matter. The high amylase starch is responsible for higher levels of enzyme resistance and slow digestion of starch (Chung et al. 2009, 2010; Parihar et al. 2016). The seed coat and cotyledon are the dietary fibre-loaded parts of pea seeds. The seed coat contains largely water-insoluble polysaccharides, primarily cellulose, whereas the cotyledon fibres comprise of polysaccharides such as hemicelluloses, pectins, and cellulose with various degrees of solubility (Tosh and Yada 2010; Guillon and Champ 2002; Reichert and MacKenzie 1982). Among the major minerals in pea, potassium and phosphorus predominate while the calcium content is relatively in least amounts. The microelements presented in pea seeds include especially iron, manganese, copper, cobalt, and zinc (Savage and Deo 1989; Sommer et al. 1994). The concentration level of other micronutrient and anti-nutritional compounds observed in the pea seeds are given in Table 2. A study in the USA, showed that field

Table 1 Nutritional composition of peas

S.N.	Component	Concentration (%)	Location/ country	References
1	Protein	21.13–27.05	Turkey	Harmankaya et al. (2010); Parihar et al. (2016)
		21.2–32.9		Dahl et al. (2012)
		20–25		Tulbek et al. (2016)
		27.9	Canada	Bhatty and Christison (1984)
		20.5–25.30	Canada	Thavarajah et al. (2010)
		29.00–29.25	Poland	Krajewski et al. (2012)
		20.2–26.7	Canada	Wang and Daun (2004)
		22.5		Katz and Weaver (2003)
		15.8–32.1	France	Burstin et al. (2007)
		25.0		Aluko et al. (2009)
		13.7–30.7		Tzitzikas et al. (2006)
		16–30	Spain	Santos et al. (2019)
		23.7–35.2	Poland	Irzykowska and Wolko (2004)
		21.2–32.9		Kumar and Pandey (2020)
2	Total carbohydrates	56–74		Dahl et al. (2012); Parihar et al. (2016)
		58.8		Katz and Weaver (2003)
3	Starch	41.6–49.0		Parihar et al. (2016)
		39.2	Canada	Bhatty and Christison (1984)
		36.9–49.0		Dahl et al. (2012)
		40–50		Tulbek et al. (2016)
		45		Kumar and Pandey (2020)
4	Amylose	20.7–33.7		Dahl et al. (2012); Parihar et al. (2016)
5	Total dietary fibre	14–26	Canada	Tosh and Yada (2010); Dahl et al. (2012)
6	Insoluble fibre	10–15	Canada	Tosh and Yada (2010); Dahl et al. (2012)
7	Soluble fibre	2–9	Canada	Tosh and Yada (2010); Dahl et al. (2012)
8	Total lipid	1.2–2.4		Dahl et al. (2012)
		2.8–3.1	Spain	Murcia and Rincon (1991)
		1.0		Katz and Weaver (2003)
9	Ash	2.3–3.4		Dahl et al. (2012); Parihar et al. (2016)
		3.0	Canada	Bhatty and Christison (1984)
10	Soluble sugar	5·3–8·7		Dahl et al. (2012); Parihar et al. (2016)

Table 2 Spectrum of minerals and anti-nutritional substances in field pea.

S.N.	Component	Concentration	Location/country	References
1	Iron	21.90–58.40 mg/kg	Turkey	Harmankaya et al. (2010)
		67.20 mg/kg	South Africa	Brand et al. (2004)
		45–49 mg kg	Canada	Gawalko et al. (2009)
		45–53 mg/kg	USA	Amarakoon et al. (2015)
		46–54 mg/kg	USA	Amarakoon et al. (2012)
		39–59 mg/kg	Canada	Bangar et al. (2017)
		23.16–105.2 ppm	USA	Kwon et al. (2012)
		54 mg/kg	USA	Thavarajah et al. (2011)
		54.0 mg/kg	USA	Cheng et al. (2015)
		47.7–58.1 mg/kg	USA	Ray et al. (2014)
		38.6–320.9 mg/kg	Turkey	Demirbas (2018)
		22–490 mg/kg	–	Kumar and Pandey (2020)
		23–105 mg/kg	–	Grusak and Cakmak (2005)
2	Zinc	21.0–57.10 mg/kg	Turkey	Harmankaya et al. (2010)
		39.0 mg/kg	South Africa	Brand et al. (2004)
		32–35 mg/kg	Canada	Gawalko et al. (2009)
		39–63 mg/kg,	USA	Amarakoon et al. (2012)
		16.10–106.63	USA	Kwon et al. (2012)
		27.4–34 mg/kg	Canada	Ray et al. (2014)
		20.4–63.5 mg/kg	–	Kumar and Pandey (2020)
		11.3–82.9 mg/kg	Turkey	Demirbas (2018)
		41.8 mg/kg	USA	Cheng et al. (2015)
		16–107 mg/kg	–	Grusak and Cakmak (2005)
3	Selenium	373–519 μg/kg	Canada	Thavarajah et al. (2010)
		457 μg /kg	Canada	Thavarajah et al. (2011)
		0.331 mg/kg	Canada	Gawalko et al. (2009)
		405–554 μg/kg	Canada	Ray et al. (2014)
4	Potassium	562.8–937.8 mg/100 g,	Turkey	Harmankaya et al. (2010)
5	Phosphorus	163.4–374.2 mg/100 g,	Turkey	Harmankaya et al. (2010)
		5.1 g/kg	South Africa	Brand et al. (2004)
6	Calcium	45.91–157.40 mg/100 g	Turkey	Harmankaya et al. (2010)
		0.7 g/kg	South Africa	Brand et al. (2004)
		786–802 mg/ kg	Canada	Gawalko et al. (2009)
		622–1219 mg/kg	USA	Amarakoon et al. (2012)
7	Magnesium	47.31–102.81 mg/100 g	Turkey	Harmankaya et al. (2010)
		1.3 g/kg	South Africa	Brand et al. (2004)
		1350–1427 mg/kg	USA	Amarakoon et al. (2012)
		1210–1270 mg/kg	Canada	Gawalko et al. (2009)
8	Sulphur	75.69–194.4 mg/100 g,	Turkey	Harmankaya et al. (2010)
9	Lutein	0.348–1.394 mg/100 g	Czech Republic	Holasová et al. (2009)
		7.2–17.6 μg/g	Canada	Ashokkumar et al. (2014)
		11.2 μg/g	Canada	Ashokkumar et al. (2015)

(continued)

Table 2 (continued)

S.N.	Component	Concentration	Location/country	References
10	β-carotene	0.1–0.2 mg/100 g	Czech Republic	Holasová et al. (2009)
		0.47 μg/g	Canada	Ashokkumar et al. (2014)
		0.5 μg/g	Canada	Ashokkumar et al. (2015)
		680 mg/100 g	USA	Amarakoon et al. (2015)
11	Zeaxanthin	0.16 μg/g	Canada	Ashokkumar et al. (2014)
		0.30 μg/g	Canada	Ashokkumar et al. (2015)
12	Violaxanthin	1.70–2.22 μg/g	Canada	Ashokkumar et al. (2014)
		0.3 μg/g	Canada	Ashokkumar et al. (2015)
13	Phytic acid	2.7–3.2 mg/g	USA	Amarakoon et al. (2015)
		4.9–7.1 mg/g	USA	Amarakoon et al. (2012)
		3–13.0 g/kg	Canada	Wang and Daun (2004)
14	Folate	41–202 mg/100 g	USA	Gupta et al. (2013)
		23–30 mg/100 g	Canada	Jha et al. (2015)
15	RFO	41.4–157.4 m/g	Canada	Gawłowska et al. (2017)

peas are a good source of micronutrients Fe, Zn, and Mg and naturally low in phytic acid (PA) or phytate (Amarakoon et al. 2012). A single serving of 100 g field pea can provide 26–78% of adult RDA of Fe, Zn, and Mg. Peas are an also good source of Selenium, and high Se may be advantageous for areas of the world where Se deficiency is prominent (Reichert and MacKenzie 1982). It is also an appreciable source of folate, vitamin C, thiamin (vitamin B1), B6, B,3 and B2 (Hedges and Lister 2006). Similar to other pulses, it contains a variety of phytochemicals such as carotenoids, including lutein and zeaxanthin and β-carotene, chlorophyll, phenolic compounds, including some flavonoids, saponins, and oxalates (Campos-Vega et al. 2010). The carotenoids are a group of yellow-orange-red pigments synthesized by plants and some microorganisms. It cannot be synthesized in the human and animal body and available exclusively either from a plant source itself or product of animal that has consumed that plant source (Fraser and Bramley 2004). The most important carotenoids in peas are the xanthophylls, lutein, and zeaxanthin, but they also contain the carotenes, α- and β-carotene (Hedges and Lister 2006). Field peas with yellow or orange cotyledons had β-carotene concentration 10 times lower than green cotyledon cultivars (Holasová et al. 2009). Beta-carotene is the most widely distributed carotenoid in plants and the one most efficiently converted to vitamin A. The concentration of lutein and zeaxanthin carotenoids in peas is many folds higher than the other food legumes (Hedges and Lister 2006).

Phenolics are a group of more than 4000 compounds existing in great amount in the plant kingdom. Among these, phenolic acids and flavonoids have dietary relevance and present in ample amounts in seed coat and cotyledon of peas. The phenolic compound and its antioxidant activity prominence in coloured seed coat (Duenas et al. 2004; Campos-Vega et al. 2010; Xu et al. 2007). Further, tannins with very high antioxidant activity are detected only in dark seed coat (Hagerman et al. 1998). In addition, a sub-group of the flavonoid category compounds well-known as isoflavones which also exist in sizeable amount in peas (Hedges and Lister 2006). Saponin is a diverse group of biologically active glycosides, which are dispersed extensively in the plant kingdom (Curl et al. 1985). A number of different saponins have been isolated in peas and of them Soyasoponin-I emerges to be predominant (Murakami et al. 2001).

Anti-nutritional Factors

Plant food substances contain many anti-nutritional apparatuses, which reduces the bio-availability of micronutrients during digestion in human gut (Welch and Graham 1999). The nutritional status of food crops can be enhanced by reducing the amount of anti-nutrient factors. Peas contain various bioactive compounds like trypsin inhibitors, lectins (phytohaemagglutinins), indigestible oligosaccharides causing flatulence, phenolic compounds, phytates, and saponins that play critical roles in metabolism in humans or animals (Campos-Vega et al. 2010; Kalač and Míka 1997). These substances, according to their effects, are regarded toxic (i.e. lectins, glycosides, alkaloids), unpalatable or indigestible (i.e. tannins, saponins), or anti-nutritive (i.e. phytates) (Enneking and Wink 2000; Champ 2002; Campos-Vega et al. 2010). These naturally available components obstruct nutrient availability consequently considered as anti-nutritional factors. For instance, protein or carbohydrate digestibility can be reduced by enzyme inhibitors, and lectins may cause reduction in nutrient absorption. Trypsin inhibitors vigorously hamper trypsin activity which reduces the digestion and absorption of dietary protein. Lectins, also recognized as phytohaemagglutinins, are capable in aggregation of red blood cells and few lectins have been designated as growth depressor in animals (Chung et al. 1998; Sandberg 2002). Similarly, phenolic compounds reduce the protein digestibility and mineral bioavailability of peas. In the same way, tannins inhibit the digestive enzymes and thus reduce the digestibility of most nutrients, particularly protein and carbohydrates. In addition, peas contain lipoxygenase in small quantities which is important in processing and/or utilization of pea fractions and contribute to both desirable and undesirable effects in foods (Owusu-Ansah and McCurdy 1991). Most interestingly, it enhances the performance of flour in baked products, but raw legumes cause off-flavours during storage. Phytic acid or phytate reduces mineral bioavailability and declines the functionality of pea protein (Sandberg 2002). Phytate (IP6), owing to negatively charged sites, performs as a strong chelator of metallic cations like potassium, iron, calcium, zinc, magnesium, and manganese (Liu et al. 2005). Consequently, a mixed salt is developed, which is mainly excreted by humans and other non-ruminant animals which lack phytase enzymes for phytate hydrolysis. Phytic acid is considered as an anti-nutrient mainly due to its ability to bind essential dietary minerals (especially Fe and Zn) as well as proteins and starch, and subsequently reduce their bioavailability in humans. Phytate is one of the major components of staple food crops that inhibit nutrients bioavailability in humans and thus responsible for the deficiencies of these minerals often referred to as "hidden hunger" (Bangar et al. 2017; UNICEF 1990). Furthermore, polyphenols reduced iron bioavailability by forming non-absorbable complexes (Petry et al. 2010). In a study evaluating the effect of phenolic compounds in common beans, some polyphenols (catechin, 3, 4-dihydroxybenzoic acid, kaempferol, and kaempferol 3-glucoside) had an enhancing effect on iron bioavailability, whereas others (myricetin, myricetin 3-glucoside, quercetin, and quercetin 3-glucoside) showed inhibitory effects (Hart et al. 2015). Mineral deficiencies or "hidden hunger" conditions arise owing to the non-availability of the bound minerals, and this happens mainly in developing countries because of their reliance on plant-based foods (Warkentin et al. 2012). Furthermore, the presence of the α-galactosides raffinose, stachyose, and verbascose collectively known as raffinose family oligosaccharides (RFO) in legume seeds is the main reason for low consumption of these crops (Wang et al. 2003; Khattab and Arntfield 2009a). Actually, in the digestive tract of humans RFOs are digested with the participation of the bacterial microflora present in the colon (Cummings and Englyst 1995; Southgate 1995). Because of sugar breakdown and subsequent fermentation of released monosaccharides, surplus amounts of carbon dioxide and hydrogen are formed, which leads to flatulence and discomfort (Coon et al. 1990; Suarez et al. 1999). On the contrary, the advantage of consuming oligosaccharides has also been noted. It is assumed that oligosaccharides may act as prebiotics in colon and encourage the growth of bifidobacteria populations, which ultimately reduce diarrhea, boost up the immune system, and amplify the resistance to infection (Muzquiz et al. 2012; Gibson et al. 2004). The content of raffinose family oligosaccharides (RFOs) in pea seeds constrains their usage in feeding humans and animals (Gawłowska et al. 2017).

Targeted Traits for Biofortification

Adoption of conventional and modern breeding techniques to develop nutritionally dense cultivars is an effective and sustainable approach towards escalating the bioavailability of minerals (Nestel et al. 2006). For this purpose, specific nutritional traits need to be identified to set target to increase or decrease their content in the field pea seeds during biofortification. Since deficiencies of Fe, Zn, Se, and I are universally seen in the rural inhabitants of developing countries. It has been deliberated that approximately 60, 33 and 15 percent human population of earth is suffering from Fe, Zn and Se deficiency, respectively (Arthur 2003; Yang et al. 2007; Hotz and Brown 2004; FAOSTAT 2007, WHO 2009). In addition, deficiencies of vitamin A, Zn, Fe, and/or I are collectively responsible for around 20 percent deaths of preschool children (Prentice et al. 2008). Pea seeds also contain some anti-nutritional substances like trypsin inhibitors, lectins (phytohaemagglutinins), oligosaccharides, gallic acid, and other phenolic acids and substances with phytoestrogenic effects (Kalač and Míka 1997). These compounds can either be toxic (i.e. lectins, glycosides, alkaloids), unpalatable or indigestible (i.e. tannins, saponins, oligosaccharides), or anti-nutritive (i.e. phytates) (Enneking and Wink 2000). Therefore, those nutrients deficiency having impact on large population worldwide among the developing countries are need to be used as target trait for increasing their concentration. However, some anti-nutritional compounds present in field pea are to be targeted for reduction of their concentration. The existing variability for these potential traits has been summarized in Table 2.

Existing Genetic Variation for Micronutrients

Abundant amounts of variability exist in field pea for protein, starch, minerals, and anti-nutritional factors (Tables 1 and 2) that may be influenced by both environmental conditions and genetic factors (Ray et al. 2014; Hood-Niefer et al. 2012; Bourgeois et al. 2011; Harmankaya et al. 2010; Wang et al. 2008). Tzitzikas et al. (2006) found that the concentration of protein in 59 pea lines ranged from 13·7 to 30·7% of seed dry matter, with an overall average of 22·3%. Likewise, Harmankaya et al. (2010) reported noteworthy variations for protein and minerals among the studied genotypes. The protein content varied from 21.13 to 27.05, potassium from 562.8 to 937.8 mg/100 g, phosphorus from 163.4 to 374.2 mg/100 g, calcium from 45.91 to 157.40 mg/100 g, magnesium from 47.31 to 102.81 mg/100 g, sulphur from 75.69 to 194.4 mg/100 g, iron from 2.19 to 5.84 mg/100 g and zinc from 2.10 to 5.71 mg/100 g. Moreover, Hood-Niefer et al. (2012) elucidated effects of environment on the concentration of protein in peas with narrow range of protein concentration (24.2–27.5%). Notably, in pea majority of proteins belongs to storage proteins, or globulins, and their amino acid outline play instrumental role in nutritional worth (Boye et al. 2011; Bourgeois et al. 2011). The protein of pulses including peas is mainly rich in lysine and marginal or deficient with respect to methionine as per human requirements (WHO 2005). During recent years being a nutritional rich pulse crop with various health benefits pea has also been acknowledged by mass in global food industry. In comparison to soybean or other plant proteins, pea protein is characterized for its better digestibility, less allergenic responses or negative health controversies (Owusu-Ansah and McCurdy 1991; Allred et al. 2004; Boye et al. 2011; Roy et al. 2010; Lam et al. 2018). In the recent past, a number of pea germplasm lines which have more than 30 percent protein in seeds were identified (Bing 2015; Shen et al. 2016; Demirbas 2018) and subsequently used in pea breeding program to enhance the amount and quality of pea protein. Several advanced breeding lines with 28–30% protein, semi-leafless, short duration, bold seeds, and good disease resistance have been developed with 20 percent high protein content, but the yield potential of the these lines is 40 percent less than the check variety. These finding clearly established that varieties with more than 30% protein and superior agronomic characteristics can be developed (Bing and Liu 2011; Bing 2015). The genes associated to starch biosynthesis have been well deliberated in pea, which pave the way for the manipulation of pea starch and composition for various applications. Shen et al. (2016) exhibited that total starch content can be reduced in peas with high protein and significantly higher amylose content.

These findings suggested that pea lines could be developed with high resistant starch and protein.

Pea is a good source of several macro- and micronutrients and substantial amount of variability for them have been observed and presented in Table 2. Initially, Reichert and MacKenzie (1982) reported potassium (1·04% of dry, dehulled weight) as the most prominent element followed by P (0·39%), Mg (0·10%), and Ca (0·08%). Pea is also an excellent source of iron (97 ppm), selenium (42 ppm), zinc (41 ppm), molybdenum (12 ppm), manganese (11 ppm), copper (9 ppm), and boron (4ppm). Interestingly, Kneen et al. (1990) identified a bronze mutant (brz) with hyperaccumulation of iron in plants. The iron concentration in mutant seeds was 163 mg/kg, compared to 65 mg/kg in wild-type seeds. Later, Gawalko et al. (2009) found that yellow peas have higher levels of Fe, Mg, and Mn, but lower levels of K, compared to green peas. He also recommended that peas produced in Canada may be advantageous for areas of the world where Se deficiency is prevalent in soils. In addition to other micronutrients field pea is an important dietary source of selenium (Se), which ranges from 373 to 519 mg/kg (Thavarajah et al. 2010). Similarly, Harmankaya et al. (2010) found significant variations in minerals among examined genotypes; potassium varied from 562.8 to 937.8 mg/100 g, phosphorus from 163.4 to 374.2 mg/100 g, calcium from 45.91 to 157.40 mg/100 g, magnesium from 47.31 to 102.81 mg/100 g, sulphur from 75.69 to 194.4 mg/100 g, iron from 2.19 to 5.84 mg/100 g, and zinc from 2.10 to 5.71 mg/100 g. Likewise, Amarakoon et al. (2012) examined six commercial field pea genotypes over seven locations in USA and interestingly these genotypes were naturally rich in Fe (46–54 mg kg^{-1}), Zn (39–63 $mgkg^{-1}$), and Mg (1350–1427 mg kg^{-1}). In the same way, Kwon et al. (2012) examined USDA core collection which contained 285 genotypes and found plenty of variation for most of the micronutrients, especially Fe ranged between 23.16 and 105.2 ppm and Zn varied from 16.1 to 106.63 ppm. In some other study, the iron content of dry pea seeds ranged from 45 to 58 mg/kg in commercial cultivars widely grown in North America (Ray et al. 2014). Additionally, these field peas were naturally also low in PA (4.9–7.1 mg g^{-1} of PA or 1.4–2 mg g^{-1} of phytic-P) despite very high total P concentrations (3.5–5 mg g^{-1}). Zinc concentrations in commercial pea varieties available in the USA fluctuated from 39 to 63 mg/kg depending on the genotype and location grown; it was superior in zinc levels previously measured in seeds of Canadian pea varieties (27–34 mg/kg) (Ray et al. 2014). As a source of iron and other micronutrients, pulses have the potential to be useful foods for alleviating nutrient deficiencies which are prevalent throughout the world. Likewise, Amarakoon et al. (2015) found that the Fe concentration ranged between 45 and 53 mg/kg with a mean concentration of 49 mg/kg. They also observed that field peas also contained substantial concentrations of Fe promoters like xanthophyll (17 mg/100 g), canthaxanthin (68 mg/100 g), beta-carotene (680 mg/100 g), kestose (1433 mg/100 g), quercetin (51.7 mg/100 g), and ferulic acid (56.1 mg/100 g). The phytic acid concentration of field peas was naturally low (2.7–3.2 mg/g) and the phytic acid:Fe molar ratio ranged between 5.0 and 5.6. Besides that they had recognized CDC Golden and DS Admiral as suitable genotypes for future Fe biofortification program. Likewise, Ma et al. (2017) examined an RIL population over two locations and enormous genotypic variability for several minerals (Fe, Zn, Ca, K, S) has been observed. In this population iron content varied from 37.3 to 71.2 (μg/g DW) and zinc content ranged between 30.7 and 64.9 μg/g DW) over the locations. Most recently, Demirbas (2018) studied micro- and macronutrients concentration in 152 landraces and 5 commercial cultivars of Turkish pea. He has found tremendous diversity for nitrogen (N) (22.3–66.7 g kg^{-1}), phosphorus (P) (1.48–8.47 g kg^{-1}), potassium (K) (6.7–18.7 g kg^{-1}), iron (Fe) (38.6–320.9 mg kg^{-1}), zinc (Zn) (11.3–82.9 mg kg^{-1}), copper (Cu) (10.5–50.8 mg kg^{-1}), and manganese (Mn) (10.2–37.9 mg kg^{-1}) in Turkish pea germplasm. The average concentrations of N, P, and Zn were detected greater in landraces while K, Fe, Cu, and Mn concentration were noticed high in commercial cultivars. This information expressed a high array of diversity in the Turkish pea germplasm for micro- and macronutrient that will be a valuable resource for the development of biofortified pea cultivars and varieties through conventional and modern breeding technologies and this variation could be used for detection of linked markers through genome-wide association studies and Quantitative trait locus (QTL) mapping. Notwithstanding, the high mineral content of peas, bioavailability may be poor due to high phytate concentrations. Sandberg (2002) reported that phytate acts as an inhibitor of Zn, Fe, and Ca absorption. Trinidad et al. (2010) observed that phytate content affected Fe but it does not influence Zn and Ca availability in pulses. In fact, it was deliberated that when Fe availability was low, Ca and Zn availability was high. It was also noticed that peas had greater Ca bioavailability as compared to other pulses. More efforts need to be put to understand the influence of food processing methods on phytate degradation. If phytate can be degraded, peas could be considered a significant source of Ca, Zn, and Fe (Sandberg 2002).

Pulses are also rich in B vitamins, specifically thiamin (vitamin B1) and folate (B9) (Sierra et al. 1998; Jha et al. 2015), although there is relatively little research efforts made to explore diversity in vitamin B concentrations among pulse crop varieties, particularly in field pea. Folates are water-soluble B vitamins and perform as cofactors in several metabolic functions in the human body. Dang et al. (2000) reported that peas contained 101 mg folate per 100 g. In another study Han and Tyler (2003) determined the concentration of folate over the locations. He reported that folate content ranged from 23·7 to 55·6 mg/100 g in yellow cotyledon genotypes and green cotyledon genotypes ranged from 24·9 to 64·8 mg/100 g. Low levels of dietary folate is responsible for anaemia and neural tube defects in humans (Dang et al. 2000; Han and Tyler 2003). Folate concentration in yellow field pea ranged from 41−55μg/100 g, and green field pea varieties varied between 50 and 202 μg/100 g (Gupta et al. 2013). Recently, Jha et al. (2015) observed that total folates concentration varied between 23 and 30 mg/100 g in pea and of them 5-methyltetrahydrofolate (5-MTHF) and tetrahydrofolate (THF) were the predominant forms in pea. Significant effects of locations and cultivar were also observed for the majority of the folates.

The most important carotenoids in peas are the xanthophylls, lutein and zeaxanthin, but they also contain the carotenes, α- and β-carotene (Hedges and Lister 2006). Holasová et al. (2009) assessed lutein and β-carotene contents in 32 genotypes of field peas comprising green, yellow, and orange cotyledons. The highest lutein concentration ranged between 0.768 and 1.394 mg/100 g was in green cotyledons varieties while yellow cotyledons contained lower amount of lutein. The highest variation in lutein content was recorded in tested breeding lines with orange cotyledons. Besides, β-carotene content in green cotyledons genotypes ranged between 0.1 and 0.2 mg/100 g, whereas yellow and orange cotyledons contain 10 times lower concentration of β-carotene. A strong positive association among lutein and chlorophyll contents was found. Bangar et al. (2017) reported that green cotyledon and yellow cotyledon pea were not significantly differing in total carotenoid concentration, but β-carotene concentration was greater in green cotyledon genotypes. Although no significant correlation was noticed between total carotenoid concentration and iron bioavailability, lutein concentration was positively correlated with iron bioavailability. The average lutein concentration ranged from 7.2 to 17.6 $\mu g\ g^{-1}$ and green cotyledon pea cultivars had approximately twice amount of many total carotenoids (16–21 $\mu g\ g^{-1}$) compared to yellow cotyledon (7–12 $\mu g\ g^{-1}$). Besides, the mean concentration of b-carotene, zeaxanthin, and violaxanthin was 0.5, 0.3, and 0.3 mg/g recorded, respectively. Interestingly cultivar had a greater effect than environment on carotenoid concentration, whereas location effects were significant for violaxanthin, lutein, and total carotenoid concentration. In addition, years had significant effects for all the carotenoids and cultivar × location interaction was significant for violaxanthin and lutein. Furthermore, carotenoid concentration was greatest in the cotyledon portion followed by embryo axis and seed coat (Ashokkumar et al. 2014, 2015).

Carbohydrates are the major components of pea available in substantial amounts of the dry matter (Dahl et al. 2012). Starch content varied between 27.6% and 56.3% of seed dry matter with an overall 46 percent average (Tzitzikas et al. 2006). Unlike other pulses starch, Pea starch has an intermediate level of amylase which is clearly witnessed by higher value of slowly digestible starch than cereal, root and tuber starches (Hoover et al. 2010). The amylose content of pea starch has been reported to vary widely among varieties and mutant lines (Guillon and Champ 2002). Similarly, Holasová et al. (2009) reported that the total starch content in pea's varieties oscillated between 53.61 and 57.23 percent and amylose content was 27.8 percent of total starch. Moreover, resistant starch content fluctuated from 2.07% to 6.31%. The flours from three pea genotypes contained 9.2–10.7, 23.3–26.5, and 10.1–14.7% of rapidly digestible starch, slowly digestible starch, and resistant starch, respectively (Chung et al. 2008a). Starch isolated from the same three genotypes consisted of 18.2–23.8, 53.7–59.0, and 8.1–12.6% of rapidly digestible starch, slowly digestible starch, and resistant starch, respectively (Chung et al. 2008b). The proportion of the starch in peas that is slowly digestible is remarkable. However, Hood-Niefer et al. (2012) noticed no effect of variety or environment on the amylose content of pea starch in field pea varieties. Dietary fibre in peas arises from both the seed coat (outer fibre), commonly referred to as the hull, and the cotyledon (inner fibre) (Martens et al. 2017; Tosh and Yada 2010). The seed coat has largely water-insoluble polysaccharides, mainly cellulose, whereas the cotyledon fibre consists of polysaccharides which have various degrees of solubility, including hemicelluloses and pectins, along with cellulose (Martens et al. 2017; Dahl et al. 2012; Tosh and Yada 2010; Guillon and Champ 2002; Reichert and MacKenzie 1982). Peas, like other legumes, contain significant amount of raffinose-family and other galactose-containing oligosaccharides (Tosh and Yada 2010) which may have prebiotic effects in the large intestine (Fernando et al. 2010). Jones et al. (1999) reported sizeable variation in content and composition of oligosaccharides in seeds of 70 pea accessions. Among them stachyose content was ranged from 0.7% to 4.1% but a verbascose content was very low amount with 3.1% of seed dry weight (g/100 g DW). Similarly, Vidal-Valverde et al. (2003) delineated a variation in 18 pea accessions and noticed that total a-D-galactosides, ranged from 22.6 to 63.4 g/kg; stachyose from 10.7 to 26.7 g/kg; verbascose from 0.0 to 26.7 g/kg; raffinose from 4.1 to 10.3 g/kg and sucrose from 11.6 to 25.4 g/kg. They also found association between a brown colour of seed coat and the lowest content of verbascose and sucrose and between seed size and amount of verbascose and total amount of oligosaccharides. Most recently, Gawłowska et al. (2017) examined 248 accessions comprised of representatives of taxa, breeding materials, and cultivars. The maximum content of total soluble carbohydrates and total RFOs were detected in accessions with wrinkled seeds (r and rb genes) and the lowest content was in wild species *P. fulvum*. It was also noticed that the content of total RFOs was most highly and positively associated with stachyose and verbascose substance. Most interestingly, all oligosaccharides contents were considerably registered lower in lines with dominant alleles of pea seed genes (R, A, and I). Then recessive mutations in mentioned genes resulted in an increased content of RFOs. Therefore, identification of mutant lines with very low concentrations of these oligosaccharides is prerequisite to develop field pea cultivars with low RFOs. The reduction in levels of raffinose oligosaccharides would prevent flatulence and discomfort-related issues and could certainly improve the popularity of this crop among all stakeholders. Raffinose synthase is a vital enzyme in raffinose biosynthesis (Peterbauer et al. 2002) and reduction of it is a most suitable target to reduce concentrations of these compounds in pea.

Peas contain a number of phytochemicals, which includes phenolic compounds, phytates, saponins, and oxalates. The major phenolic constituents in pulses are tannins, phenolic acids, and flavonoids (Campos-Vega et al. 2010). Phenolic compounds behave as antioxidants, and the highest concentrations of most phenolics exist in the pea seed coat, particularly in dark-seeded varieties (Campos-Vega et al. 2010; Troszynska and Ciska 2002; Duenas et al. 2004). Likewise, Xu et al. (2007) also delineated that the antioxidant activity was associated significantly with seed coat colour in pea. Besides that, the assessment of the seed coat and cotyledon in two dark-coloured pea varieties revealed that the seed coat contained glycosides of quercetin, luteolin and apigenin, in addition to a sort of simple phenolics and proanthocyanidins. The cotyledon mostly contained hydroxybenzoic and hydroxycinnamic compounds and some of the glycosides detected in the seed coat (Duenas et al. 2004). Peas contain other phytochemicals including saponins and phytates, which may demonstrate hypocholesterolaemic and anticarcinogenic actions (Campos-Vega et al. 2010). Overall, field pea is a good source of carbohydrates, protein, Fe, Zn, Se, carotenoids and also has genotypes with low level of anti-nutritional factors. The data from most of these reports have limited growing season or genotypes. Therefore, in future multi-location field studies are required to understand the true genetic and genetic × environment interaction effects with more number of genotypes. However, these reports do provide basic information for biofortification research efforts directed at field pea.

Strategies to Alleviate Nutritional Deficiency

In the recent past, sincere efforts have been made to overcome the problems of malnutrition using different approaches like dietary supplementation, food fortification, food processing, dietary diversification and biofortification (Frossard et al. 2000; Kumar et al. 2016). Of them, biofortification is the most effective way to uplift intake of nutrients among resource-poor people by providing nutrient-dense biofortified staple foods through daily diet (Bouis et al. 2011). Biofortification of food crops can be done by adopting the following approaches: firstly, agronomic approach (White and Broadley 2009; Thavarajah et al. 2015; Smrkolj et al. 2006; Turakainen 2007). Secondly, genetic approach in which nutritionally dense high-yielding varieties of food crops are developed by changing the genetic makeup of high-yielding varieties using the classical plant breeding and modern genomic approaches. This approach is well-known as genetic biofortification, which offers a sustainable and cost-effective way of providing the essential micronutrients to the people in both developing and developed countries compared to agronomic approach that involves some technology and costs (Graham et al. 2007; White and Broadley 2009). However, very limited efforts have been made in case of field pea for biofortification (Frossard et al. 2000; Gomez-Galera et al. 2010; Mayer et al. 2008; Welch and Graham 2004; Amarakoon et al. 2012). Therefore, this section of the chapter has been focused on briefing about the current efforts made towards nutritional enrichment of field pea using different approaches to cater the nutritional requirements of poor people in developing countries.

Dietary Diversification and Supplementation

Dietary diversification is a food-based conventional strategy that involves consumption of a wide range of different foods, especially different plant-based foods such as vegetables, fruits, and whole grains in daily diets (White and Broadley 2009). In dietary supplementation, micronutrients are being provided in the form of tablets, powders, capsules and syrups, particularly where daily diet insufficient to provide required amount of micronutrients. For example, the enrichment in folate, Vitamin A and Zn amount in diets has been accomplished with the use of their supplements

(Blancquaert et al. 2013; Hefni et al. 2010; Black et al. 2008). However, folic acid, iron, and zinc supplements have been supportive for children and pregnant women; but this approach is not cost-effective, especially for low-income consumers (Bailey et al. 2015; Wiltgren et al. 2015). In addition, food supplementation needs access to medical amenities, adequate awareness programs, and management of supply vs. demand chain with sufficient storage space with all facilities (Bailey et al. 2015; Stoltzfus 2011). In the case of pea, beta-carotene was microencapsulated in pea protein isolate wall system with and without maltodextrin using emulsification technology and spray drying. It was found that pea protein or pea protein combined with maltodextrin could be used as good microencapsulating agents for food ingredients, nutraceuticals, and pharmaceuticals (Qi 2004). It can be used as temporary method to enhance nutritional health; however, it is unsustainable for a large population (Jha and Warkentin 2020). Most importantly, it is only accessible to a resourceful population that has enough money to purchase food supplements, and therefore this approach is unaffordable to resource-poor population.

Food-Fortification

Fortification is another strategy to uplift the nutritional level of food products by addition of essential micronutrients including minerals and vitamins (Jha and Warkentin 2020; Haas and Miller 2006). Many food supporting programs by the world food program (WFP) are in position using partially pre-cooked and milled cereals and pulses fortified with micronutrients to conquer nutritional deficiencies and provide health benefits with nominal risk. It is considered as a sustainable and cost-effective technique to defeat iron deficiency through the fortification of food items using various high bioavailability iron compound i.e. ferrous sulfate, ferrous fumarate, ferric pyrophosphate, and electrolytic iron powder are being used usually (WHO 2006). Among them, ferrous sulphate is water soluble and strongly interacts with food components, which leads to off-flavours, colour changes, or fat oxidation (Zimmermann and Hurrell 2007). Accordingly, the iron fortification of flours is performed with less water soluble forms such as elemental iron powders, which are less soluble compared to ferrous sulphate. By adopting dual fortification of salt with iodine and iron, a decrease in the prevalence of anaemia and iron deficiency was noticed in school-going children in south India (Andersson et al. 2008). Similarly, salt iodization (fortification with iodine) has been successfully achieved to reduce the occurrence of goiter (Gomez-Galera et al. 2010). In the same way, food can be also fortified with folic acid to boost levels of folates in diets (Blancquaert et al. 2013; Hefni et al. 2010).

Food-Processing Techniques

Food processing methods, for instance, dehulling, soaking, germination, cooking/boiling, and roasting, enhance the taste and deliciousness of peas-based food products, in conjunction with boosting the bioavailability of nutrients by disengaging anti-nutritional substances (Khattab and Arntfield 2009a, b). The improvement in digestibility of protein along with partial or complete elimination of anti-nutritional components (polyphenols, tannins, phatic acid, and trypsin inhibitor) using different processing techniques i.e. water soaking, boiling, roasting, microwave cooking, autoclaving, fermentation, and micronization have been reported (Khattab and Arntfield 2009a). In addition, the seeds also experience significant physicochemical changes such as gelatinization and swelling of starch, denaturation of protein, solubilization of some of the polysaccharides, and softening of structure, which could result in a desirable texture with transformed functional properties (Ma et al. 2011; Ning et al. 2003). The pea starches have lower digestibility compared with cereals but have more digestibility as compared to potato and other high amylose starches (Liljeberg Elmståhl 2002). Many processing techniques are capable of enhancing the in vitro digestibility of pea starch to a different level (Eyaru et al. 2009; Ma et al. 2015). An interesting aspect is that the boiling of frozen peas increases β-carotene concentration than raw peas, as freezing and boiling processes break down cell structure and discharge the compounds that were previously bound to other components (Hedges and Lister 2006).

Additionally, different food processing methods like fermentation, sprouting/germination, and soaking are usually exercised to reduce phytate content in cereals and legumes by activating endogenous phytase (Sandberg and Svanberg 1991). Sprouting subsidizes phytate content in pigeon pea by 35% to 39% (Duhan 2002). Sprouting also amplifies the activity of phytase and degraded phytate in rye (79%), barley (80%), and rice (71%) (Larsson and Sandberg 1992). Urbano et al. (2006) reported that soaking of pea seeds before to germination reduced Zn content by 49% followed by minor losses during germination. The Mg content was declined by 6% owing to the soaking of seeds and by 20–28% during germination. Sprouting for 2 and 4 days enhanced the bio-availability of Zn and Mg from pea seed. Most interestingly, the presence or absence of light at the time of germination process does not affect the results. Overall sprouting of peas for 4 days is the most effective treatment to enhance the bioavailability of Zn and Mg in pea seeds. Likewise, field pea seed soaking and germination for 18 and 48 h reduced the concentration of polyphenols by 52% and 88%, respectively. Interestingly, seeds soaking with de-hulling and pressure cooking leached out 76% of the polyphenols (Bishnoi et al. 1994). Grain processing also influences mineral bioavailability in field pea. For instance, milling and boiling of grains cause Se losses (Thavarajah et al. 2008; Poblaciones et al. 2014b) but also improve Zn bioavailability by declining phytate substance (Brigide et al. 2014). Cooking legume grains usually results in significant decrease in concentrations of phytate, potassium, and Zn (Wang et al. 2008, 2009) and increase in concentrations of protein, Ca, and copper (Vijayakumari et al. 1998; Wang et al.

2009). The processing of field pea grain (24 h freezing and 5 min cooking at 1008C) resulted in a decrease in grain concentrations of Se (by 7.4%), Zn (by 19%), and phytate (by 3%); consequently, phytate: Zn ratio increased by 13% on average suggesting that cooking lowered bioavailability of Zn to humans. The consumption of 100 g of cooked field peas biofortified with the highest combined doses (0.06 Se and 0.5 Zn) would provide 50% of the recommended dietary allowance of Zn with good bioavailability and 45% of the total Se recommended for human daily intake (Poblaciones and Rengel 2017; Fairweather-Tait et al. 2011). Contrarily, Wang et al. (2009) found an increase in Ca and decrease in Fe and Mg grain concentrations upon cooking. Therefore, it is important to elucidate the effects of processing (freezing and boiling typically used in field peas) on bioavailability of micronutrients. Likewise, Moore et al. (2018) reported that the iron content in microwaved immature peas was more bioavailable than in boiled mature peas since cooking destabilized the ferritin-iron. They also demonstrated that the phytic acid is the main inhibitor of iron uptake from mature peas in vitro. However, iron from immature peas is more bioavailable because of lower phytic acid levels compared to mature peas. Recently, Ma et al. (2017) subjected field pea to different processing treatment and recorded significantly higher in vitro protein and starch digestibility along with significant reduction in trypsin inhibitor activity and tannin content. These reports would provide elementary information to help to better comprehend the functionality of field peas as ingredients, and particularly in regard to agri-food industry to enhance the process competence of field peas with improved nutritional and techno-functional qualities.

Biofortification

"Biofortification" or "biological fortification" refers to nutritionally rich food crops with improved bioavailability to the human population that are developed and cultivated by adopting modern biotechnology techniques, conventional plant breeding, and agronomic practices (Garg et al. 2018; Amarakoon et al. 2015; Welch and Graham 2002). Several approaches are being adopted to alleviate micronutrient deficiencies, for instance, pharmaceutical preparation and food fortification, but these strategies have been proved to be unrealistic and restrained because of various reasons (Frossard et al. 2000; Li et al. 2017). Biofortification is the best strategy for enrichment of food crops to overcome malnutrition problems (deficiency of Fe, Zn, and other micronutrients) in resource-less peoples of both developing and developed countries (Welch and Graham 2005; Bouis et al. 2011). Biofortification primarily targets low-income population, but is also attractive to people with higher income who want highly nutritious plant-based foods. Biofortified food crops can be developed through the following approaches. Firstly, agronomic approach, which involves increase in concentration of nutrients in the edible parts of crop plants by applying the micronutrient fertilizers containing Fe-chelates and by using intercropping and crop rotations, as well as soil microorganisms for improving solubilization and mobilization of Fe in the soil (White and Broadley 2009). Secondly, genetic approach, in which nutritionally dense high-yielding varieties of food crops are developed by changing the genetic constitution of high-yielding varieties using the classical plant breeding and modern genomic approaches. This approach is known as genetic biofortification, which provides a sustainable and lucrative means of supply of the essential micronutrients to the people in both developing and developed countries as compared to agronomic approach that involves some technology and costs (White and Broadley 2009; Graham et al. 2007). The enhancement of nutritional worth of staple food crops through biofortification of pulses might bring significant impact owing to their high consumption rate globally.

Biofortification Through Agronomic Approaches

Notwithstanding legumes being staple food for billions of people, very few reports on agronomic biofortification are available. The crops planted in Zn and Se deficit soils has to suffer from Zn deficiency and also contain low Zn and Se content in edible parts. In such specific conditions agronomic biofortification may be the most effective way to increase concentrations of Zn and Se in the edible parts of various crops (Cakmak et al. 2010; Zou et al. 2012; Gomez-Coronado et al. 2016; Poblaciones and Rengel 2016). In agronomic biofortification, level of micronutrients is enhanced by using inorganic fertilizers (Prasad et al. 2014) and improving solubilization and mobilization of micronutrients in the soil adopting intercropping and crop rotations, as well as by escalating the activities of soil microorganisms (Rengel et al. 1999; Zuo et al. 2000; White and Broadley 2009; Bouis et al. 2011). Different agronomic biofortification methods i.e. seed priming, seed coating, and soil or/and foliar fertilization, contain the potential to enhance micronutrient level in the grain. Increased concentration of micronutrients can be achieved without concomitant loss of yield (Harris et al. 2007; Singh 2007; Masuthi et al. 2009; Shivay et al. 2016). Biofortification of food crops with Fe through agricultural approaches is a widely applied strategy (Pfeiffer and McClafferty 2007; Borg et al. 2009). Limited attempts have been made to investigate role of foliar-application of micronutrients in improving shoot and grain Fe concentration in pea. Gupta (1991) reported the significant boost in grain yield owing to foliar spray of $ZnSO_4$ or $FeSO_4$, than Zn or Fe concentration in grain. Further, transport of Fe is faster than Zn in field pea grains when foliar application was done, and around 40% of the applied iron was recovered in field pea grains, while the recovery of zinc was only 5–9 percent. This clearly shows that the regulation mechanism of micronutrient loading into grain is different for iron and zinc (Pearson and Rengel 1995). Many researchers found that the use of FeSO4 as supplement increased grain yield of corn and sorghum grown on Fe deficient soil (Chad et al. 2003; Patel et al. 2004). Similarly, Kabir et al. (2016) found iron (Fe) foliar sprays effective for boosting grain Fe content under Fe deficiency conditions. They used EDDHA [ethylenediamine-N,N′-bis(2-hydroxyphenylacetic acid)] followed by $FeSO_4$ (73.7 mg/l Fe) treatment. The

Fe content of grains significant increased with all foliar sprays at the time of grain filling in Fe-deficient plants. Among them, $FeSO_4$ (73.7 mg/l Fe) was the most efficient in enrichment of Fe in mature grain under Fe deficiency in peas. Also pinpoint that flowering stage is the most suited for foliar application of iron sprays to improve Fe in mature grains. The results of experiments conducted field pea also showed beneficial effect of Zn in terms of seed yield and Zn concentration and uptake by the grain. Application of $ZnSO_4$/ha gave the highest grain yield at Kota, Rajasthan and Shillongani. Field pea responded well to soil application of ammonium molybdate up to 1.5 kg/ha (Khamparia et al. 2010). Poblaciones et al. (2013) suggested peas as highly suitable to introduce Se in the diet of humans because of its ability to accumulate a great amount of Se in the grain. He had found that the sodium selenate as foliar application was much more effective than sodium selenite. A strong positive association between the total Se content in grain and the dose rates of the application was observed and a dose of 10 g Se ha−1 was sufficient to increase Se levels close to daily recommendation. Similarly, the foliar zinc applications alone or in combination with soil zinc applications help in enrichment of field pea grain for zinc (Poblaciones and Rengel 2016). The combined foliar Se and Zn fertilisation in field pea (*Pisum sativum* L.) applied individually and in all combinations (0, 0.03% or 0.06% (w/v) $NaSeO_4$, and 0, 0.25% or 0.5% (w/v) $ZnSO_4 \cdot 7H_2O$) at early grain filling stage demonstrated positive association between total Se or Zn concentration in raw or cooked grains and respective Se or Zn application dose. Notably, the grain Zn accrual was positively influenced by the combined application of Se and Zn. The grain cooking decreases more Zn concentration (19%) as compared to Se (7.4%); nevertheless, cooking improved Zn bioavailability. The eating of 100 g of cooked, biofortified field peas would provide ~50% of recommended daily intake of Zn and 45% of Se. The foliar application of combined 0.06 Se and 0.5 Zn on field pea resulting in grain nutrient concentrations such that eating of 100 g of cooked grains would supply ~5% of the recommended daily allowance of Ca and 35% of Mg, both with low bioavailability, and 90% of Fe with satisfactory bioavailability. Therefore, this successfully biofortified field peas with Fe, Zn, and Se using combined foliar application of Se and Zn would be the best agro-fortification option to strike against malnutrition (Poblaciones and Rengel 2017). Soil application of $ZnSO_4$, ZnO, $FeSO_4$ and other micronutrient containing fertilizers have the potential to improve crop yield along with micronutrient concentration in mature grains. Foliar fertilization with EDTA of $FeSO_4$, $ZnSO_4$ may be a more effective measure (Shivay et al. 2016). In future concentrated efforts are needed for assessment of various ferti-fortification strategies to enhance micronutrients concentration in grains of pulses under field conditions. Other important factors that affecting nutrient enrichment/density in grain are cultivars, environmental conditions, soil type, soil fertility status, crop species, etc., may also be considered while making any such recommendation.

Genetic Biofortification

In genetic biofortification approach nutritionally intense high-yielding varieties of food crops are developed by altering the genetic constitution of high-yielding genotypes using the classical plant breeding and modern genomic approaches (White and Broadley 2009; Bouis et al. 2011).

Conventional Breeding-Enabled Biofortification

Field pea is one of the most important pulse crops for human health particularly in developing countries. Nutritional enrichment of food crops through plant breeding has been widely discussed and accepted in the fields of nutrition and public health worldwide (Bouis 2002). Biofortification using conventional breeding has been considered as a most sustainable, cost-effective alternative to transgenic- and agronomic-based strategies to combat against global human mineral micronutrient deficiencies (Garg et al. 2018). The sufficient genotypic diversity for the targeted nutritional traits is necessary for successful implementation of conventional breeding program for biofortification, and subsequently breeder can access available variation to enhance the levels of trait of interest i.e. minerals and vitamins in high-yielding genotypes. Most importantly, the scientist must focus not only on developing high-yielding genotypes but also on micronutrients-rich foods (Bouis and Welch 2009). In conventional plant breeding, trait specific donors are used for hybridization with recipient genotypes having good agronomic base followed by selection in subsequent generations for developing biofortified genotypes. In case substantial variability is not available for targeted traits, then breeders need to explore other methods like distant hybridization and mutagenesis. Keeping in mind these things, Kneen et al. (1990) identified hyperaccumulation mutants, bronze (brz) and degenerate leaves (dgl), which displayed great increase in iron uptake. The iron concentration in dgl mutant seeds was 163 mg/kg, compared to 65 mg/kg in wild-type seeds. However, although brz plants showed increased iron uptake, there was no increase in seed-iron content and iron over accumulated in other parts of the plant, causing phytotoxicity. Recently several international organizations have started different programs, for instance, "The Health Grain Project (2005–2010)" and "HarvestPlus" program, to enhance the nutritional content of different crops through breeding programs involving many partners from different countries to provide nutritional security to undernourished populations (Garg et al. 2018; Sarker and Agrawal 2015; Bouis and Welch 2010). Due to better acceptability, large numbers of crops have been targeted for biofortification using conventional crop breeding. In case of pulses not much attention has been paid towards biofortification, but in some pulses like lentils, cowpea, and common bean efforts have been made recently. In collaboration with the HarvestPlus program, a number of lentil varieties i.e. five in Bangladesh (Barimasur-4, Barimasur-5, Barimasur-6, Barimasur-7, and

Barimasur-8), seven in Nepal (ILL 7723, Khajurah-1, Khajurah-2, Shital, Sisir Shekhar, Simal), two in India (L4704, Pusa Vaibhav), one in Ethiopia (Alemaya), and two in Syria (Idlib-2, Idlib-3), have been released with high iron and zinc by ICARDA (Garg et al. 2018; Shivay et al. 2016; HarvestPlus 2014). Lentil varieties have been screened for variation in Se content (Thavarajah et al. 2008). Similarly, cow pea has been biofortified for iron content by conventional breeding, and four varieties have been developed i.e. Pant Lobia-1, Pant Lobia-2, Pant Lobia-3, and Pant Lobia-4 by GB Pant University, Pantnagar, India in association to HarvestPlus (Shivay et al. 2016; Garg et al. 2018). In common bean several biofortified varieties for high iron and zinc have been released in Rwanda and Democratic Republic of Congo under the HarvetPlus program (Garg et al. 2018). In case of field pea, very less efforts have been made towards the screening of existing released varieties and germplasm for macro- and micronutrient content. However, ample amount of genetic variability exists for micronutrients in field pea. It is naturally rich in Fe (46–54 mg kg^{-1}), Zn (39–63 mg kg^{-1}), and Mg (1350–1427 mg kg^{-1}) with low phytic acid (PA) or phytate. A single serving of field pea could provide 28–68% of the recommended daily allowance (RDA) for Fe, 36–78% of the RDA for Zn, and 34–46% of the RDA for Mg. Field pea is not a good source of Ca (622–1219 mg kg^{-1}; 6–12% of RDA). In addition, these field peas are naturally low in PA (4.9–7.1 mg g^{-1} of PA or 1.4–2 mg g^{-1} of phytic-P) despite very high total P concentrations (3.5–5 mg g^{-1}). Overall, field pea is a good food source of Fe, Zn, and Mg, and selection of genetic material to enrich micronutrients in combination with growing location may further enhance mineral concentrations (Amarakoon et al. 2012). Field pea is an exceptional source of complex carbohydrates, protein, dietary fibre, vitamins, and minerals (Gawalko et al. 2009; Wang and Daun 2004; Wang et al. 2010). The western Canadian grown field pea Fe content ranged from 45.2 to 48.9 mg/kg, Zn from 32.3 to 35.0 mg/kg, Ca from 786 to 802 mg/kg, Mg from 1210 to 1270 mg/kg, and Se from 0.413 to 429 mg/kg (Gawalko et al. 2009). Field pea micronutrient level could be further enhanced by selection of location specific genotypes and their utilization in conventional breeding for developing site specific biofortified varieties of field pea as a food-based solution to global micronutrient malnutrition. In addition to enhancing seed micronutrient concentrations, improving the bioavailability of micronutrients could be achieved through breeding to achieve lower levels of anti-nutritional factors, such as phytate, and enhanced levels of absorption-promoting compounds, such as xanthophyll, ascorbate, and betacarotene, which are known to promote iron absorption (Hurrell and Egli 2010; Lockyer et al. 2018).

However, phytate could subsidize essential micronutrient bioavailability; therefore the reduction of phytate concentration is quintessential for field pea biofortification. The biochemical pathway of phytate has been altered using mutagenesis followed by conventional breeding. Chemical mutagenesis was used to develop two low-phytate mutants (1-150-81 and 1-2347-144) of field pea from CDC Bronco cultivar at the Crop Development Centre, University of Saskatchewan (Warkentin et al. 2012). The Genotype x environment interaction was also ascertained for the two low-phytate pea lines (1-150-81 and 1-2347-144) along with their progenitor

(CDC Bronco) and two check varieties (Cutlass and CDC Golden) by planting at three diverse environments (Delgerjav 2012; Warkentin et al. 2012). The GE interaction effect was significant for concentration of phytate phosphorus, inorganic phosphorus, and concentration of iron. The concentration of phytate phosphorus was significantly reduced in 1-2347-144 (1.13 mg g^{-1}) and 1-150-81 (1.20 mg g^{-1}) as compared to the other cultivars, which ranged from 2.94 to 2.99 mg g^{-1}. As a result of low phytate phosphorus concentration, there was a proportionate raise in inorganic phosphorus. Iron concentration in low phytate pea lines was 42.1 mg kg^{-1} as compared to 39.4 mg kg^{-1} in CDC Bronco. Environment had influenced iron concentration which varied between 35.1 and 57.0 mg kg^{-1} (Shunmugam et al. 2015). The low phytate concentration in field pea is controlled by single recessive genes (Rehman et al. 2012). The high carotenoid concentration in pulse crop seeds is part of a biofortification strategy. Therefore, Ashokkumar et al. (2014) evaluated the concentration and distribution of carotenoids in the seeds of twelve pea (*Pisum sativum* L.) cultivars planted over multi-locations consecutively for two years in Saskatchewan, Canada. The Lutein was the main carotenoid with average concentration oscillated between 7.2 and 17.6 μg g^{-1}. The green cotyledon pea cultivars had around twice as many total carotenoids (16–21 μg g^{-1}) than yellow cotyledon pea cultivars (7–12 μg g^{-1}). The genotypic effect was greater than environment on carotenoid content. Further, the environment effects were significant for violaxanthin, lutein, and total carotenoid concentration. Contrarily the year effect was significant for all carotenoids in pea. The cultivar × location interaction was significant for violaxanthin and lutein. The carotenoid concentration was greatest in the cotyledon followed by the embryo axis and seed coat. The results of this investigation should be useful for improving nutritional quality in pulse crops. Iron bioavailability was improved by 50–100% in lpa lines compared to controls, in experiments that used simulated digestion and absorption into Caco-2 cells (Liu et al. 2015). Bangar et al. (2017) reported that iron concentration is positively correlated with iron bioavailability and phytate concentration is negatively correlated with iron bioavailability. The β-carotene concentration was greater in green cotyledon genotypes. Although no significant correlation was detected between total carotenoid concentration and iron bioavailability, lutein concentration was positively correlated with iron bioavailability. Similarly, the relationship between phytate concentration and iron bioavailability was further supported by Moore et al. (2018), which explained that lower phytate levels in immature peas correlated with better iron bioavailability compared to mature peas. The protease inhibitors (Bowman–Birk) that decrease the digestibility of protein are controlled by two genes (TI1 and TI2) in pea. A wild pea (*P. elatius*) line was identified that had mutations within both genes and drastically reduces levels of protease inhibitor activity (Clemente et al. 2015), which leads to improved amino acid bioavailability. Most importantly, several promising genotypes possessing high iron and zinc levels have been identified, which are listed in Table 3. These genotypes are being used in conventional and molecular breeding for development of high-yielding nutritionally rich genotypes, development of mapping populations, and identification of associated genes/QTLs.

Table 3 Field pea promising genotypes identified with high content of protein, Fe, and Zn

S.N.	Name of genotypes	Countries	Protein (%)	Fe (ppm)	Zn (ppm)	References
1	Agassiz	USA	–	52.0	63.0	Amarakoon et al. (2012)
2	CDC Golden	USA	–	52.0	39.0	Amarakoon et al. (2012)
3	DS Admiral	USA	–	46.0	45.0	Amarakoon et al. (2012)
4	Manuell	Turkey	-	44.7	24.2	Harmankaya et al. (2010)
5	PS 3048	Turkey	–	44.1	36.6	Harmankaya et al. (2010)
6	PS3045	Turkey	27.05	–	–	Harmankaya et al. (2010)
7	PS4053-1	Turkey	–	58.40	32.30	Harmankaya et al. (2010)
8	PS3029-2	Turkey	–	43.80	57.10	Harmankaya et al. (2010)
9	Cruiser	USA	–	54.0	42.0	Amarakoon et al. (2012)
10	CDC Striker	USA	–	53.0	46.0	Amarakoon et al. (2012)
11	MI3391	Canada	32.0	–	–	Bing (2015)
12	CDC647-1	Canada	26.0	–	–	Bing (2015)
13	SW Marquee	Canada	–	58.1	31.5	Ray et al. (2014)
14	SW Midas	Canada	–	56.1	33.9	Ray et al. (2014)
15	Tudor	Canada	–	56.8	27.6	Ray et al. (2014)
16	Cooper	Canada	–	55.9	32.8	Ray et al. (2014)
17	CDC Centennial	Canada	–	55.2	30.4	Ray et al. (2014)
18	Eclipse	Canada	–	52.5	34.0	Ray et al. (2014)
19	Tekirdağ2	Turkey	–	320.9	60.1	Demirbas (2018)
20	Tokat1	Turkey	–	119.10	48.0	Demirbas (2018)
21	Konya3	Turkey	–	157.5	72.6	Demirbas (2018)
22	İzmir4	Turkey	–	183.5	51.4	Demirbas (2018)
23	Giresun	Turkey	–	247.8	44.2	Demirbas (2018)
24	Elazığ	Turkey	–	154.8	82.9	Demirbas (2018)
25	Adıyaman2	Turkey	–	125.10	62.5	Demirbas (2018)

Genomics-Enabled Biofortification

The exploitation of molecular marker-assisted selection in pulse crop biofortification has been initiated around the world.The combination of plant breeding with new approaches of biotechnology has resulted in the development of staple crops enriched with nutrients (Nestel et al. 2006; Bouis et al. 2011). Biofortified maize, rice, and barley have been produced with increased concentration of iron, zinc, or provitamin A (Raboy et al. 2000; Larson et al. 1998, 2000). The genomics embrace great promise to accelerate the progress of breeding nutritious legume crops (Bohra et al. 2014). In the current genomic scenario, the identified markers associated with genes/QTLs controlling level of micronutrients can be utilized in marker-assisted breeding program to develop biofortified varieties (Grusak 2002; Bouis 2003; Bohra et al. 2014). During the past years sincere efforts have been made to develop genomic resources in field pea (Tayeh et al. 2015; Smýkal et al. 2012; Krajewski et al. 2012; Burstin et al. 2007; Loridon et al. 2005; Irzykowska and Wolko 2004).

However, these genomic resources have not been used extensively in identification of genes/QTL for biofortified traits as compared to other legume crops. Nevertheless, recently few attempts have been made to map and tag the gene(s)/QTL controlling micronutrient status. Some of the researchers elucidated the genetic basis of the iron content in seeds from current germplasm stocks and got success in finding genetic markers and quantitative trait loci to aid in breeding programmes (Kwon et al. 2012; Diapari et al. 2015; Ma et al. 2017; Gali et al. 2018). Significant marker trait association for 12 markers including RAPD and SSR with different minerals has been established in core collection contained 285 accessions of pea (Kwon et al. 2012). Diapari et al. (2015) discovered total nine SNPs significantly associated with iron, and two SNPs with zinc concentration in seeds using a panel of 94 genotypes; however, none of the markers was associated with seed Se concentration. Likewise, Cheng et al. (2015) detected five SNPs marker on LG -4 significantly associated mineral nutrients (Calcium & magnesium). Furthermore, 26 SNPs had association with low molecular weight carbohydrate glucose in dry seed concentrations which were situated on all LG, and one SNP from LG V was found associated with inositol. Bangar et al. (2017) used a RIL population (PR-07) derived from the cross Carrera/CDC Striker segregated for iron concentration, and QTLs were detected on LG3, LG4, and LG7, which collectively demonstrated 51% of the phenotypic variance. Pea being one of the oldest domesticated crops in the world remains behind many other crops in the availability of genomic and genetic resources. To further improve mineral nutrient levels in pea seeds requires the development of genome-wide tools. For this, Ma et al. (2017) used a RIL population derived from "Kiflica" and "Aragorn" and generated linkage map of size 1310.1 cM. Comparative mapping with other legumes demonstrated that the highest level of synteny was observed between pea and the genome of *Medicago truncatula.* Overall, 46 seed mineral concentration QTLs, 37 seed mineral content QTLs, and 6 seed weight QTLs were discovered. The QTLs explained 2.4% to 43.3% of the phenotypic variance. The genome-wide SNPs and the genetic linkage map developed here permitted QTL identification for pea seed mineral nutrients that will serve as important resources to enable marker-assisted selection (MAS) for nutritional quality traits in pea breeding programs.

Future Prospective

It is clearly evident that mineral malnutrition is becomeing a significant global challenge. Several traditional approaches like dietary supplementation, fortification of foods, and agro-fortification are being used to increase the availability of minerals in daily diet. However, the final solution is dietary diversification; but this is not instantly possible. Therefore, biofortification of edible crops including field pea is advocated through mineral fertilization and/or plant breeding. As explained, field pea is an important supplementary food in many developed and developing countries. It is being considered as an integral part of daily diet of resource-poor vegetarian rural population in developing countries. Additionally, it is an economic and nutritive crop and is often regarded as "poor man's meat" due to its high protein, vitamin and minerals, and prebiotic carbohydrate substance which are available at affordable price for poorer consumers. As more than three billion people worldwide have mineral deficiencies and hence even a small increase in the nutritive value of field pea seed may be highly noteworthy for betterment of nutritional status of resource-poor vegetarian population. In addition to other approaches, conventional plant breeding might provide a more sustainable and cost-effective solution in the long run, delivering minerals to the entire population. There is ample natural genetic variation in field pea for for macro- and micronutrient among available germplasm and cultivars (Table 3). These variations could be used as foundation in breeding program for development of new cultivars of field pea with high nutritional level. Besides, such variation can also be explored to identify quantitative trait loci (QTL) associated with minerals. Additionally, knowledge of the genes that have an impact on mineral accumulation and biosynthetic pathways producing anti-nutrients and promoters can be used to adopt targeted strategies, such as mutant screening or genetic engineering, to manipulate the amounts and bioavailability of minerals in the edible portions of plants.

References

Allred CD, Allred KF, Ju YH, Goeppinger TS, Doerge DR, Helferich WG (2004) Soy processing influences growth of estrogen-dependent breast cancer tumors. Carcinogenesis 25(9):1649–1657. https://doi.org/10.1093/carcin/bgh178

Aluko RE, Mofolasayo OA, Watts BM (2009) Emulsifying and foaming properties of commercial yellow pea (Pisum Sativum L.) seed flours. J Agric Food Chem 57:9793–9800

Alves-Rodrigues A, Shao A (2004) The science behind lutein. Toxicol Lett 150:57–83. https://doi.org/10.1016/j.toxlet.2003.10.031

Amarakoon D, McPhee K, Thavarajah P (2012) Iron-, zinc-, and magnesium-rich field peas (Pisum sativum L.) with naturally low phytic acid: a potential food-based solution to global micronutrient malnutrition. J Food Compos Anal 27(1):8–13

Amarakoon D, Gupta DS, McPhee K, DeSutter T, Thavarajah P (2015) Genetic and environmental variation of seed iron and food matrix factors of North-Dakota-grown field peas (Pisum sativum L.). J Food Compos Anal 37:67–74

Andersson M, Thankachan P, Muthayya S, Goud RB, Kurpad AV, Hurrell RF, Zimmermann MB (2008) Dual fortification of salt with iodine and iron: a randomized, double-blind, controlled trail of micronized ferric pyrophosphate and encapsulated ferrous fumarate in southern India. Am J Clin Nutr 88:1378–1387

Arthur JR (2003) Selenium supplementation: does soil supplementation help and why? Proc Nutr Soc 62:393–397. https://doi.org/10.1079/PNS2003254

Ashokkumar K, Tar'an B, Diapari M, Arganosa G, Warkentin TD (2014) Effect of cultivar and environment on carotenoid profile of pea and chickpea. Crop Sci 54:2225–2235

Ashokkumar K, Diapari M, Jha AB, Tar'an B, Arganosa G, Warkentin TD (2015) Genetic diversity of nutritionally important carotenoids in 94 pea and 121 chickpea accessions. J Food Compos Anal 43:49–60

Bailey RL, West KP Jr, Black RE (2015) The epidemiology of global micronutrient deficiencies. Ann Nutr Metab 66:2233

Bangar P, Glahn RP, Liu Y, Arganosa GC, Whiting S, Warkentin TD (2017) Iron bioavailability in field pea seeds: correlations with iron, phytate, and carotenoids. Crop Sci 57(2):891–902
Batra J, Seth PK (2002) Effect of iron deficiency on developing rat brain. Indian J Clin Biochem 17(2):108–114
Bhatty RS, Christison GI (1984) Composition and nutritional quality of pea (Pisum sativum L.), faba bean (Vicia faba L. spp. minor) and lentil (*Lens culinaris* Medik.) meals, protein concentrates and isolates. Plant Foods Hum Nutr 34(1):41–51
Bing DJ (2015) Breeding field pea cultivars with improved protein content. In: Eucarpia international symposium on protein crops, Pontevedra, 5–7 Apr, Abstracts pp 31–32
Bing DJ, Liu Q (2011) Investigation of relationships of yield, seed size, seed protein and starch content and development of varieties with improved protein content of field pea (*Pisum sativum* L.). Can J Plant Sci 91:381
Bishnoi S, Khetarpaul N, Yadav RK (1994) Effect of domestic processing and cooking methods on phytic acid and polyphenol contents of pea cultivars (*Pisum sativum*). Plant Foods Hum Nutr 45:381–388
Biswas S, Talukder G, Sharma A (1999) Prevention of cytotoxic effects of arsenic by short-term dietary supplementation with selenium in mice in vivo. Mutat Res/Genet Toxicol Environ Mutagen 441(1):155–160
Black RE, Allen LH, Bhutta ZA, Caulfield LE, de Onis M, Ezzati M, Mathers C, Rivera J (2008) Maternal and child undernutrition: global and regional exposures and health consequences. Lancet 371:243–260
Blair MW, Astudillo C, Rengifo J, Beebe SE, Graham R (2011) QTL for seed iron and zinc concentrations in a recombinant inbred line population of Andean common beans (Phaseolus vulgaris L.). Theor Appl Genet 122:511–521
Blancquaert D, Storozhenko S, Van Daele J, Stove C, Visser R, Lambert W, Van Der Straeten D (2013) Enhancing pterin and paraaminobenzoate content is not su_cient to successfully biofortify potato tubers and Arabidopsis thaliana plants with folate. J Exp Bot 64:3899–3909
Bohra A, Sahrawat KL, Kumar S, Joshi R, Parihar AK, Singh U, Singh D, Singh NP (2014) Genetics- and genomics-based interventions for nutritional enhancement of grain legume crops: status and outlook. J Appl Genet. https://doi.org/10.1007/s13353-014-0268-z
Borg S, Brinch-Pedersen H, Tauris B, Holm PB (2009) Iron transport, deposition and bioavailability in the wheat and barley grain. Plant Soil 325:15–24
Bouis HE (2002) Plant breeding: a new tool for fighting micronutrient malnutrition. J Nutr 132:491S–494S
Bouis HE (2003) Micronutrient fortification of plants through plant breeding: can it improve nutrition in man at low cost? Proc Nutr Soc 62:403e411
Bouis HE, Welch RM (2009) Biofortification-A sustainable agricultural strategy for reducing micronutrient in global south. Crop Sci 50:S20–S32
Bouis HE, Welch RM (2010) Biofortification: a sustainable agricultural strategy for reducing micronutrient in the global south. Crop Sci 50:S20–S32. https://doi.org/10.2135/cropsci2009.09.0531
Bouis HE, Hotz C, McClafferty B, Meenakshi JV, Pfeiffer WH (2011) Biofortification: a new tool to reduce micronutrient malnutrition. Food Nutr Bull 32:31S–40S
Bourgeois M, Jacquin F, Casseculle F et al (2011) A PQL (protein quantity loci) analysis of mature pea seed proteins identifies loci determining seed protein composition. Proteomics 11:1581–1594
Boye J, Zare F, Pletch A (2011) Pulse proteins: processing, characterization, functional properties and applications in food and feed. Food Res Int 43:414–431
Bramley PM (2000) Is lycopene beneficial to human health? Phytochemistry 54:233–236. https://doi.org/10.1016/S0031-9422(00)00103-5
Brand TS, Brandt DA, Cruywagen CW (2004) Chemical composition, true metabolisable energy content and amino acid availability of grain legumes for poultry. South Afr J Animal Sci 34(2):116–122
Brigide P, Canniatti-Brazaca SG, Silva MO (2014) Nutritional characteristics of biofortified common beans. Food Sci Technol 34:493–500. https://doi.org/10.1590/1678-457x.6245

Burstin J, Marget P, Huart M et al (2007) Developmental genes have pleiotropic effects on plant morphology and source capacity, eventually impacting on seed protein content and productivity in pea. Plant Physiol 144:768–781

Cakmak I, Pfeiffer WH, McClafferty B (2010) Biofortification of durum wheat with zinc: agronomic or genetic biofortification. Cereal Chem 87:1–17

Campos-Vega R, Loarca-Piña G, Oomah BD (2010) Minor components of pulses and their potential impact on human health. Food Res Int 43(2):461–482

Chad BG, John PS, Alan JS, Randal KT, Curtis RT, Ronald JG (2003) Correcting iron deficiency in corn with seed row–applied iron sulfate. Agron J 95:160–166

Champ MMJ (2002) Non-nutrient bioactive substances of pulses. Br J Nutr 88(S3):307–319

Cheng P, Holdsworth W, Ma Y, Coyne CJ, Mazourek M, Grusak MA, Fuchs S, McGee RJ (2015) Association mapping of agronomic and quality traits in USDA pea single-plant collection. Mol Breed 35(2):1–13

Chung KT, Wei CI, Johnson MG (1998) Are tannins a double edged sword in biology and health? Trends Food Sci Technol 9:168–175

Chung H-J, Liu Q, Hoover R et al (2008a) In vitro starch digestibility, expected glycemic index, and thermal and pasting properties of flours from pea, lentil and chickpea cultivars. Food Chem 111:316–321

Chung H-J, Liu Q, Donner E et al (2008b) Composition, molecular structure, properties and in vitro digestibility of starches from newly released Canadian pulse cultivars. Cereal Chem 85:471–479

Chung HJ, Liu Q, Hoover R (2009) Impact of annealing and heat-moisture treatment on rapidly digestible, slowly digestible and resistant starch levels in native and gelatinized corn, pea and lentil starches. Carbohydr Polym 75(3):436–447

Chung HJ, Liu Q, Hoover R (2010) Effect of single and dual hydrothermal treatments on the crystalline structure, thermal properties, and nutritional fractions of pea, lentil, and navy bean starches. Food Res Int 43(2):501–508

Clemente A, Arques MC, Dalmais M et al (2015) Eliminating antinutritional plant food proteins: the case of seed protease inhibitors in pea. PLoS ONE 10:e0134634

Coon CN, Leske KL, Akavanichan O, Cheng TK (1990) Effect of oligosaccharide-free soybean meal on true metabolizable energy and fiber digestion in adult roosters. Poult Sci 69(5):787–793

Cummings JH, Englyst HN (1995) Gastrointestinal effects of food carbohydrate. Am J Clin Nutr 61(4):938S–945S

Curl CL, Price KR, Fenwick GR (1985) The quantitativeestimation of saponin in pea (Pisumsativum L.) andsoya (glycine-max). Food Chem 18:241–250

Dahl WJ, Foster LM, Tyler RT (2012) Review of the health benefits of peas (Pisum sativum L.). Br J Nutr 108(S1):S3–S10

Dang J, Arcot J, Shrestha A (2000) Folate retention in selected processed legumes. Food Chem 68:295–298

Delgerjav O (2012) Genotype by environment analysis of the performance of the two low phytate pea lines. M.S. thesis, University of Saskatchewan, Saskatoon

Demirbas A (2018) Micro and macronutrients diversity in Turkish pea (Pisum sativum) germplasm. Int J Agric Biol 20(4):701–710

Depar N, Rajpar I, Memon MY, Imtiaz M, Hassan Z (2011) Micronutrient nutrient densities in some domestic and exotic rice genotypes. Pak J Agri Agril Engg Vet Sc 27:134–142

Diapari M, Sindhu A, Warkentin TD, Bett K, Tar'an B (2015) Population structure and market-trait association studies of iron, zinc and selenium concentrations in seed of field pea (*Pisum sativum* L.). Mol Breed 35:30–42. https://doi.org/10.1007/s11032-015-0252-2

Duenas M, Estrella I, Hernandez T (2004) Occurrence of phenolic compounds in the seed coat and the cotyledon of peas (Pisum sativum L.). Eur Food Res Technol 219:116–123

Duhan A (2002) Changes in phytates and HCl extractability of calcium, phosphorus, and iron of soaked, dehulled, cooked, and sprouted pigeon pea cultivar (UPAS-120). Plant Foods Hum Nutr 57:275–284. https://doi.org/10.1023/A:1021814919592

Elmadfa I (2009) The European nutrition and health report. Forum of Nutrition, Vienna
Enneking D, Wink M (2000) Towards the elimination of anti-nutritional factors in grain legumes. In: Linking research and marketing opportunities for pulses in the 21st century. Springer, Dordrecht, pp 671–683
Eyaru R, Shrestha AK, Arcot J (2009) Effect of various processing techniques on digestibility of starch in Red kidney bean (Phaseolus vulgaris) and two varieties of peas (Pisum sativum). Food Res Int 42(8):956–962
Fairweather-Tait SJ, Bao Y, Broadley MR, Collings R, Ford D, Hesketh JE, Hurst R (2011) Selenium in human health and disease. Antioxid Redox Signal 14:1337–1383. https://doi.org/10.1089/ars.2010.3275
FAO (2019) Food and agriculture organization statistics. http://www.fao.org/faostat/en/#data/QC
FAOSTAT (2007). http://faostat@fao.org/
Fernando WMU, Hill JE, Zello GA et al (2010) Diets supplemented with chickpea or its main oligosaccharide component raffinose modify fecal microbial composition in healthy adults. Benefic Microbes 1:197–207
Food and Agriculture Organization (2011) FAO Statisitics. Food Security Data and Definitions 2005–2007. Food Deprivation. Number of Undernourished Persons. http://www.fao.org/economic/ess/ess-fs/fs-data/ess-fadata/en/. Accessed June 2011
Fordyce FM (2013) Selenium deficiency and toxicity in the environment. In: Essentials of medical geology. Springer, Dordrecht, pp 375–416
Fraser PD, Bramley PM (2004) The biosynthesis and nutritional uses of carotenoids. Prog Lipid Res 43:228–265
Frossard E, Bucher M, Mächler F, Mozafar A, Hurrell R (2000) Potential for increasing the content and bioavailability of Fe, Zn and Ca in plants for human nutrition. J Sci Food Agric 80(7):861–879
Gali KK, Liu Y, Sindhu A et al (2018) Construction of high-density linkage maps for mapping quantitative trait loci for multiple traits in field pea (Pisum sativum L.). BMC Plant Biol 18:172
Garg M, Sharma N, Sharma S et al (2018) Biofortified crops generated by breeding, agronomy, and transgenic approaches are improving lives of millions of people around the world. Front Nutr 5:12
Gawalko E, Garrett RG, Warkentin T, Wang N, Richter A (2009) Trace elements in Canadian field peas: a grain safety assurance perspective. Food Addit Contam A 26:1002–1012
Gawłowska M, Święcicki W, Lahuta L, Kaczmarek Z (2017) Raffinose family oligosaccharides in seeds of Pisum wild taxa, type lines for seed genes, domesticated and advanced breeding materials. Genet Resour Crop Evol 64(3):569–578
Gibson GR, Probert HM, Loo JV, Rastall RA, Roberfroid MB (2004) Dietary modulation of the human colonic microbiota: updating the concept of prebiotics. Nutr Res Rev 17(2):259–275
Gomez-Coronado F, Poblaciones MJ, Almeida AS, Cakmak I (2016) Zinc (Zn) concentration of bread wheat grown under Mediterranean conditions as affected by genotype and soil/foliar Zn application. Plant Soil 401:331–346. https://doi.org/10.1007/s11104-015-2758-0
Gomez-Galera S, Rojas E, Sudhakar D, Zhu CF, Pelacho AM, Capell T et al (2010) Critical evaluation of strategies for mineral fortification of staple food crops. Transgenic Res 19:165–180. https://doi.org/10.1007/s11248-009-9311-y
Graham RD, Welch RM, Saunders DA, Ortiz-Monasterio I, Bouis HE, Bonierbale M, de Haan S, Burgos G, Thiele G, Liria R, Meisner CA, Beebe SE, Potts MJ, Kadian M, Hobbs PR, Gupta RK, Twomlow S (2007) Nutritious subsistence food systems. Adv Agron 92:1–74
Grusak M (2002) Enhancing mineral content in plant food products. J Am Coll Nutr 21:178S–183S
Grusak MA, Cakmak I (2005) Methods to improve the crop-delivery of minerals to humans and livestock. In: Broadley MR, White PJ (Eds.), Plant Nutritional Genomics. Blackwell Publishing, Oxford, pp 265–286
Guillon F, Champ MJ (2002) Carbohydrate fractions of legumes: uses in human nutrition and potential for health. Br J Nutr 88(S3):293–306
Gupta UC (1991) Iron status of crops in Prince Edward Island and effect of soil pH on plant iron concentration. Can J Soil Sci 71:197–202

Gupta DS, Thavarajah D, Knutson P, Thavarajah P, McGee RJ, Coyne CJ, Kumar S (2013) Lentils (*Lens culinaris* L.) a rich source of folates. J Agric Food Chem 61:7794–7799

Haas JH, Miller DD (2006) Overview of experimental biology 2005 symposium: food fortification in developing countries. Am Soc Nutr J Nutr 136:1053–1054

Hagerman AE, Riedl KM, Jones A et al (1998) Highmolecular weight plant polyphenolics (tannins) asantioxidants. J Agric Food Chem 46:1887–1892

Han JY, Tyler RT (2003) Determination of folate concentrations in pulses by a microbiological method employing trienzyme extraction. J Agric Food Chem 51:5315–5318

Harmankaya M, Özcan MM, Karadaş S, Ceyhan E (2010) Protein and mineral contents of pea (Pisum sativum L.) genotypes grown in Central Anatolian region of Turkey. South West J Hortic Biol Environ 1(2):159–165

Harris D, Rashid A, Miraj G et al (2007) 'On–farm' seed priming with zinc sulphate solution–A cost–effective way to increase the maize yields of resource–poor farmers. Field Crop Res 102(2):119–127

Hart JJ, Tako E, Kochian LV, Glahn RP (2015) Identification of Black Bean (*Phaseolus vulgaris* L.) Polyphenols that Inhibit and Promote Iron Uptake by Caco-2 Cells. J Agric Food Chem 63:5950–5956

HarvestPlus (2014) Biofortification progress briefs: iron and zinc lentils. Brief 9, p 19. Available online at www.HarvestPlus.org

Hedges LJ, Lister CE (2006) The nutritional attributes of legumes. Crop Food Res Confidential Rep 1745:50

Hefferon K (2019) Biotechnological approaches for generating zinc-enriched crops to combat malnutrition. Nutrients 11(2):253

Hefni M, Öhrvik V, Tabekha M, Witthöft C (2010) Folate content in foods commonly consumed in Egypt. Food Chem 121(2):540–545

Holasová M, Dostálova R, Fieldlerová V, Horáček J (2009) Variability of lutein content in peas (*Pisum sativum* L.) in relation to the variety, season and chlorophyll content. Czech J Food Sci 27:S188–S191

Holmberg RE Jr, Ferm VH (1969) (1969) Interrelationships of selenium, cadmium, and arsenic in mammalian teratogenesis. Arch Environ Health Int J 18(6):873–877

Hood-Niefer SD, Warkentin TD, Chibbar RN, Vandenberg A, Tyler RT (2012) Effect of genotype and environment on the concentrations of starch in, and protein and the physicochemical properties of starch from, field pea and fababean. J Sci Food Agric 92(1):141–150. https://doi.org/10.1002/jsfa.4552

Hoover R, Hughes T, Chung HJ et al (2010) Composition, molecular structure, properties, and modification of pulse starches: a review. Food Res Int 43:399–413

Hotz C, Brown KH (2004) Assessment of the risk of zinc deficiency in populations and options for its control. Food Nutr Bull 25:S94–S204

Hurrell R, Egli I (2010) Iron bioavailability and dietary reference values. Am J Clin Nutr 91:1461S–1467S

Irzykowska L, Wolko B (2004) Interval mapping of QTLs controlling yield-related traits and seed protein content in Pisum sativum. J Appl Genet 45:297–306

Jha AB, Warkentin TD (2020) Biofortification of pulse crops: Status and future perspectives. Plants 9(1):73

Jha AB, Ashokkumar K, Diapari M et al (2015) Genetic diversity of folate profiles in seeds of common bean, lentil, chickpea and pea. J Food Compos Anal 42:134–140

Jones DA, Dupont MS, Ambrose MJ, Frias J, Hedley CL (1999) The discovery of compositional variation for the raffinose family of oligosaccharides in pea seeds. Seed Sci Res 9(04):305–310

Kabir AH, Paltridge N, Stangoulis J (2016) Chlorosis correction and agronomic biofortification in field peas through foliar application of iron fertilizers under Fe deficiency. J Plant Interact 11(1):1–4

Kalač P, Míka V (1997) Přirozené škodlivé látky v ros-tlinných krmivech. 1. vydání. Praha, ÚZPI, 317pp

Katz SH, Weaver WW (2003) Encyclopedia of food and culture. Scribner

Khamparia RS, Singh MV, Sharma BL, Kulhare PS, Sharma GD (2010) Research publication No. 9, AICRP micro and secondary nutrients and pollutant elements in soil and plants. Indian Inst Soil Sci Bhopal Madhya Pradesh 6:1–113

Khattab RY, Arntfield SD (2009a) Nutritional quality of legume seeds as affected by some physical treatments 2. Antinutritional factors. LWT-Food Sci Technol 42(6):1113–1118

Khattab RY, Arntfield SD (2009b) Nutritional quality of legume seeds as affected by some physical treatments 2. Antinutritional factors. Lebensm Wiss Technol 42(6):1113–1118

Kneen BE, Larue TA, Welch RM et al (1990) Pleiotropic effects of brz: a mutation in Pisum sativum (L) cv. 'Sparkle' conditioning decreased nodulation and increased iron uptake and leaf necrosis. Plant Physiol 93:717–722

Krajewski P, Bocianowski J, Gawłowska M, Kaczmarek Z, Pniewski T, Święcicki W, Wolko B (2012) QTL for yield components and protein content: a multienvironment study of two pea (Pisum sativum L.) populations. Euphytica 183(3):323–336

Krinsky NI, Russett MD, Handelman GJ, Snodderly DM (1990) Structural and geometric isomers of carotenoids in human plasma. J Nutr 120:1654–1662

Kumar S, Pandey G (2020) Biofortification of pulses and legumes to enhance nutrition. Heliyon 6(3):e03682

Kumar J, Gupta DS, Kumar S, Gupta S, Singh NP (2016) Current knowledge on genetic biofortification in lentil. J Agric Food Chem 64(33):6383–6396

Kwon S-J, Brown A, Hu J, McGee R, Watt C, Kisha T, Timmerman-Vaughan G, Grusak M, McPhee K, Coyne C (2012) Genetic diversity, population structure and genome-wide marker-trait association analysis emphasizing seed nutrients of the USDA pea (Pisum sativum L.) core collection. Genes Genom 34(3):305–320

Lam ACY, Karaca AC, Tyler RT, Nickerson MT (2018) Pea protein isolates: structure, extraction and functionality. Food Rev Int 34(2):126–147. https://doi.org/10.1080/87559129.2016.1242135

Larson SR, Young KA, Cook A, Blake TK, Raboy V (1998) Linkage mapping of two mutations that reduce phytic acid content of barley grain. Theor Appl Genet 97:141–146. https://doi.org/10.1007/s001220050878

Larson SR, Rutger JN, Young KA, Raboy V (2000) Isolation and genetic mapping of a non-lethal rice (Oryza sativa L.) low phytic acid mutation. Crop Sci 40:1397–1405

Larsson M, Sandberg AS (1992) Phytate reduction in oats during malting. J Food Sci 57:994–997. https://doi.org/10.1111/j.1365-2621.1992.tb14340.x

Li H, Lian C, Zhang Z et al (2017) Agro-biofortification of iron and zinc in edible portion of crops for the global south. Adv Plants Agric Res 6(2):52–54. https://doi.org/10.15406/apar.2017.06.00210

Liljeberg Elmståhl H (2002) Resistant starch content in a selection of starchy foods on the Swedish market. Eur J Clin Nutr 56(6):500–505

Liu ZH, Cheng FM, Cheng WD, Zhang GP (2005) Positional variations in phytic acid and protein content within a panicle of japonica rice. J Cereal Sci 41(3):297–303

Liu X, Glahn RP, Arganosa GC et al (2015) Iron bioavailability in low Phytate Pea. Crop Sci 55:320

Lockyer S, White A, Buttriss JL (2018) Biofortified crops for tackling micronutrient deficiencies – what impact are these having in developing countries and could they be of relevance within Europe? Nutr Bull 43:319–357

Loridon K, McPhee K, Morin J, Dubreuil P, Pilet Nayel ML, Aubert G et al (2005) Microsatellite marker polymorphism and mapping in pea (Pisum sativum L.). Theor Appl Genet 111:1022–1031. https://doi.org/10.1007/s00122-005-0014-3

Lyons G, Ortiz-Monasterio I, Stangoulis J, Graham R (2005) Selenium concentration in wheat grain: Is there sufficient genotypic variation to use in breeding? Plant Soil 269:369–380

Ma Z, Boye JI, Simpson BK, Prasher SO, Monpetit D, Malcolmson L (2011) Thermal processing effects on the functional properties and microstructure of lentil, chickpea, and pea flours. Food Res Int 44(8):2534–2544

Ma Z, Boye JI, Azarnia S, Simpson BK (2015) Volatile flavour profile of Saskatchewan grown pulses as affected by different thermal processing treatments. Int J Food Prop 19(10):1–21

Ma Y, Coyne CJ, Grusak MA, Mazourek M, Cheng P, Main D, McGee RJ (2017) Genome-wide SNP identification, linkage map construction and QTL mapping for seed mineral concentrations and contents in pea (Pisum sativum L.). BMC Plant Biol 17(1):43

Martens LG, Nilsen MM, Provan F (2017) Pea hull fibre: novel and sustainable fibre with important health and functional properties. EC Nutr 10:139–148

Masuthi DA, Vyakaranahal BS, Deshpande VK (2009) Influence of pelleting with micronutrients and botanical on growth, seed yield and quality of vegetable cowpea. Karnataka J Agric Sci 22:898–900

Mayer JE, Pfeiffer WH, Beyer P (2008) Biofortified crops to alleviate micronutrient malnutrition. Curr Opin Plant Biol 11(2):166–170

Meydani M, Martin A, Ribaya Mercado JD, Gong J, Blumberrg JB, Russel RM (1994) Beta-carotene supplementation increases antioxidant capacity of plasma in older women. J Nutr 124:2397–2403

Moeller SM, Jacques PF, Blumberg JB (2000) The potential role of dietary xanthophylls in cataract and age-related macular degeneration. J Am Coll Nutr 19:522S–527S. https://doi.org/10.1080/07315724.2000.10718975

Monsen ER, Hallberg L, Layrisse M, Hegsted DM, Cook JD, Mertz W, Finch CA (1978) Estimation of available dietary iron. Am J Clin Nutr 31:134–141

Moore KL, Rodríguez-Ramiro I, Jones ER et al (2018) The stage of seed development influences iron bioavailability in pea (Pisum sativum L.). Sci Rep 8:6865

Murakami T, Kohno K, Matsuda H, Yoshikawa M (2001) Medicinal foodstuffs. XXII. Structures of oleanane-type triterpene oligoglycosides, pisumsaponins I and II, and kaurane-type diterpene oligoglycosides, pisumosides A and B, from green peas, the immatureseeds of Pisum sativumL. Chem Pharm Bull 49(1):73–77

Murcia MA, Rincon F (1991) Fatty acid composition of pea (Pisum sativum L., var. Citrina) during seed growth. Grasas Aceites 42(6):444–449

Muzquiz M, Varela A, Burbano C, Cuadrado C, Guillamón E, Pedrosa M (2012) Bioactive compounds in legumes: pronutritive and antinutritive actions. Implications for nutrition and health. Phytochem Rev 11(2–3):227–244

Nestel P, Bouis HE, Meenakshi JV, Pfeiffer W (2006) Biofortification of staple food crops. J Nutr 136(4):1064–1067

Ning W, Daun JK, Malcolmson LJ (2003) Relationship between physicochemical and cooking properties, and effects of cooking on antinutrients, of yellow field peas (Pisum sativum). J Sci Food Agric 83(12):1228–1237

Olmedilla B, Granado F, Blanco I, Vaquero M, Cajigal C (2001) Lutein in patients with cataracts and age-related macular degeneration: a long-term supplementation study. J Sci Food Agric 81:904–909. https://doi.org/10.1002/jsfa.905

Owusu-Ansah YJ, McCurdy SM (1991) Pea proteins: a review of chemistry, technology of production, and utilization. Food Rev Int 7(1):103–134

Parihar AK, Bohra A, Dixit GP (2016) Nutritional benefits of winter pulses with special emphasis on Peas and Rajmash. In: Biofortification of food crops. Springer, New Delhi, pp 61–71

Patel GJ, Ramakrishnayya BV, Patel BK (2004) Effect of soil and foliar application of ferrous sulphate and of acidulation of soil on iron chlorosis of paddy seedlings ingoradu soil nurseries in India. Plant Soil 46:209–219

Pearson JN, Rengel Z (1995) Uptake and distribution of 65Zn and 54Mn in wheat grown in sufficient and deficient levels of Zn and Mn II. During grain development. J Exp Bot 46(7):841–845. https://doi.org/10.1093/jxb/46.7.841

Peterbauer T, Mach L, Mucha J et al (2002) Functional expression of a cDNA encoding pea (Pisum sativum L.) raffinose synthase, partial purification of the enzyme from maturing seeds, and steady-state kinetic analysis of raffinose synthesis. Planta 215:839–846

Petry N, Egli I, Zeder C, Walczyk T, Hurrell R (2010) Polyphenols and phytic acid contribute to the low iron bioavailability from common beans in young women. J Nutr 140:1977–1982. https://doi.org/10.3945/jn.110.125369

Pfeiffer WH, McClafferty B (2007) HarvestPlus: breeding crops for better nutrition. Crop Sci 47:S88–S105

Poblaciones MJ, Rengel Z (2016) Soil and foliar zinc biofortification in field pea (*Pisum sativum* L.): grain accumulation and bioavailability in raw and cooked grains. Food Chem 212:427–433. https://doi.org/10.1016/j.foodchem.2016.05.189

Poblaciones MJ, Rengel Z (2017) Combined foliar selenium and zinc biofortification in field pea (Pisum sativum): accumulation and bioavailability in raw and cooked grains. Crop Pasture Sci 68(3):265–271

Poblaciones MJ, Rodrigo SM, Santamaria O (2013) Evaluation of the potential of peas (*Pisum sativum* L.) to be used in selenium biofortification programs under Mediterranean conditions. Biol Trace Elem Res 151:132–137. https://doi.org/10.1007/s12011-012-9539-x

Poblaciones MJ, Santamaria O, Garcia-White T, Rodrigo SM (2014a) Selenium biofortification in bread-making wheat under Mediterranean conditions: Influence on grain yield and quality parameters. Crop Pasture Sci 65:362–369. https://doi.org/10.1071/CP14061

Poblaciones MJ, Rodrigo S, Santamaría O, Chen Y, McGrath SP (2014b) Selenium accumulation and speciation in biofortified chickpea (*Cicer arietinum* L.) under Mediterranean conditions. J Sci Food Agric 94:1101–1106. https://doi.org/10.1002/jsfa.6372

Prasad R, Shivay YS, Kumar D (2014) Agronomic biofortification of cereal grains with iron and zinc. Adv Agron 125:55–91

Prentice AM, Gershwin ME, Schaible UE, Keusch GT, Victoria LG, Gordon JI (2008) New challenges in studying nutrition disease interactions in the developing world. J Clin Invest 118:1322–1329

Qi J (2004) Microencapsulation of beta-carotene in pea protein wall system. Thesis submitted to the Faculty of Graduate Studies in partial fulfillment of the requirements for the degree of Master of Science, Department of Food Science, University of Manitoba, Winnipeg

Raboy V, Gerbasi PF, Young KA, Stoneberg SD, Pickett SG, Bauman AT, Murthy PPN, Sheridan WF, Ertl DS (2000) Origin and seed phenotype of maize low phytic acid 1-1 and low phytic acid 2-1. Plant Physiol 124:355–368

Ray H, Bett K, Tar'an B, Vandenberg A, Thavarajah D, Warkentin T (2014) Mineral micronutrient content of cultivars of field pea, chickpea, common bean, and lentil grown in Saskatchewan, Canada. Crop Sci 54(4):1698–1708

Rehman AU, Shunmugam A, Arganosa G, Bett KE, Warkentin TD (2012) Inheritance of the low-phytate trait in pea. Crop Sci 52:1171–1175

Reichert RD, MacKenzie SL (1982) Composition of peas (Pisum sativum) varying widely in protein content. J Agric Food Chem 30:312–317

Reifen R (2002) Vitamin A as an anti-inflammatory agent. Proc Nutr Soc 3:397–400. https://doi.org/10.1079/PNS2002172

Reilly C (1996) Biological role of selenium. In: Selenium in food and health. Blackie, London

Rengel Z, Batten GD, Crowley DE (1999) Agronomic approaches for improving the micronutrient density in edible portions of field crops. Field Crop Res 60(1–2):27–40

Roy F, Boye JI, Simpson BK (2010) Bioactive proteins and peptides in pulse crops: pea, chickpea and lentil. Food Res Int 43:432–442

Sandberg AS (2002) Bioavailability of minerals in legumes. Br J Nutr 88(S3):281–285

Sandberg AS, Svanberg U (1991) Phytate hydrolysis by phytase in cereals: effects on in vitro estimation of iron availability. J Food Sci 56:1330–1333. https://doi.org/10.1111/j.1365-2621.1991.tb04765.x

Santos CS, Carbas B, Castanho A, Vasconcelos MW, Vaz Patto MC, Domoney C, Brites C (2019) Variation in pea (Pisum sativum L.) seed quality traits defined by physicochemical functional properties. Foods 8(11):570

Sarker A, Agrawal SK (2015) Combating Micronutrient Malnutrition with Biofortified Lentils. Amman Jordan the International Center for Agriculture Research in the Dry Areas. The International Center for Agriculture Research in the Dry Areas

Savage GP, Deo S (1989) The nutritional value of peas *(Pisum sativum)*. A literature review. Nutr Abstr Rev (Ser A) 59:65–88

Shen S, Hou HW, Ding CB, Bing DJ, Lu ZX (2016) Protein content correlates with starch morphology, composition and physicochemical properties in field peas. Can J Plant Sci 96(3):404–412

Shivay YS, Singh U, Prasad R, Kaur R (2016) Agronomic interventions for micronutrient biofortification of pulses. Indian J Agron 61(4th IAC Special Issue):S161–S172

Shunmugam ASK, Liu X, Stonehouse R, Bett KE, Tar'an B, Warkentin TD (2015) Mapping of seed phytate concentration and iron bioavailability in pea recombinant inbred line population. Crop Sci 55:828–836. https://doi.org/10.2135/cropsci2014.08.0544

Sierra I, Vidal-Valverde C, Kozlowska H (1998) Effect of ripening stage on thiamin and riboflavin levels in lupin, pea and faba bean seeds. Zeitschrift für Lebensmitteluntersuchung und Forschung A 206:126–129

Singh MV (2007) Efficiency of seed treatment for ameliorating zinc deficiency in crops. In: Proceedings of zinc crops: improving crop production and human health, Istanbul, 24–26 May 2007

Singh J, Nadarajan N, Basu PS, Srivastava RP, Kumar L (2013) Pulses for Human Health and Nutrition, pp 1–35. Technical Bulletin No. 7/2013, Indian Institute of Pulses Research, Kanpur

Singh U, Praharaj CS, Chaturvedi SK, Bohra A (2016) Biofortification: introduction, approaches, limitations and challenges. In: Singh U et al (eds) Biofortification of food crops. Springer, New Delhi, pp 3–18. https://doi.org/10.1007/978-81-322-2716-8_1

Smrkolj P, Germ M, Kreft I, Stibilj V (2006) Respiratory potential and Se compounds in pea (*Pisum sativum* L.) plants grown from Se-enriched seeds. J Exp Bot 57(14):3595–3600

Smýkal P, Aubert G, Burstin J, Coyne CJ, Ellis NT, Flavell AJ, Ford R, Hýbl M, Macas J, Neumann P, McPhee KE (2012) Pea (Pisum sativum L.) in the genomic era. Agronomy 2(2):74–115

Snodderly DM (1995) Evidence for protection against age-related macular degeneration by carotenoids and antioxidant vitamins. Am J Clin Nutr 62:1448S–1461S

Sommer A, Čerešňáková Z, Frydrych Z, Králík O, Králíková Z, Krása A, Pajtáš M, Petrikovič P, Pozdíšek J, Šimek M, Třináctý J, Vencl B, Zeman L (1994) Potřeba živin a tabulky výživné hodnoty krmiv pro přežvýkavce. VÚVZ Pohořelice, 196pp

Southgate DA (1995) Digestion and metabolism of sugars. Am J Clin Nutr 62(1 Suppl):203S–210S

Spallholz JE, Mallory Boylan L, David Robertson J, Smith L, Rahman MM, Hook J, Rigdon R (2008) Selenium and arsenic content of agricultural soils from Bangladesh and Nepal. Toxicol Environ Chem 90(2):203–210

Stephens D, Jackson PL, Gutierrez Y (1996) Subclinical vitamin A deficiency: A potentially unrecognized problem in the United States. Pediatr Nurs 22:377–389

Stewart CP, Dewey KG, Ashoran P (2010) The undernutrition epidemic: an urgent health priority. Lancet 375:282

Stoltzfus RJ (2011) Iron interventions for women and children in low-income countries. J Nutr 141:756S–762S

Suarez FL, Springfield J, Furne JK, Lohrmann TT, Kerr PS, Levitt MD (1999) Gas production in human ingesting a soybean flour derived from beans naturally low in oligosaccharides. Am J Clin Nutr 69(1):135–139

Tayeh N, Aubert G, Pilet Nayel ML, Lejeune Hénaut I, Warkentin TD, Burstin J (2015) Genomic tools in pea breeding programs: status and perspectives. Front Plant Sci 6:1037. https://doi.org/10.3389/fpls.2015.01037

Thavarajah D, Ruszkowski J, Vandenberg A (2008) High potential for selenium biofortifi cation of lentils (*Lens culinaris* L.). J Agric Food Chem 56:10747–10753

Thavarajah D, Warkentin T, Vandenberg A (2010) Natural enrichment of selenium in Saskatchewan field peas (*Pisum sativum* L.). Can J Plant Sci 90:383–389. https://doi.org/10.4141/CJPS09154

Thavarajah D, Thavarajah P, Wejesuriya A, Rutzke M, Glahn RP, Combs Jr GF and Vandenberg A (2011) The potential of lentil (Lens culinaris L) as a whole food for increased selenium, iron and zinc intake: preliminary results from a 3 year study. Euphytica 180:123–128

Thavarajah D, Thavarajah P, Vial E, Gebhardt M, Lacher C, Kumar S, Combs GF (2015) Will selenium increase lentil (LensculinarisMedik) yield and seedquality? Front Plant Sci 6:356

Tömösközi S, Lásztity R, Haraszi R, Baticz O (2001) Isolation and study of the functional properties of pea proteins. Nahrung 45:399–401

Tosh SM, Yada S (2010) Dietary fibres in pulse seeds and fractions: Characterization, functional attributes, and applications. Food Res Int 43(2):450–460

Trinidad TP, Mallillin AC, Loyola AS et al (2010) The potential health benefits of legumes as a good source of dietary fibre. Br J Nutr 103:569–574

Troszynska A, Ciska E (2002) Phenolic compounds of seed coats of white and coloured varieties of pea (Pisum sativum L.) and their total antioxidant activity. Czech J Food Sci 20(1):15–22

Tulbek MC, Lam RSH, Wang YC, Asavajaru P, Lam A (2016) Pea: a sustainable vegetable protein crop. In: Nadathur SR, Wanasundara JPD, Scanlin L (eds) Sustainable protein sources. Academic, San Diego, pp 145–164

Turakainen M (2007) Selenium and its effects on growth, yield and tuber quality in potato. Julkaisuja/Helsingin yliopisto, Soveltavan biologian laitos, 4 May 2007

Tzitzikas EN, Vincken JP, de Groot J, Gruppen H, Visser RGF (2006) Genetic variation in pea seed globulin composition. J Agric Food Chem 54(2):425–433. https://doi.org/10.1021/jf0519008

Umbreit J (2005) Iron deficiency: a concise review. Am J Hematol 78:225–231

UNICEF (1990) First call for children. World declaration and 1990–2000 plan of action on the survival, protection and development of children. UNICEF, New York

Urbano G, López-Jurado M, Aranda C, Vilchez A, Cabrera L, Porres JM, Aranda P (2006) Evaluation of zinc and magnesium bioavailability from pea (Pisum sativum, L.) sprouts. Effect of illumination and different germination periods. Int J Food Sci Technol 41(6):618–626

Vidal-Valverde C, Frias J, Hernández AM, Martín-Alvarez PJ, Sierra I, Rodríguez C, Blazquez I, Vicente G (2003) Assessment of Nutritional Compounds and Antinutritional Factors in Pea (Pisum Sativum) Seeds. J Sci Food Agric 83:298–306

Vijayakumari K, Sidduraju P, Pugalenthi M, Janardhanan K (1998) Effect of soaking and heat processing on the levels of antinutrients and digestible proteins in seeds of *Vigna sinensis*. Food Chem 111:132–138

Wang N, Daun JK (2004) Effect of variety and crude protein content on nutrient and certain antinutrients in field peas (Pisum sativum). J Sci Food Agric 84:1021–1029

Wang TL, Domoney C, Hedley CL, Casey R, Grusak MA (2003) Can weimprove the nutritional quality of legume seeds? Plant Physiol 131:886–891

Wang N, Hatcher DW, Gawalko EJ (2008) Effect of variety and processing of nutrients and certain anti-nutrients in field peas (*Pisum sativum*). Food Chem 111:132–138. https://doi.org/10.1016/j.foodchem.2008.03.047

Wang N, Hatcher DW, Toews R, Gawalko EJ (2009) Influence of cooking and dehulling on nutritional composition of several varieties of lentils (*Lens culinaris*). Lebensmittel-Wissenschaft +Technologie 42:842–848. https://doi.org/10.1016/j.lwt.2008.10.007

Wang N, Hatcher DW, Warkentin TD, Toews R (2010) Effect of cultivar and environment on physicochemical and cooking characteristics of field pea (Pisum sativum). Food Chem 118:109–115

Warkentin TD, Delgerjav O, Arganosa G, Rehman AU, Bett KE, Anbessa Y, Rossnagel B, Raboy V (2012) Development and Characterization of Low-Phytate Pea. Crop Sci 52:74–78

Welch RM (2003) Farming for nutritious foods: agricultural technologies for improved human health. IFA-FAO Agriculture Conference on Global food security and the role of sustainable fertilization, Rome, 26–28 Mar, pp 2–24

Welch RM, Graham RD (1999) A new paradigm for world agriculture: meeting human needs: productive, sustainable, nutritious. Field Crop Res 60:1–10

Welch RM, Graham RD (2002) Breeding crops for enhanced micronutrient content. Plant Soil 245:205–214. https://doi.org/10.1023/A:1020668100330

Welch RM, Graham RD (2004) Breeding for micronutrients in staple food crops from a human nutrition perspective. J Exp Bot 55:353–364. https://doi.org/10.1093/jxb/erh064

Welch RM, Graham RD (2005) Agriculture: the real nexus for enhancing bioavailable micronutrients in food crops. J Trace Elem Med Biol 18:299–307. https://doi.org/10.1016/j.jtemb.2005.03.001
Wessells KR, Brown KH (2012) Estimating the global prevalence of zinc deficiency: results based on zinc availability in national food supplies and the prevalence of stunting. PLoS One 7(11):e505–e568
White PJ, Broadley MR (2009) Biofortification of crops with seven micronutrient elements often lacking in human diets – iron, zinc, copper, calcium, magnesium, selenium and iodine. New Phytol 182:49–84
WHO (2009) The World Health Report. World Health Organization, Geneva
Wiltgren AR, Booth AO, Kaur G, Cicerale S, Lacy KE, Thorpe MG, Keast RS, Riddell LJ (2015) Micronutrient supplement use and diet quality in university students. Nutrients 7:1094–1107
World Health Organization (2005) Protein and Amino Acid Requirements in Human Nutrition. Report of a Joint WHO/FAO/UNU Expert Consultation. WHO Technical Report Series no. 935. WHO, Geneva
World Health Organization (WHO) (2012) The World Health Report. World Health Organization, Geneva
World Health Organization and Food and Agriculture Organization of the United Nations (2006) Guidelines on food fortification with micronutrients; Allen L, de Benoist B, Dary O, Hurrell R (eds). WHO, Geneva
World Health Organization WHO (2008) Worldwide prevalence of Anaemia 1993–2005: WHO Global Database on Anaemia; de Benoist B, McLean E, Egli I, Cogswell M (eds). World Health Organization Press, Geneva
Xu BJ, Yuan SH, Chang SKC (2007) Comparative analyses of phenolic composition, antioxidant capacity, and color of cool season legumes and other selected food legumes. J Food Sci 72:S167–S175
Yang XE, Chen WR, Feng Y (2007) Improving human micronutrient nutrition through biofortification in the soilplant system: China as a case study. Environ Geochem Health 29:413–428
Zimmermann MB, Hurrell R (2007) Nutritional iron deficiency. Lancet 370:511–520
Zou CQ, Zhang YQ, Rashid A, Ram H, Savasli E, Arisoy RZ, Ortiz-Monasterio I, Simunji S, Wang ZH, Sohu V, Hassan M, Kaya Y, Onder O, Lungu O, Yaqub Mujahid M, Joshi AK, Zelenskiy Y, Zhang FS, Cakmak I (2012) Biofortification of wheat with zinc through zinc fertilization in seven countries. Plant Soil 361:119–130. https://doi.org/10.1007/s11104-012-1369-2
Zuo YM, Zhang FS, Li XL et al (2000) Studies on the improvement in iron nutrition of peanut by intercropping with maize on a calcareous soil. Plant Soil 220(1):13–25

Genetic Potential of Lentil as a Nutritionally Rich Food Legume Crop

Jitendra Kumar, Debjyoti Sen Gupta, and Shiv Kumar

Abstract Micronutrient deficiency affects more than two billion population worldwide, especially in sub-Saharan Africa and South Asia. Their deficiency in human body is commonly known as "hidden hunger" and causes many health hazards, including low birth weight, anemia, learning disabilities, increased morbidity and mortality rates, low work productivity, and high healthcare costs. Biofortification of food crop varieties with essential micronutrients is one of the means to combat micronutrient deficiencies through classical plant genetic improvement. Lentil, which is rich source of protein and other minerals including iron, zinc, selenium, folates, carotenoids, and vitamins, has been shown to have genetic variability among the lentil germplasm for these traits. Therefore, lentil crop has been identified as an ideal crop for micronutrient biofortification and a possible whole food solution to the global micronutrient malnutrition. The present chapter discusses the current efforts made toward the genetic biofortification in lentil using different tools of classical plant breeding and modern genomics.

Keywords *Lens culinaris* · Iron · Zinc · Protein · Raffinose family oligosaccharides · Phytic acid · Genetic biofortification

Introduction

Insufficient availability of proteins, carbohydrates, vitamins, and essential minerals and the presence of antinutritional compounds in our daily diet cause nutritional malnutrition. Among these, deficiency of micronutrients is a more serious problem among the world populations due to more intake of carbohydrate-rich cereal-based diet, which is low in these essential micronutrients (Stewart et al. 2010; Bouis et al.

J. Kumar (✉) · D. Sen Gupta · S. Kumar
Division of Crop Improvement, ICAR-Indian Institute of Pulses Research, Kanpur, Uttar Pradesh, India
e-mail: jitendra.kumar@icar.gov.in

D. S. Gupta et al. (eds.), *Breeding for Enhanced Nutrition and Bio-Active Compounds in Food Legumes*, https://doi.org/10.1007/978-3-030-59215-8_4

2011). Thus, such populations need essential micronutrients despite enough food intake. As a result, it causes several health problems, including low birth weight, anemia, learning disabilities, increased morbidity and mortality rates, low work productivity, and high healthcare costs especially in developing nations (Batra and Seth 2002; Welch and Graham 1999). Deficiencies of iron (Fe), zinc (Zn), selenium (Se), and iodine (I) are commonly seen among the people of rural areas in Southeast Asia. Studies showed that about 60% of world population is deficient in Fe (Yang et al. 2007), 33% in Zn (Hotz et al. 2004), and 15% in Se (FAOSTAT 2007). About 20% of children aged less than 5 years are dying due to the deficiencies of vitamin A, Zn, Fe, and/or I (Prentice et al. 2008). In preschool children and pregnant women, deficiencies of essential micronutrients, especially Zn and Fe, are more commonly observed (Welch and Graham 1999; White and Broadley 2009; WHO 2012). An estimate of the World Health Organization revealed that ~ 25% of the world's population suffers from anemia (WHO 2008), while 17.3% of people worldwide suffer from inadequate intake of Zn (Wessells and Brown 2012) leading to an annual death of 433,000 children under the age of 5 (WHO 2009). All over the world, more than one billion people who have Se deficiency mainly suffer from Keshan (cardiomyopathy) and Kashin–Beck (osteoarthropathy) diseases (Reilly 1996). This problem is more serious in those countries (i.e., New Zealand, Australia, Thailand, Africa, Denmark, Finland, central Siberia, northeast to south central China, Turkey, parts of India, Nepal, and Bangladesh) where soils have low level of bioavailable Se (100–2000 μg kg−1) (Swaine 1955; Fordyce 2005; Lyons et al. 2005; Spallholz et al. 2004, 2008). Studies showed that Se prevents cytotoxic effect of arsenic (Biswas et al. 1999) because both Se and As work as mutual detoxification (Holmberg Jr and Ferm 1969; Levander 1977). Therefore, it is an important mineral. Also, deficiency of folate occurs globally among millions of people in both developed and developing countries, and it causes several health problems (Gupta et al. 2013).

To overcome the problems of malnutrition, different strategies have been adopted during the past years. Efforts are being made to overcome these by adopting different approaches. These strategies include providing diet enriched with minerals and vitamins. One of the strategies is to provide mineral- and vitamin-enriched food supplements. However, this approach is only applicable to those people who can afford to buy the costly food supplements, and thus this approach is unaffordable to poor people. Another strategy is to provide the nutrient-dense biofortified staple foods through daily diet. This is the most effective way to increase intake of nutrients among resource-poor people (Bouis et al. 2011).

Biofortified food crops can be developed by following two approaches: First is the agronomic approach, which increases the concentration of nutrients in the edible parts of crop plants by applying the micronutrient fertilizers containing Fe-chelates and by using intercropping and crop rotations, as well as soil microorganisms for making improvement in solubilization and mobilization of Fe in the soil (White and Broadley 2009). For example, Se application increases Se concentration in lentil (Thavarajah et al. 2015) and also in other crops such as potato tubers, tea leaves, and field pea seeds (Hu et al. 2003; Smrkolj et al. 2006; Turakainen 2007). Though it is a rapid method to develop the micronutrient dense staple food crop, it is not always a successful and sustainable approach because they increase the cost of production, especially in developing countries (Graham and Rengel 1993). Moreover, there is a need to be careful in the application of dose of inorganic fertilizers because of the narrow window between toxic and beneficial levels especially in the application of Se fertilizers (Terry et al. 2000). Second is the genetic approach in which nutritionally dense high-yielding varieties of food crops are developed by changing the genetic makeup of genotypes using the classical plant breeding and modern genomic approaches. This approach is known as genetic biofortification, which offers a sustainable and cost-effective way of providing the essential micronutrients to the people of both developing and developed countries compared to agronomic approach that involves some technology and costs (Graham et al. 2007; White and Broadley 2009). Earlier several reviews have been published on general and specific aspects of biofortification. The general aspects focused on food system strategies for preventing micronutrient malnutrition (Miller and Welch 2013) and progressing toward a more nourishing future (Saltzman et al. 2013), while specific aspect covered strategies to increase selenium, zinc, and iron content (Hawkesford and Zhao 2007; Velu et al. 2014) and the use of genomic for Fe and Zn biofortification in wheat (Borrill et al. 2014) and in common bean (Blair 2013; Petry et al. 2015). In the recent past years, genetic potential for increasing the concentrations of Fe and Zn and reducing antinutrients such as PA in/from grains of various food crops such as maize (*Zea mays* L.), rice (*Oryza sativa* L.), wheat (*Triticum sativum* L.), and common bean (*Phaseolus vulgaris* L.) and field pea has been reviewed in detail (Frossard et al. 2000; Gomez-Galera et al. 2010; Mayer et al. 2008; Welch and Graham 2004; Amarakoon et al. 2012). This chapter focuses on reviewing the current efforts made to study the genetic potential of lentil as a nutritionally rich food legume crop for poor people of developing countries.

Nutritional Composition of Lentil Seeds

Lentil (*Lens culinaris* Medikus subsp. *culinaris)* is one of the important cool-season pulses. Globally, it is cultivated in 52 countries, and among these countries, India (36%), western Canada (18%), southeastern Turkey (15%), and Australia (4%) are major lentil-growing countries (FAOSTAT 2011). Naturally, lentils are rich source of proteins, iron, zinc, selenium, folates, carotenoids, vitamins, and other nutritional components (Thavarajah et al. 2011a, b; Johnson et al. 2013a, b; Gupta et al. 2013), and several previous studies have described its nutritional value (Table 1). Studies showed that intake of 100 g lentil grain can be enough to fulfill the recommended daily allowance (RDA) (Fe, 41–113%; Zn, 40–68%; Se, 77–122%) based on given nutritional value in Table 1. Naturally, lentils are also found rich in beta-carotene (2–12 μg/g) and low in PA-phosphorus (0.7–1.2 mg/g) in the lentils grown in the USA. The level of PA has been observed comparatively lower than from those genotypes of other crops that contain low PA (rice, 1.22–2.23 mg/g; soybean, 1.77–4.86 mg/g; wheat, 1.24–2.51 mg/g; maize, 3.3–3.7 mg/g; common bean, 0.52–1.38 mg/g). Also, its faster booking ability (<10 min) helps to save time and energy required for cooking (Thavarajah et al. 2011a). Thus, naturally lentil is suitable for developing the nutrient-dense pulse crop that can help to overcome problem of global malnutrition among poor populations (Thavarajah et al. 2011a). The concentration level of other nutritional and antinutritional compounds observed in the lentil seeds is also given in Table 2.

Table 1 Nutritional value of lentil seeds

Protein	20–25%
Carbohydrate	50–60%
Fat	0.7–0.8%
Ca	60–70 mg/100 g
Fe	7–9 mg/100 g
Zn	4–5 mg/100 g
Se	42–67 μg/100 g
Folate	261–290 μg/100 g

Table 2 Nutritional traits for genetic biofortification in lentil

Targeted nutritional traits	Concentration
Nutritional traits for increasing their concentration	
Protein	15.9–32%
Starch	34.7–65.0%
Dietary fibers	5.1–26.6%
Fatty acids	0.3–3.5 g/100 g
Micronutrients	
Iron	73–90 mg/kg
Zinc	44–54 mg/kg
Antinutritional traits for decreasing their concentration	
Total phenolics	6.24–27.73 mg GAE/g defatted sample
Total flavonoids	1.15–4.94 mg CE/g defatted sample
Condensed tannin content	3.14–12.97 mg CE/g defatted sample
Phytoestrogens	8.9–12.3 μg/100 g dry matter
Phytate	3.9–11.9 mg/g
Saponins	0.07–0.13 g/100 g
Protease inhibitor	25–55 TIA/mg of protein
α-Amylase inhibitor	–
Lectins	–
Vicilin protein	–
Low molecular weight carbohydrates as nutritional traits increasing their concentration	
Sorbitol	1250–1824 mg/100 g
Mannitol	57–132 mg/100 g
Galactinol	46–89 mg/100 g
Sucrose	1750–2355 mg/100 g
Raffinose + stachyose	3314–4802 mg/100 g
Verbascose	1907–2453 mg/100 g
Nystose	8–450 mg/100 g

Source: modified from Kumar et al. (2016)
GAE gallic acid equivalent, *CE* catechin equivalent, *TIA* trypsin inhibitor activity

Targeted Traits for Genetic Biofortification in Lentil

The use of conventional and modern breeding approaches for developing the nutritionally dense cultivars is an effective and a long-term sustainable solution for increasing the bioavailability of minerals (Nestel et al. 2006). For this, several nutritional traits are required to target for increasing or decreasing their concentration in the lentil seeds through breeding. Those nutritional deficiencies, which affect a large population worldwide, especially among the people of developing countries, are used as target trait for increasing their concentration. However, some phytochemicals present in lentil are to be toxic to the human body and hence they are targeted for reduction of their concentration by introgressing the genes that control production of antinutritional phytochemicals in less concentration in the plants. These potential traits have been summarized in Table 2. These traits have been discussed recently in a review (Kumar et al. 2016).

Genetic Variability Available in Gene Pool

Cultivated Gene Pool

Identification of natural variants having favorable alleles for a nutritional trait in the cultivated gene pool is a simple approach for developing the biofortified genotypes of lentil. Such natural variants can be used as donor for transferring the useful genes in the background of cultivated gene pool through breeding or can be directly released as biofortified variety, if identified variant is already a high-yielding variety. In the past years, cultivated gene pool of different countries, including Turkey, Syria, Canada, and Pakistan, has been screened for identification of genetic variability for nutritional traits especially folate and macro-/micronutrient traits (see Kumar et al. 2016, Table 2). The International Center for Agricultural Research in the Dry Areas (ICARDA) under the HarvestPlus Challenge Program screened a collection of >1600 accessions belonging to cultivated gene pool (i.e., local landraces, breeding lines, released cultivars) for iron (42–132 ppm) and zinc (23–78 ppm) content (Sarker et al. 2007; HarvestPlus 2014). However, in another study, little genetic variation has been observed for Fe and Zn concentrations among 19 genotypes belonging to different market classes in Canada. In these genotypes, Fe was ranged from 73 to 90 mg kg-1, and Zn was ranged from 44 to 54 mg kg-1 (Thavarajah et al. 2009a, b). For folate concentration, significant genetic variation ranging from 216 to 290 μg/100 g has been observed among 10 lentil cultivars of the USA, and compared to other pulse crops (chickpea, 42–125 μg/100 g; yellow field pea, 41–55 μg/100 g; green field pea, 50–202 μg/100 g), this concentration was higher (Gupta et al. 2013). Turkish landraces also showed a range diversity not only for micronutrients but also for macronutrients (Karakoy et al. 2012). The effect of genetic constitution has been observed on Fe and Zn concentration among 41 elite lines of lentil (Kumar et al. 2014a). Organic Se, selenomethionine, is also found in sufficient amount in lentil seeds (Thavarajah et al. 2007, 2008), and its concentration does not change during cooking (Thavarajah et al. 2008). For this compound, significant genotypic differences were observed among the genotypes of different countries (Thavarajah et al. 2011b; Rahman et al. 2013). Among 23 genotypes of cultivated gene pool of lentil, genetic variability has been observed for Fe and phytic acid (PA) concentration, and these two traits showed significant positive correlation to each other (DellaValle et al. 2013). Lentils are a rich source of beta-carotene, and significant variability for this trait (2–12 μg/g) has also been reported among of the USA germplasm lines (Thavarajah et al. 2011b). Genetic variation has been observed among lentil varieties for prebiotic carbohydrates. It is one of the important components of healthy diets supporting healthful hindgut microflora, and hence its concentration can be improved through breeding (de Almeida Costa et al. 2006; Tahir et al. 2011; Wang et al. 2009). In another study, genetic variability for prebiotic carbohydrates such as raffinose family of oligosaccharides (RFO), sugar alcohols, fructooligosaccharides (FOS), and resistant starch (RS) has been observed among 10 commercial lentil varieties belonging to cultivated gene pool that may be possible to enhance through breeding and location sourcing (Johnson et al. 2013b; Table 3).

Wild Gene Pool

Wild gene pool is a rich reservoir of useful alien genes that are no longer available within the cultivated gene pool (Hawkes 1977; Doyle 1988; Tanksley and McCouch 1997). In lentil, 587 accessions representing six wild *Lens* species from 26 countries have been collected and conserved by ICARDA. Many wild species have shown their cross compatibility with cultivated species (Abbo and Ladizinsky 1991, 1994; Fratini et al. 2004; Fratini and Ruiz 2006; Muehlbauer et al. 2006). Wild gene pool carries useful genes that can be used to generate new variability through recombination breeding, and in other legume crops, accessions of wild gene pool have been identified as donors for high concentration of minerals (Blair and Izquierdo 2012; Monasterio and Graham 2000; Cakmak et al. 2000; Ortiz-Monasterio et al. 2007). However, in lentil, not much efforts have been made to identify the wild relatives having their richness for nutritional traits. A study determined protein content, phenol content, and antioxidant activity among 10 accessions of lentil wild species. In this study, average protein content, phenol content, and higher total antioxidant activity were observed, 29.7%, 8.9 mg/100 g, and 16.17 μmole TE/g, respectively, that were higher than the accessions belonging to cultivated species (personal communication with Dr. Jagdish Singh, Indian Institute of Vegetable Research, Varanasi, India). Recently, biofortification potential of wild gene pool in lentil was determined, and considerable variation (mg/100 g) was observed for Na (30–318), K (138.29–1578), P (37.50–593.75), Ca (4.74–188.75), Mg (15–159), Fe (2.82–14.12), Zn (1.29–12.62), Cu (0.5–7.12), Mn (1.22–9.99), Mo (1.02–11.89), Ni (0.16–3.49),

Table 3 Genetic variability studied among germplasm lines of lentil

Type of genetic material	No. of accessions	Compound with range of concentration	Country	References
Landraces, wild types, and breeding lines	1600	Fe: 43–132	ICARDA, Syria	Sarker et al. (2007)
		Zn: 22–78	do	Baum et al. (2008)
Lentil germplasm		Fe: 41–109		
		Zn: 22–78		
Breeding lines, germplasm, and modern high-yielding genotypes	900	Fe: 73–90	Canada	Thavarajah et al., (2009a, b)
		Zn: 44–54		
Varieties	19	Se: 425–673	Canada	Thavarajah et al. (2008)
Landraces, cultivars	46	Fe: 49–81	Turkey	Karakoy et al. (2012)
		Zn: 42–73		
Landraces, breeding material	96	Fe: 37–157		
		Zn: 26–65	India	Kumar et al. (2018a)
Exotic lines		Se: 240–630		
Varieties	19	Fe: 73–90	Canada	Thavarajah et al. (2009a, b)
		Zn: 44–54 mg		
Varieties		Folate: 216–290 μg/100 g	USA	Gupta et al. (2013)
Elite breeding lines	41	Fe: 50.85–136.9 mg	India	Kumar et al. (2014a)
		Zn: 40.26–81.5		
Varieties	7	Se: 74–965 μg kg^{-1}	Bangladesh	Rahman et al. (2013)
Varieties	23	Fe: 43–92 ppm	Canada	DellaValle et al. (2013)
		PA: 3.8–15.9 mol/g		
Germplasm line	–	Beta-carotene: 2–12 μg/g	USA	Thavarajah et al. (2011b)
Varieties	9	Prebiotic carbohydrate	USA	Johnson et al. (2013b)
Cultivars, breeding lines	192	Se: 6–254 μg/	Syria, Nepal Morocco, USA, Australia, Turkey	Thavarajah et al. (2011b)

Pb (0.01–0.58), Cd (0–0.03), Co (0–0.63), and As (0–0.02). In this study, accessions of wild species belonging to Turkey and Syria showed maximum variability for different minerals (Kumar et al. 2018b). Therefore, wild gene pool can be useful genetic resources for minerals and other nutritional traits and can be used for transferring favorable alien genes for these traits.

Location-Specific Breeding for Biofortified Traits

Growing conditions related to soil and environments (i.e., pH, temperature, radiation, precipitation, organic matter, and soil texture) influence the accumulation of micronutrients to seeds (Tisdale and Nelson 1975; Cakmak 2008; Joshi et al. 2010). Therefore, knowledge of optimal lentil-growing conditions becomes important for harvesting highest concentration of a compound during mass cultivation of biofortified crops. Environmental conditions, particularly growing locations, also complicate the breeding for nutritional traits. As the result, knowledge of genotype × environment interactions has become important for developing stable biofortified cultivars or for designing location-specific breeding program for biofortified traits. In lentil, accumulation of PA, Fe, and Zn in the seeds is known to vary with growing weather (rainfall and temperature), location, and soil conditions (Thavarajah et al. 2009a, b). More recently, genotype × environment interactions have affected the concentration of Fe and Zn in lentil (Kumar et al. 2018a). It has been shown that concentration of micronutrients varies from one geographical region to other region when studied international lentil samples (see Thavarajah et al. 2011a). For example, concentration of Fe has been observed high in seed sample of Syria (63 mg/kg), Turkey (60 mg/kg), USA (56 mg/kg), and Nepal (50 mg/kg) and low in the seed sample of Australia (46 mg/kg) and Morocco (42 mg/kg). Likewise, Zn was high in the seeds of lentil genotypes grown in Syria (36 mg/kg), Turkey (32 mg/kg), and USA (28 mg/kg) and low in the seeds of genotypes grown in Australia (18 mg/kg) and Morocco (27 mg/kg). Also concentration of Se has been observed high in the seeds of those genotypes that belonged to Nepal (180 μg/kg) and Australia (148 μg/kg) compared to the seeds of those genotypes that pertained to Syria (22 μg/kg), Morocco (28 μg/kg), and Turkey (47 μg/kg) (Thavarajah et al. 2011a). Concentration of Ca in the seeds of Turkish landraces was high (0.48–1.28 g/kg), while concentration of Fe (37–156 mg/kg) and Zn (26–65 mg/kg) was high in the seeds of lentil samples grown in Indian conditions (Karakoy et al. 2012; Kumar et al. 2016). Studies showed that concentration of Fe is highly fluctuated with environmental fluctuations as compared to the concentration of zinc (HarvestPlus 2014; Kumar et al. 2014a). In Bangladesh, significant genotype and location differences were observed for selenium concentration but no genotype × location interaction was observed in a study when seven lentil genotypes were evaluated over four locations along with a farmers' field survey (Rahman et al. 2013). Similar results have also been obtained in another study when 12 genotypes were evaluated over seven locations in Australia (Rahman et al. 2013, 2014). For folate concentration, a significant year × location interaction was observed among 10 lentil cultivars of the USA studied over 2 years (Gupta et al. 2013). In another study, temperature influences phytic acid (PA), iron (Fe), and zinc (Zn) concentration among mature seeds of 11 lentil genotypes when these genotypes have been studied under simulated long-term temperature regimes representing Saskatoon, Canada (decreasing temperatures), and Lucknow, India (increasing temperatures). In this study, PA and Zn concentrations in lentil seeds have been observed significantly higher in the rising temperature regime (8.8 mg/g and 69 mg/kg, respectively) compared to the decreasing temperature regime (6.7 mg/g and 61 mg/kg, respectively). Fe concentrations followed the same trend (116 vs. 113 mg/kg). Thus, for developing the lentil cultivars with lower concentration of PA, the cooler temperatures of temperate summers might be an important factor, and it can be a biofortification strategy with an aim of lowering the PA content in staple crop (Thavarajah et al. 2010). Local environments highly influenced the nutritional traits such as folate, Se, and Zn, and hence location-specific biofortified cultivars can be developed by utilizing the genetic variability of these traits (Gupta et al. 2013). It is more important under the global warming conditions where it has been forecasted to increase winter temperature patterns because it can increase the level of antinutrients such as PA. Therefore, success in global micronutrient malnutrition management will depend on using the genetics for designing the biofortification strategies toward the development of cultivars having a high level of Fe and Zn and a low level of antinutrients.

Genomics for Biofortification

The current genomic approaches help to identify the markers associated with gene/QTL controlling concentration of micronutrients that can be used in marker-assisted breeding program for developing the biofortified cultivars (Grusak 2002; Bouis 2003). During the past years, significant progress has been made in the development of genomic resources in lentil (Kumar et al. 2014b, 2015). However, these genomic resources are not used widely for identification of genes/QTL controlling biofortified traits in lentil. Few studies have been conducted to map and tag the gene(s)/QTL controlling micronutrient traits. In lentil, those genes that control Fe uptake have been mapped in a RIL population (ILL 8006–BM (Barimasur-4) × CDC Milestone) through QTL analysis. Phenotyping data recorded on RILs over three different locations (Izmir, Adana, and S Urfa) of Turkey ranged between 37 and 149 mg/kg. Out of 181 markers including 150 AFLPs, 27 SSRs, and 4 SNPs, 149 markers were mapped on 16 linkage groups covering 496.5 cM. This information was used along with phenotyping data in QTL analysis leading to the identification of four QTLs. These QTLs can be used in molecular breeding toward the development of biofortified cultivars (Aldemir et al. 2014). Association mapping analysis identified SSR markers associated with Fe concentration explaining 9–11% of phenotypic variation and Zn concentration explaining 14–21% of phenotypic variation. These loci have been shown to be associated with Fe and Zn concentration stably across the locations (Singh et al. 2017). In a biparental mapping population, 21 QTLs explaining 5.9–14.0% of the phenotypic variation distributed over six linkage groups (LG1, LG2, LG4, LG5, LG6, and LG7) have been identified for Fe uptake in lentil (Aldemir et al. 2017). In another study, six QTLs explaining 15.3–24.1% of the phenotypic variation have been identified for manganese uptake in lentil (Ates et al. 2018). These QTLs can be used for the development of micronutrient-enriched lentil genotypes. Moreover, lentil genotypes varying for grain Fe and Zn concentration have been characterized on the basis of molecular markers (Kumar et al. 2014a; Singh et al. 2018). Based on molecular diversity and multilocation phenotyping of Fe and Zn, diverse parents have been identified and used in hybridization for obtaining the transgressive segregants and for developing mapping populations in lentil (Kumar et al. 2014a; Singh et al. 2018).

Conventional Breeding Approach Used in the Development of Biofortified Cultivars

The efforts have been made toward the screening of existing released varieties for Fe and Zn content under the HarvestPlus Challenge Program in different countries (i.e., Bangladesh, Ethiopia, Nepal, Morocco, Turkey, Syria, Lesotho, and Portugal). As a result, several varieties possess high Fe and Zn levels along with good agronomic performance have been identfied as biofortified variety in lentil (Table 4). These varieties are being disseminated to farmers on a fast-track mode through national program. For example, in Bangladesh, the government has taken a massive dissemination program to promote promising lentil varieties having high Fe and Zn – Barimasur 5 and Barimasur 6 in traditional and nontraditional areas. Similarly in Nepal, lentil varieties such as Sishir, Khajurah-2, Khajurah-1, and Shital are spreading fast in the Terai region. In India, the variety Pusa Vaibhav rich in Fe is being grown by farmers in its northwest plain zone (ICARDA 2012). In the future, more varieties of lentil would be released as biofortified varieties in different countries. For example, in Nepal, ILL 7723 has been recommended by the National Variety Release Committee and can be released shortly for farmers' cultivations (HarvestPlus 2014). Moreover, breeding lines developed for high yielding have also screened for Fe and Zn concentration in their seeds. This led to the identification of a high-yielding breeding line IPL 220 having a high concentration of Fe (73–114 ppm) and Zn (51–65 ppm). This breeding line has been released as biofortified variety in India for cultivation in northeastern regions of the country.

Table 4 Lentil variety rich in Fe and Zn identified through analysis of released cultivars under the HarvestPlus Challenge Program of CGIAR

Country	Name of variety	Content (ppm)	
		Fe	Zn
Bangladesh	Barimasur -4	86.2	
	Barimasur-5	86	59
	Barimasur-6	86	63
	Barimasur-7	81	
Nepal	Shishir	98	64
	Khajurah-2	100.7	59
	Khajurah-1	–	58
	Sital	–	59
	Shekhar	83.4	–
	Simal	81.6	–
India	Pusa Vaibhav	102	–
	L 4704	125	74
	IPL 220	73–114	51–65
Syria/Lebanon	Idlib-2	73	–
	Idlib-3	72	–
Ethiopia	Alemaya	82	66

Future Directions

As discussed above, lentil is an indispensable supplementary food in many countries, particularly in South Asia, West Asia, Northeast and North Africa. In South Asia, particularly in Pakistan, Nepal, Bhutan, India, and Bangladesh, red lentil is an integral part of daily diet, most particularly among rural population. Red lentil is very popular in Turkey and other Mediterranean countries owing to its abundant nutritional and functional components. As more than three billion people worldwide have mineral deficiencies, even a small increase in the nutritive value of lentil seed may be highly significant for improvement of human nutrition. The studies showed considerable variation for macro- and micronutrient among the landraces and cultivars of lentil. These variations provide a useful foundation for using them in breeding program for the development of new cultivars of lentil with high mineral content. In addition to this, such variation can also be used to identify quantitative trait loci (QTL) associated with mineral uptake and transport.

References

Abbo S, Ladizinsky G (1991) Anatomical aspects of hybrid embryo abortion in the genus Lens L. Bot Gaz 152(3):316–320

Abbo S, Ladizinsky G (1994) Genetical aspects of hybrid embryo abortion in the genus Lens L. Heredity 72:193–200

Aldemir SB, Sever T, Ates D, Yagmur B, Kaya HB, Temel HY, Kahriman A, Ozkan H, Tanyolac MB (2014) QTL mapping of genes controlling to fe uptake in lentil seed (*Lens culinaris* L.) using recombinant inbred lines. Plant & Animal Genome XXII Conference, January 11–15, 2014, San Diego, CA, USA. Available at https://pag.confex.com/pag/xxii/webprogram/Paper9689.html

Aldemir S, Ateş D, Temel HY, Yağmur B, Alsaleh A, Kahriman A, Özkan H, Vandenberg A, Tanyolac MB (2017) QTLs for iron concentration in seeds of the cultivated lentil (*Lens culinaris* Medic.) via genotyping by sequencing. Turk J Agric For 41(4):243–255

Amarakoon D, Thavarajah D, McPhee K, Thavarajah P (2012) Iron-, zinc-, and magnesium-rich field peas (Pisum sativum L.) with naturally low phytic acid: a potential food-based solution to global micronutrient malnutrition. J Food Compos Anal 27(2012):8–13

Ates D, Aldemir S, Yagmur B, Kahraman A, Ozkan H, Vandenberg A, Tanyolac MB (2018) QTL mapping of genome regions controlling manganese uptake in lentil seed. G3: Genes, Genomes, Genetics 8:1409–1416. g3-200259

Batra J, Seth PK (2002) Effect of iron deficiency on developing rat brain. Indian J Clin Biochem 17(2):108–114

Baum M, Hamweih A, Furman B, Sarker A (2008) Erskine W (2008) Biodiversity, genetic enhancement and molecular approaches in lentils. In: Sharma AK, Sharma A (eds) Plant genome, biodiversity and evolution. Science Publishers, Enfield, pp 135–152

Biswas S, Talukder G, Sharma A (1999) Prevention of cytotoxic effects of arsenic by short-term dietary supplementation with selenium in mice in vivo. Mutat Res Genet Toxicol Environ Mutagen 441(1):155–160

Blair MW (2013) Mineral biofortification strategies for food staples: the example of common bean. Agric Food Chem 61:8287–8294

Blair MW, Izquierdo P (2012) Use of the advanced backcross- QTL method to transfer seed mineral accumulation nutrition traits from wild to Andean cultivated common beans. Theor Appl Genet 125:1015–1031

Borrill P, Connorton JM, Balk J, Miller AJ, Sanders D, Uauy C (2014) Biofortification of wheat grain with iron and zinc: integrating novel genomic resources and knowledge from model crops. Front Plant Sci 5:53. https://doi.org/10.3389/fpls.2014.00053

Bouis HE (2003) Micronutrient fortification of plants through plant breeding: can it improve nutrition in man at low cost? Proc Nutr Soc 62:403e411

Bouis HE, Hotz C, McClafferty B, Meenakshi JV, Pfeiffer WH (2011) Biofortification: a new tool to reduce micronutrient malnutrition. Food Nutr Bull 32:31S–40S

Cakmak I (2008) Enrichment of cereal grains with zinc: agronomic or genetic biofortification? Plant Soil 302:1e17

Cakmak I, Ozkan H, Braun H-J, Welch RM, Romheld V (2000) Zinc and iron concentrations in seeds of wild primitive and modern wheats. Food Nutr Bull 21:e401–e403

de Almeida Costa GE, da Silva Queiroz-Monici K, Pissini Machado Reis SM, de Oliveira AC (2006) Chemical composition, dietary fibre and resistant starch contents of raw and cooked pea, common bean, chickpea and lentil legumes. Food Chem 94(3):327–330

DellaValle DM, Thavarajah D, Thavarajah P, Vandenberg A, Glahn RP (2013) Lentil (*Lens culinaris* L.) as a candidate crop for iron biofortification: is there genetic potential for iron bioavailability. Field Crop Res 144:119–125

Doyle JJ (1988) 5S ribosomal gene variation in the soybean and its progenitor. Theor Appl Genet 75:621–624

FAOSTAT (2007). http://faostat.fao.org/

FAOSTAT (2011). http://faostat.fao.org/

Fordyce FM (2005) Selenium deficiency and toxicity in the environment. In: Essentials of medical geology 2013. Springer, Dordrecht, pp 375–416

Fratini R, Ruiz ML (2006) Interspecific hybridization in the genus lens applying in vitro embryo rescue. Euphytica 150:271–280

Fratini R, Ruiz ML, Perez de la Vega M (2004) Intra-specific and inter-sub-specific crossing in lentil (*Lens culinaris* Medik). Can J Plant Sci 84:981–986

Frossard E, Bucher M, Mächler F, Mozafar A, Hurrell R (2000) Potential for increasing the content and bioavailability of Fe, Zn and Ca in plants for human nutrition. J Sci Food Agric 80(7):861–879

Gomez-Galera S, Rojas E, Sudhakar D, Zhu CF, Pelacho AM, Capell T et al (2010) Critical evaluation of strategies for mineral fortification of staple food crops. *Transgenic Res* 19:165–180. https://doi.org/10.1007/s11248-009-9311-y

Graham RD, Rengel Z (1993) Genotypic variation in zinc uptake and utilization. In: Robson AD (ed) Zinc in soils and plants. Kluwer Academic Publishers, Dordrecht, pp 107–118

Graham RD, Welch RM, Saunders DA, Ortiz-Monasterio I, Bouis HE, Bonierbale M, de Haan S, Burgos G, Thiele G, Liria R, Meisner CA, Beebe SE, Potts MJ, Kadian M, Hobbs PR, Gupta RK, Twomlow S (2007) Nutritious subsistence food systems. Adv Agron 92:1–74

Grusak M (2002) Enhancing mineral content in plant food products. J Am Coll Nutr 21:178Se183S

Gupta DS, Thavarajah D, Knutson P, Thavarajah P, McGee RJ, Coyne CJ, Kumar S (2013) Lentils (*Lens culinaris* L) a rich source of folates. J Agric Food Chem 61:7794–7799

HarvestPlus (2014) Biofortification progress briefs: iron and zinc lentils. Brief 9, pp. 19. Available online at www.HarvestPlus.org

Hawkes JG (1977) The importance of wild germplasm in plant breeding. Euphytica 26:615–621

Hawkesford MJ, Zhao FJ (2007) Strategies for increasing the selenium content of wheat. J Cereal Sci 46(3):282–292

Holmberg RE Jr, Ferm VH (1969) Interrelationships of selenium, cadmium, and arsenic in mammalian teratogenesis. Arch Environ Health Int J 18(6):873–877

Hotz C, Brown KH, Rivera JA et al (2004) Assessment of the risk of zinc deficiency in populations and options for its control. Food Nutr Bull 25(Suppl 2):S94–S204

Hu Q, Xu J, Pang G (2003) Effect of selenium on the yield and quality of green tea leaves harvested in early spring. J Agric Food Chem 51(11):3379–3381

ICARDA (2012). Biofortified lentils to enhance nutritional security in South Asia. ICARDA-South Asia and China Regional Program, New Delhi.

Johnson C, Thavarajah D, Thavarajah P (2013a) The influence of phenolic and phytic acid food matrix factors on iron bioavailability potential in 10 commercial lentil genotypes (Lens culinaris L). J Food Compos Anal 31:82–86. https://doi.org/10.1016/j.jfca.2013.04.003

Johnson CR, Thavarajah D, Combs GF Jr, Thavarajah P (2013b) Lentil (Lens culinaris L.): a prebiotic-rich whole food legume. Food Res Int 51(2013):107–113

Joshi AK, Crossa I, Arun B, Chand R, Trethowan R, Vargas M, Ortiz- Monasterio I (2010) Genotype × environment interaction for zinc and iron concentration of wheat grain in eastern Gangetic plains of India. Field Crop Res 116:268–277

Karakoy T, Halil E, Baloch FS, Toklu F, Eker S, Kilian B, Ozkan H (2012) Diversity of macro- and micronutrients in the seeds of lentil landraces. Sci World J 2012., Article ID 710412:1–9. https://doi.org/10.1100/2012/710412

Kumar H, Singh A, Jain N, Kumari J, Singh AM, Singh D, Sarker A, Prabhu KV (2014a) Characterization of grain iron and zinc in lentil (*Lens culinaris* Medikus culinaris) and analysis of their genetic diversity using SSR markers. Aust J Crop Sci 8:1005–1012

Kumar S, Hamweih A, Manickavelu A, Kumar J, Sharma TR, Baum M (2014b) Advances in lentil genomics. In: Gupta S, Nadarajan N, Gupta DS (eds) Legumes in omics era. Springer, New York, pp 111–130

Kumar S, Rajendran K, Kumar J, Hamwieh A, Baum M (2015) Current knowledge in lentil genomics and its application for crop improvement. Front Plant Sci 6:1–13. https://doi.org/10.3389/fpls.2015.00078

Kumar J, Thavarajah D, Kant R, Sarker A, Singh NP (2016) Genetic variability for mineral nutrients in lentil. J Food Legum 29(2):94–101

Kumar J, Kumar S, Sarker A, Singh NP (2018a) Analysis of genetic variability and genotype× environment interactions for iron and zinc content among diverse genotypes of lentil. J Food Sci Technol 55(9):3592–3605

Kumar S, Choudhary AK, Rana KS, Sarker A, Singh M (2018b) Bio-fortification potential of global wild annual lentil core collection. PLoS One 13(1):e0191122

Levander OA (1977) Metabolic interrelationships between arsenic and selenium. Environ Health Perspect 19:159

Lyons G, Ortiz-Monasterio I, Stangoulis J, Graham R (2005) Selenium concentration in wheat grain: is there sufficient genotypic variation to use in breeding? Plant Soil 269:369–380

Mayer JE, Pfeiffer WH, Beyer P (2008) Biofortified crops to alleviate micronutrient malnutrition. Curr Opin Plant Biol 11(2):166–170

Miller DD, Welch RM (2013) Food system strategies for preventing micronutrient malnutrition. Food Policy 42(2013):115–128

Monasterio I, Graham RD (2000) Breeding for trace minerals in wheat. Food Nutr Bull 21:393e396

Muehlbauer FJ, Cho S, Sarker A, McPhee KE, Coyne CJ, Rajesh PN, Ford R (2006) Application of biotechnology in breeding lentil for resistance to biotic and abiotic stress. Euphytica 147:149–165

Nestel P, Bouis HE, Meenakshi JV, Pfeiffer W (2006) Biofortification of staple food crops. J Nutr 136(4):1064–1067

Ortiz-Monasterio I, Palacios-Rojas N, Meng E, Pixley K, Trethowan R, Pena RJ (2007) Enhancing the mineral and vitamin content of wheat and maize through plant breeding. J Cereal Sci 46:293e307

Petry N, Boy E, Wirth JP, Hurrell RF (2015) Review: the potential of the common bean (*Phaseolus vulgaris*) as a vehicle for iron biofortification. Nutrients 7:1144–1173

Prentice AM, Gershwin ME, Schaible UE, Keusch GT, Victoria LG, Gordon JI (2008) New challenges in studying nutrition disease interactions in the developing world. J Clin Invest 118:1322e1329

Rahman M, Erskine W, Zaman MS, Thavarajah P, Thavarajah D, Siddique K (2013) Selenium biofortification in lentil (*Lens culinaris* Medikus subsp. culinaris): farmers' field survey and genotype × environment effect. *Food Res Int* 54:1596–1604. https://doi.org/10.1016/j.foodres.2013.09.008

Rahman MM, Erskine W, Siddique KHM, Thavarajah P, Thavarajah D, Zaman MS, Materne MA, Mcmurray LM (2014) Selenium biofortification of lentil in Australia and Bangladesh. 6th International Food Legume Research Conference, July 7–11, 2014, TCU Place, Sakatoon, Saskatchewan, Canada

Reilly C (1996) Selenium in food and health. Blackie, London

Saltzman A, Birol E, Bouis HE, Boy E, DeMoura FF, Islam Y, Pfeiffer WH (2013) Biofortification: progress toward a more nourishing future. Glob Food Sec 2:9–17

Sarker A, El-Askhar F, Uddin MJ, Million E, Yadav NK, Dahan R, Wolfgang P (2007) Lentil improvement for nutritional security in the developing world. In: Paper presented at the ASA-CSSA-SSSA international annual meetings, New Orleans, LA, 4–8 Nov 2007

Singh A, Sharma V, Dikshit HK, Aski M, Kumar H, Thirunavukkarasu N, Patil BS, Kumar S, Sarker A (2017) Association mapping unveils favorable alleles for grain iron and zinc concentrations in lentil (*Lens culinaris* subsp. *culinaris*). PLoS One 12(11):e0188296

Singh A, Sharmaand V, Dikshit HK (2018) Molecular characterization of elite lentil (Lens culinaris Medikus) lines varying for grain Fe and Zn concentration. Vegetos Int J Plant Res 31(special):9–18

Smrkolj P, Germ M, Kreft I, Stibilj V (2006) Respiratory potential and Se compounds in pea (*Pisum sativum* L.) plants grown from Se-enriched seeds. J Exp Bot 57(14):3595–3600

Spallholz JE, Boylan LM, Rhaman MM (2004) Environmental hypothesis: is poor dietary selenium intake an underlying factor for arsenicosis and cancer in Bangladesh and West Bengal, India? Sci Total Environ 323(1–3):21–32

Spallholz JE, Mallory Boylan L, David Robertson J, Smith L, Rahman MM, Hook J, Rigdon R (2008) Selenium and arsenic content of agricultural soils from Bangladesh and Nepal. Toxicol Environ Chem 90(2):203–210

Stewart CP, Dewey KG, Ashoran P (2010) The undernutrition epidemic: an urgent health priority. Lancet 375:282

Swaine DJ (1955) The trace-element content of soils, Commonwealth Bureau of Soil Science, Technical Communication no. 48. Rothamsted Experimental Station, Harpenden, p 157

Tahir M, Vandenberg A, Chibbar RN (2011) Influence of environment on seed soluble carbohydrates in selected lentil cultivars. J Food Compos Anal 24(4–5):596–602

Tanksley SD, McCouch SR (1997) Seed banks and molecular maps unlocking genetic potential from the wild. Science 277:1063–1066

Terry N, Zayed AM, De Souza MP, Tarun AS (2000) Selenium in higher plants. Annu Rev Plant Biol 51(1):401–432

Thavarajah D, Vandenberg A, George GN, Pickering IJ (2007) Chemical form of selenium in naturally selenium-rich lentils (*Lens culinaris* L.) from Saskatchewan. *J Agric Food Chem* 55:7337–7341. https://doi.org/10.1021/jf070681i

Thavarajah D, Ruszkoski J, Vandenberg A (2008) High potential for selenium bio-fortification of lentils (*Lens culinaris* L.). J Agric Food Chem 56:10747–10753

Thavarajah D, Thavarajah P, Sarker A, Vandenberg A (2009a) Lentils (*Lens culinaris* Medikus subsp. *culinaris*): a whole food for increased iron and zinc intake. J Agric Food Chem 57:5413–5419

Thavarajah P, Thavarajah D, Vandenberg A (2009b) Low phytic acid lentils (*Lens culinaris* L.): a potential solution for increased micronutrient bioavailability. J Agric Food Chem 57:9044–9049

Thavarajah D, Thavarajah P, See CT, Vandenberg A (2010) Phytic acid and Fe and Zn concentration in lentil (*Lens culinaris* L.) seeds is influenced by temperature during seed filling period. Food Chem 122(2010):254–259

Thavarajah D, Thavarajah P, Wejesuriya A, Rutzke M, Glahn RP, Combs GF Jr, Vandenberg A (2011a) The potential of lentil (*Lens culinaris* L) as a whole food for increased selenium, iron and zinc intake: preliminary results from a 3 year study. Euphytica 180:123–128

Thavarajah D, Thavarajah P, Sarker A, Michael M, George V, Renuka S, Omar I, Omar H, Beybin B, Albert V (2011b) A global survey of effects of genotype and environment on selenium concentration in lentils (Lens culinaris L.): implications for nutritional fortification strategies. Food Chem 125(2011):72–76

Thavarajah D, Thavarajah P, Vial E, Gebhardt M, Lacher C, Kumar S, Combs GF (2015) Will selenium increase lentil (Lens culinaris Medik) yield and seed quality? Front Plant Sci 6:356

Tisdale SL, Nelson WL (1975) Accounting principles Fifth Canadian Edition. In: Soil fertility and fertilizer, vol 1, 3rd edn. McMillan Publishing Company, New York

Turakainen M (2007) Selenium and its effects on growth, yield and tuber quality in potato. Julkaisuja/Helsingin yliopisto, Soveltavan biologian laitos. 2007 May 4

Velu G, Ortiz-Monasterio I, Cakmak I, Hao Y, Singh RP (2014) Biofortification strategies to increase grain zinc and iron concentrations in wheat. J Cereal Sci 59:365e372

Wang N, Hatcher DW, Toews R, Gawalko EJ (2009) Influence of cooking and dehulling on nutritional composition of several varieties of lentils (Lens culinaris). LWT Food Sci Technol 42(4):842–848

Welch MR, Graham DR (1999) A new paradigm for world agriculture: meeting human needs productive sustainable nutritious. Field Crop Res 60:1–10

Welch RM, Graham RD (2004) Breeding for micronutrients in staple food crops from a human nutrition perspective. J Exp Bot 55:353–364. https://doi.org/10.1093/jxb/erh064

Wessells KR, Brown KH (2012) Estimating the global prevalence of zinc deficiency: results based on zinc availability in national food supplies and the prevalence of stunting. PLoS One 7(11):e505-68

White PJ, Broadley MR (2009) Biofortification of crops with seven micronutrient elements often lacking in human diets – iron, zinc, copper, calcium, magnesium, selenium and iodine. New Phytology 82:49–84

WHO (2008) The world health report. World Health Organization, Geneva
WHO (2009) The world health report. World Health Organization, Geneva
WHO (2012) The world health report. World Health Organization, Geneva
Yang XE, Chen WR, Feng Y (2007) Improving human micronutrient nutrition through biofortification in the soil plant system: China as a case study. Environ Geochem Health 29:413–428

Breeding for Low Phytates and Oligosaccharides in Mungbean and Blackgram

J. Souframanien, V. J. Dhole, and K. S. Reddy

Abstract Mungbean or Green gram (*Vigna radiata* [L.] Wilczek) and blackgram (*V. mungo* [L.] Hepper) are important legume crops in Asia, where it is a major source of dietary protein for its predominantly vegetarian population. Both mungbean and blackgram are consumed in various forms in their diet, as they are a rich source of protein, carbohydrates, and minerals. These crops also accumulate certain antinutritional factors such as phytic acid (PA) and oligosaccharides in their seeds along with others. Breeding efforts are underway to breed varieties with reduced content of PA and oligosaccharides. Genetic variation for PA and oligosaccharide content in mungbean and blackgram ranged from 6.17 to 12 $mgg^{-1}$and 6.97 to 7.50 mgg^{-1}, respectively. Low PA content was reported in mungbean VC-6379 (5.74 mg g^{-1}), YBSM (5.85 mg g^{-1}), blackgram KUG-365 (1.7 mg g^{-1}), Shekhar-2 (3.7 mg g^{-1}), and KUG-230 (4.0 mg g^{-1}) genotypes. In mungbean, PA accumulation was reported to be controlled by dominant alleles at two independent loci of major genes showing duplicate recessive epistasis. Two major QTLs, viz., SDPAP4.1 and SDPAP11.1, were also reported to be present on linkage group 4A and 11A in interval markers CEDG139-MBSSR179 and BM141-VR222. Genes and enzymes involved in the biosynthesis of PA and oligosaccharides are characterized in other legume crops which can help in the genetic manipulation of these traits toward the development of cultivars with reduced content without affecting their biological consequences.

Keywords *Vigna mungo* · *Vigna radiata* · Iron · Zinc · Protein · Raffinose family oligosaccharides · Phytic acid · Genetic biofortification

J. Souframanien (✉) · V. J. Dhole · K. S. Reddy
Nuclear Agriculture and Biotechnology Division, Bhabha Atomic Research Centre, Mumbai, Maharashtra, India
e-mail: souf@barc.gov.in

D. S. Gupta et al. (eds.), *Breeding for Enhanced Nutrition and Bio-Active Compounds in Food Legumes*, https://doi.org/10.1007/978-3-030-59215-8_5

Introduction

Pulses are a rich source of nutrients particularly proteins, carbohydrates, and minerals for the vegetarian population. Among the pulses, mungbean and blackgram are the most important pulse crops of India after chickpea and pigeon pea. Blackgram and mungbean are grown throughout the year in some or other parts of India as a kharif, rabi, and summer/spring crop. Mungbean and blackgram became an integral part of daily diet of Indians and supplementing cereal-based diet due to their rich lysine content. Both mungbean and blackgram are consumed as dal, sprouts, used in preparation of papad, fermented food like wada, dosa, and idli. In India, per capita availability of pulses declined (42 g/day in 1990–1991 to 33 g/day in 2005), while prices have raised because production is not increased in proportion to population growth. In the future, challenge is not only to increase the production but also to improve the quality and bioavailability of nutrients. Mungbean and blackgram seeds are a rich source of easily digestible protein, carbohydrates, vitamin C, folic acid, thiamin, iron, zinc, potassium, magnesium, copper, manganese, and phosphorous (Mubarak 2005; Taunk et al. 2012; Dahiya et al. 2013). Plants which produce seeds, rich in energy supplies (carbohydrates, lipids, proteins), usually accumulates potent chemical defense compounds. Protein-rich grain legumes often contain substantial amounts of "antinutritive" factors (ANF), such as trypsin inhibitors, lectins, phytic acid, tannins, saponins, oligosaccharides, non-protein amino acids (NPAAs), alkaloids, cyanogenic glycosides, pyrimidine glycosides, and isoflavones (Bardocz et al. 1996; Enneking and Wink 2000). Since many of the ANFs are toxic, unpalatable, or indigestible, elimination or reduction strategies (germination, boiling, leaching, fermentation, extraction, etc.) are followed to minimize the impact. Reduction of ANFs through genetic manipulation is of prime objective of plant breeders though selection from natural variation or from mutant population and genetic engineering of biosynthetic pathways. However, the biological consequences and economic constraints of changing ANFs need to be considered to breed for ANF free crops or to reduce ANFs with option for processing. In this chapter, we discuss the strategy to improve/reduce them, respectively, with special reference to phytic acid and oligosaccharides in mungbean and blackgram.

Phytic Acid

Apart from nutrients, mungbean and blackgram also contain certain antinutritional factors such phytic acid, oligosaccharides, etc. Phytic acid (myo-inositol hexaphosphoric acid) (Fig. 1) is an abundant plant constituent, comprising 1–5% by weight of edible legumes, cereals, oil seeds, pollens, and nuts (Cheryan 1980). Phytic acid (PA) is the main storage organic form of phosphorous (P) in plants. The PA, a major antinutritional factor, is of prime concern for human nutrition and its effect on health, because it is present in more or less concentration in all plant-based diets.

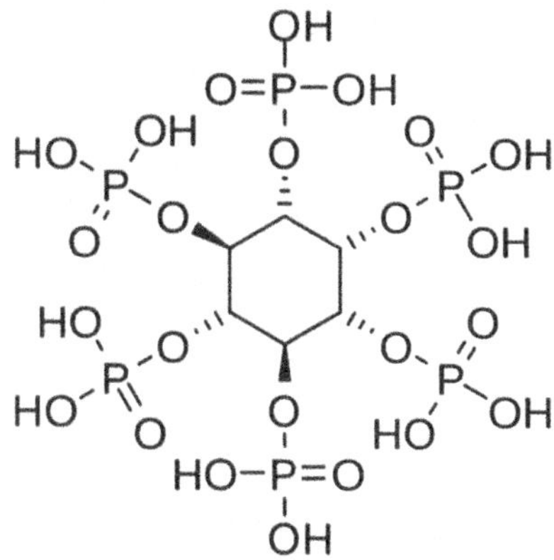

Fig. 1 Structure of phytic acid

Being an effective chelator of positively charged cations, PA will bind to nutritionally important mineral cations such as calcium, iron, and zinc and is also responsible for the inhibition of trypsin (Singh and Krikorian 1982). Phytates also form complexes with proteins at both low and high pH values. These complex formations alter the protein structure, which may result in decreased protein solubility, enzymatic activity, and proteolytic digestibility (O'Dell and de Boland 1976; Ravindran et al. 1995). PA also forms phytate-carbohydrate complexes making carbohydrate less degradable. Amylase activity is inhibited by phytate-Ca^{++} complexes, which reduce degradation of carbohydrate (Rickard and Thompson 1997; Selle et al. 2000), while "lipophytin" complexes may lead to metallic soaps in gut lumen, resulting in lower lipid availability (Matyka et al. 1990; Leeson 1993; Vohra and Satyanarayana 2003). Apart from this, humans as well as other nonruminants were lacking enzyme phytase; hence, they are unable to digest PA and excrete a large fraction of these salts. This phenomenon can contribute to human mineral deficiency, particularly with respect to iron and zinc, and also cause eutrophication of waterways (Erdman 1981; Cromwell and Coffey 1991; Brown and Solomons 1991). Iron deficiency is the most prevalent micronutrient disorder worldwide. Iron deficiency limits oxygen delivery to cells, leading to fatigue, poor work performance, decreased immunity, and death (Jones 1997). A large proportion of the population in developing countries consume less than the recommended dietary allowance (RDA) of iron, which for adult women is approximately 0.06 g day^{-1} with a low iron bioavailability (5%) diet and 0.02 g day^{-1} with a high iron bioavailability (15%) diet (WHO 2004). Zinc is a component of more than 300 enzymes involved in carbohydrate metabolism, DNA synthesis, protein synthesis, digestion, and bone metabolism. The RDA for zinc is 0.011 g day^{-1} for male adults and 0.008 g day^{-1} for female adults (IMFNB 2001).

Vegetarian population of developing countries like India has a greatest risk for mineral deficiencies caused by dietary PA particularly children and child-bearing women in rural communities that depend on cereals and pulses as staple foods (Raboy 2002). PA can reduce bioavailability of Zn, Ca, Mg, and Fe up to 5–15% (Das et al. 2012). Due to reduced bioavailability, deficiency of micronutrients like zinc and iron occurs in humans and causes various physiological disorders, e.g., zinc deficiency causes impaired growth, immune dysfunction, increased morbidity and mortality, adverse pregnancy outcomes, and abnormal neurobehavioral development. The daily intake of phytate for humans on vegetarian diets, on an average, is 2000–2600 mg for inhabitants of rural areas in developing countries, while on mixed diets, it is 150–1400 mg (Reddy 2002).

Phytic Acid and Phytate

PA (inositol hexakisphosphate) is designated as phytate, when it forms a salt with mono- and divalent cations K^+, Mg^{2+}, and Ca^{2+}. PA is mostly present as salts of the phytate that accumulates in the seeds during the ripening period and is the main storage form of both phosphate and inositol in plant seeds and grains (Loewus 2002). It is the principal storage form of phosphorus in many plant tissues. Its molecular formula is $C_6H_{18}O_{24}P_6$, and its molecular mass is 660.04 g mol^{-1} (Kumar et al. 2010). Precipitation of phytate started when at least four out of the six phosphate groups are able to bind with the Fe^{3+}, leading to iron phytate with 2–4 iron atoms per molecule of PA (Mali et al. 2006). In the most idealized version of the salt, the ferric ions will be organized in a grid, where each phosphate group is bound to two iron atoms and each iron atom is bound to three phosphate groups shared between two PA molecules (Thompson and Erdman 1982).

Phytic Acid Synthesis in Plants

In plants, biosynthesis of PA takes place through two different ways. One is a "lipid-dependent" pathway, which operates in all plant tissues, while second is a "lipid-independent" pathway, which functions in seeds (Sparvoli and Cominelli 2015). PA biosynthesis is initiated by conversion of D-glucose-6-phosphate to *myo*-inositol 3-phosphate (Ins(3)P1) with the help of enzyme D-*myo*-inositol 3-phosphate synthase (MIPS) (Fig. 2). *Myo*-inositol 3-phosphate is then dephosphorylated to free "Ins" by a specific Mg^{2+}-dependent inositol monophosphate phosphatase (IMP). The reaction catalyzed by IMP can be reversed by the action of *myo*-inositol kinase (MIK). This reverse reaction could provide more substrate diversity for the generation of inositol bisphosphate to feed the lipid-independent pathway by producing multiple inositol monophosphates (Shi et al. 2005).

Inositol triphosphates (InsP3) are generated through different ways in the "lipid-dependent" and "lipid-independent" pathways. In the "lipid-dependent" pathway, Ins is converted to phosphatidylinositol (PtdIns) by a phosphatidylinositol synthase (PtdIS). After that, PtdIns(4,5)P2 is produced by phosphorylation of the PtdIns with the help of phosphatidylinositol kinases. PtdIns(4,5)P2 is the substrate of a PtdIns-specific phospholipase C activity that releases Ins(1,4,5)P3, a molecule central to

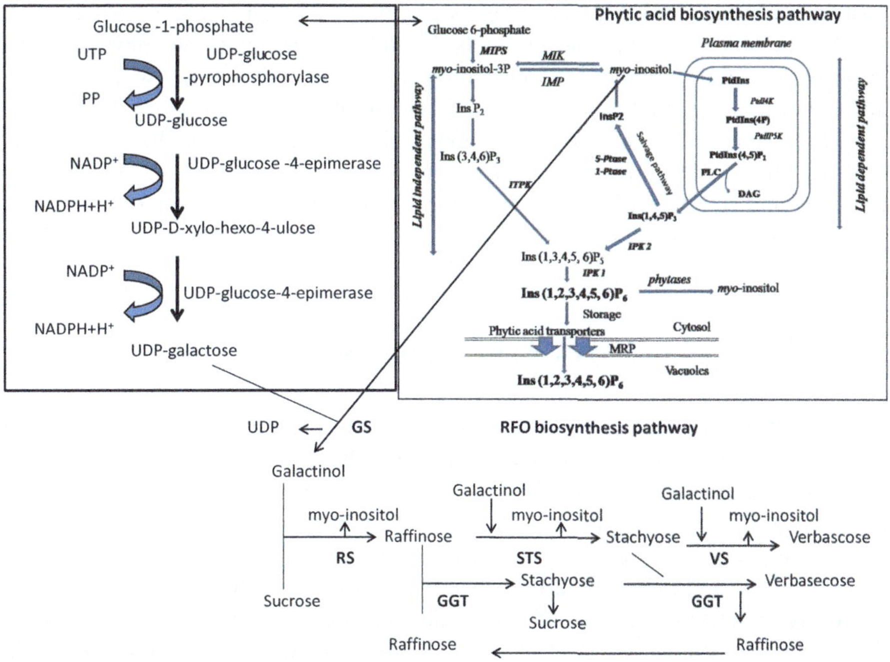

Fig. 2 Schematic representation of phytic acid biosynthetic pathway. MIPS, *myo*-inositol-3-phosphate synthase; IMP, bifunctional enzyme: *myo*-inositol-phosphate monophosphatase and galactose-1-phosphate phosphatase; MIK, *myo*-inositol kinase; IPK2, inositol 1,4,5-tris-phosphate kinase; ITPK, inositol 1,3,4-triphosphate 5/6-kinase; IPK1, inositol 1,3,4,5,6 penta*kis*phosphate 2-kinase; PtdIS, phosphatidylinositol phosphate synthase; PtdI4K, phosphatidylinositol 4-kinase; PtdIP5K, phosphatidylinositol 4-phospate 5-kinase; PtdIns, phosphatidylinositol; PtdInsP2, phosphatidylinositol bisphosphate; PLC, phospholipase C; MRP, multidrug-resistance-associated protein ATP-binding cassette. Schematic representation of phytic acid biosynthetic pathway. GS, galactinol synthase (EC 2.4.1.123); RS, raffinose synthase (EC 2.4.1.82); STS, stachyose synthase (EC 2.4.1.67); VS, verbascose synthase; GGT, galactan:galactan galactosyltransferase

signal transduction (Odom et al. 2000). Then Ins(1,3,4,5,6)P5 is formed from Ins(1,4,5)P3 with the help of inositol 1,4,5-tris-phosphate kinase (IPK2). In "lipid-independent" pathway, InsP6 is formed by sequential phosphorylation of the Ins ring through the action of a number of specific inositol phosphate kinases. Thus, Ins(1,3,4,5,6)P5 is formed in both the pathways finally converted into Ins(1,2,3,4,5,6) P6 by inositol 1,3,4,5,6 penta*kis*phosphate 2-kinase (IPK1).

After synthesis, PA along with coprecipitated cations are stored in electron-dense spherical particles called globoids inside the storage vacuoles where it is actively transported by a specific InsP6 transporter, a multidrug-resistance-associated protein (MRP) (Sparvoli et al. 2014). The location of globoids is mainly the aleurone layer in wheat and barley or the embryo in maize (O'Dell et al. 1972). The globoids are compartmentalized inside protein storage vacuoles in the seeds which can be distinguished morphologically in three distinct regions. First is a matrix that contains most of the soluble storage proteins, second is crystalloids composed of proteins in lattice structure, and third is globoids of PA or oxalate crystals (Lott 1980). The size of the phytate globoids depends on the amount of phytate in the grain. A low phytic acid (*lpa*) wheat mutant has smaller globoids with the same amount of P in the grains but lowered phytate concentration (Joyce et al. 2005). PA stored in seeds serves as important sources of P and cations for the germinating embryo (Raboy 1997). During germination, phytin is hydrolyzed to phosphate by phytases and a series of lower phosphoric esters of *myo*-inositol (Loewus and Murthy 2000).

Genes Involved in Phytic Acid Synthesis

Phytic acid synthesis takes place with the help of a series of enzymes as described in biosynthetic pathway (Fig. 2). Genes encoding these enzymes were identified and characterized in various plants like *Arabidopsis,* rice, wheat, soybean, and common bean. The number of loci of these genes present in plants varies from species to species. Each gene family can express differentially in different tissues of the same plant (Sparvoli and Cominelli 2015). Hence, information of these genes will be very useful for the characterization of *lpa* mutants/genotypes. The gene families coding for different enzymes involved in phytic acid biosynthetic pathway are given in Table 1.

***MIPS* genes:** The *MIPS* genes are an important gene family in phytic acid biosynthesis. The *MIPS* genes present in variable loci numbers in different species. In some crops like barley, only one locus was identified, while two loci in rice and common bean and multiple loci were identified in *Arabidopsis*, maize, and soybean (Suzuki et al. 2007; Fileppi et al. 2010). Expression of *MIPS* gene was found to be started at very early during seed development and then decline after the accumulation of phytic acid in both monocot and dicot crops (Suzuki et al. 2007, Fileppi et al. 2010). In mungbean, the maximal transcript levels of gene *Vigna radiata* d-myo-inositol-3-phosphate synthase (VrMIPS1) were reported from 7 to 9 days after flowering (Wongkaew et al. 2010). **IMP and MIK genes:** The IMP genes form a small gene family and present in a different number of loci coding for IMP enzymes, i.e., one *IMP* gene in barley and in common bean (Fileppi et al. 2010, Fu et al. 2008), one *IMP* and one *IMPL* gene in rice (Suzuki et al. 2007), one *IMP* and two *IMPL* genes in *Arabidopsis* (Torabinejad et al. 2009), and three *IMP* genes in tomato (Gillaspy et al. 1995). The expression of IMP gene is found to be very high at early stages of seed development and then declining to undetectable levels before the start of PA accumulation in seeds (Fileppi et al. 2010). Only one *MIK* gene has been found in plant genomes, which is expressed at high levels during seed development stage in *Arabidopsis* and common bean (Kim and Tai 2011; Fileppi et al. 2010).

***PGK* and *IPK2* gene:** The 2-PGK gene was identified in rice and *Arabidopsis* (Odom et al. 2000; Kim and Tai 2014). This gene is expressed in shoot, root, and panicle of rice, while two genes of this family (*At5g60760* and *At3g45090*) were found to be highly expressed during silique development in *Arabidopsis* and

Table 1 Classes of genes coding for enzymes for phytic acid synthesis

Sr. no	Classes of genes	Full form	Stage of expression	Crops	References
1	MIPS	*Myo*-inositol-3-phosphate synthase	Early seed development	Common bean soybean	Fileppi et al. (2010) and Chappell et al. (2006)
2	IMP	*Myo*-inositol-phosphate monophosphatase	Early seed development	Common bean	Fileppi et al. (2010)
3	MIK	*Myo*-inositol kinase	Seed development	Common bean	Fileppi et al. (2010)
4	PGK	Phosphoglycerate kinase	Shoot, root, and panicle/silique development	*Arabidopsis*	Kim and Tai (2011)
5	IPK2	Inositol 1,4,5-tris-phosphate kinase (*specific for the lipid-dependent pathway)	Seed development	*Arabidopsis*	Kim and Tai (2011)
6	ITPK	Inositol 1,3,4-triphosphate 5/6-kinase	Embryo-specific expression	Soybean	Stiles et al. (2008)
7	IPK1	Inositol 1,3,4,5,6 penta*kis*phosphate 2-kinase	Immature ears, seeds at 12 DAF, middle-stage endosperm and maturing embryos, roots	Soybean common bean	Yuan et al. (2012) and Fileppi et al. (2010)
8	MRP	Multidrug-resistance-associated protein ATP-binding cassette	Almost all tissues except seed	Soybean common bean	Gillman et al. (2009) and Panzeri et al. (2011)

required for InsP6 synthesis (Kim et al. 2008; Kim and Tai 2014). The *IPK*2 gene is specifically involved in the lipid-dependent pathway, which is not the major route to PA in the seed. In *Arabidopsis*, two genes of *IPK*2 family were found. The first gene *AtIPK2*α plays a role in pollen germination and root growth (Xu et al. 2005), while *AtIPK2*β is involved in auxiliary shoot branching through the auxin signaling pathway (Zhang et al. 2007). The IPK2 is expressed in later steps of phytic acid synthesis during seed development.

***ITPK* and *IPK*1 genes:** In *ITPK* gene family, six different ITPKs have been reported in rice (Suzuki et al. 2007), four each in *Arabidopsis* (Sweetman et al. 2007, Wilson and Majerus1997, Chen et al. 2003), soybean (Stiles et al. 2008), and wheat (Bhati et al. 2014), at least three in common bean (Fileppi et al. 2010), and one in maize (Raboy et al. 2000; Shi et al. 2003). *ITPK* genes are expressed at seed development stage in aleurone, embryo, and silique. A single *IPK1* gene was identified in many crops like rice and wheat, while two genes were identified in maize and in *Arabidopsis* and three genes in soybean (Yuan et al. 2012). These genes were found to be expressed in tissues like immature ears, middle-stage endosperm, maturing embryos, roots, and aleurone (Fileppi et al. 2010; Bhati et al. 2014). ***MRP* gene:** The MRP gene family is not involved in PA synthesis, but it is very important for the transportation of phytic acid from its site of synthesis to the storage vacuoles present in seeds. The MRPs are found as single copy genes in *Arabidopsis*, maize, and rice. These genes were found to be expressed in different organs including seeds (Klein et al. 2006; Shi et al. 2007; Xu et al. 2009; Wanke and Kolukisaoglu 2010; Kang et al. 2011).

Role of Phytic Acid in Plant Growth and Development

The role of PA in plant growth and development is not fully understood. It is a reserve form of phosphorous, which is translocated from vegetative plant parts to the developing seed soon after anthesis (Abernethy et al. 1973). During germination, phytase sequesters orthophosphate groups from the inositol ring of PA to produce free organic P, which is utilized for plant growth and development (Ockenden et al. 2004; Debnath et al. 2005). Apart from P reserve, PA is also involved in stress responses, membrane biogenesis, intracellular signaling, DNA repair, chromatin remodeling, endocytosis, and nuclear messenger RNA export (York et al. 1999; Hanakahi et al. 2000; Loewus and Murthy 2000; Hoy et al. 2002; Steger et al. 2003). *Myo*-inositol synthesis and catabolism impact metabolites involved in many critical plant biochemical pathways, such as (i) the production of compatible solutes, like galactinol, raffinose family of oligosaccharides, pinitol, and cell wall polysaccharides; (ii) the generation of inositol polyphosphates (InsPs), PA, and inositol polyphosphate pyrophosphates (PP-InsPs); and (iii) the synthesis of phosphoinositides and the production of Ins(1,4,5)P3 (Sparvoli and Cominelli 2015). Studies on *lpa* mutants in several crops elucidated the role of PA on plant growth and development.

Signal Transduction and Vesicle Trafficking

Ins(1,4,5)P3 (InsP3) and InsP6, the products of inositol metabolism, act as signaling molecule in a wide range of plant developmental and physiological processes, such as response to diverse stimuli (light, gravitropism, abiotic and biotic stresses), downstream responses to ABA and sugars, and auxin-mediated processes (Sparvoli and Cominelli 2015), while InsP7 and InsP8 molecules have been considered as unique signaling molecules involved in energy sensing and metabolism (Laha et al. 2015; Fassetti et al. 2011; Desai et al. 2014; Shears 2015; Williams et al. 2005). In seed development stage, expression of *MIPS* genes is very high indicating the importance of *myo*-inositol in seed and embryo development. MIPS-mediated *myo*-inositol synthesis is necessary for the normal functioning of endo-membrane trafficking and for maintaining endo-membrane structure (Luo et al. 2011). The various phosphorylated forms of PtdIns have critical roles in cytoskeletal rearrangements, membrane trafficking, and organelle labeling.

Biotic and Abiotic Stress Response

Decrease in PA content may also impact plant defense response. Decreased PtdIns availability for sphingolipid production resulted in the elevated ceramides and hydroxyceramides, and changes in *myo*-inositol, galactinol, and ascorbic acid induce spontaneous cell death due to altered oxidative stress sensitivity (Donahue et al. 2010). Recently, the crystal structure of TIR1 (transport inhibitor response 1) reveals PA molecule tightly bound to the protein at a functionally important position of the hormone receptor involved in expression of stress resistance genes (Macbeth et al. 2005; Calderon-Villalobos et al. 2010). The PA interacts with the auxin binding pocket in TIR 1 LLR domain. The high affinity and the binding mode of PA at the core of the auxin receptor strongly suggest that it is a functional cofactor of TIR1 (Tan et al. 2007). During germination, phytase degraded PA to produce free organic P, which is utilized for plant growth and development (Ockenden et al. 2004, Debnath et al. 2005). Specific pool of PA regulates defense against phytopathogens (Murphy et al. 2008). Yield penalties in some *lpa* mutants are reported due to low stress tolerance (Bregitzer and Raboy 2006; Ertl et al. 1998). Decreased levels of PA in transgenic potato plants constitutively expressing an antisense gene sequence for myo-inositol 3-phosphate synthase and *lpa* mutant plants of *Arabidopsis thaliana* showed increased susceptibility to various viral, bacterial, and fungal diseases (Murphy et al. 2008). In mungbean, high PA content was found in genotypes resistant to yellow mosaic disease (YMD) and powdery mildew, while susceptible genotypes contain less PA (Dhole and Reddy 2016). Transgenic plants overexpressing *MIPS* gene in sweet potato showed significantly enhanced salt and drought tolerance and stem nematode resistance (Zhai et al. 2015). Accumulation of cyanogenic glycosides and PA in mungbean seeds during seed maturation plays an important role in defense against bruchid (Lattanzio et al. 2005). High PA content was reported in wild species *Vigna radiata* var. *sublobata*, which showed resistance to bruchid (Dhole and Reddy 2016).

Genetic Variability for Phytic Acid in Mungbean and Blackgram

Inositols with 4, 5, or 6 phosphate groups are common in the seeds of many grain legumes and can reach concentration higher than 10% of dry matter (Bisbyet al.1994). In mungbean and blackgram, reports on phytic acid studies are very limited. In germplasm, phytic acid content in seeds of mungbean ranged from 6.17 to

Table 2 Phytic acid content in mungbean and blackgram in comparison to cereals, oilseeds, and other pulses

Crops	Phytic acid (mgg^{-1})	References
Pulses		
Chickpea	2.9–11.7	Kumar et al. (2010)
Pea	1.8–11.5	Kumar et al. (2010)
Mungbean	6.17–12	Chitra et al. (1995), Raboy (1997), Tajoddin et al. (2011), Sompong et al. (2012), Dahiya et al. (2013) and Dhole and Reddy (2015)
Blackgram	6.47–13.7	(Duhan et al. (1989), Chitra et al. (1995) and Suneja et al. (2011)
Cereals		
Rice	12.7–21.6	Kumar et al. (2010)
Wheat	4.97–15.02	Shitre et al. (2015)
Maize		
Sorghum	5.9–11.8	Kumar et al. (2010)
Oilseeds		
Groundnut	5.29–11.19[a]	Hande et al. (2013)
Sesame	39.3–57.2	Kumar et al. (2010)
Soybean	9.2–16.7	Kumar et al. (2010)

[a]Figures converted from PAP to PA (mg g^{-1})

12 mgg^{-1}(Chitra et al. 1995; Raboy 1997; Sompong et al. 2010; Tajoddin et al. 2011; Sompong et al. 2012; Dahiya et al. 2013; Dhole and Reddy 2015), while in blackgram it was ranged from 6.47 to 13.7 mgg^{-1}(Duhan et al. 1989; Chitra et al. 1995; Suneja et al. 2011). Phytic acid content was reported to be very narrowly ranged, i.e., 6.97 to 7.50 mgg^{-1}, in grains of blackgram and mungbean amphidiploids (Kataria et al. 1989). Phytic acid content in mungbean and blackgram in comparison to cereals, oilseeds, and other pulses is given in Table 2.

Sources of Low Phytic Acid in Mungbean and Blackgram

Germplasm of mungbean and blackgram are needed to be screened for *lpa* source. So far, few reports are available on PA studies in germplasm of these crops. All available germplasm and landrace database on PA content will clear the picture whether source for *lpa* is available or not. Low PA content was reported in mungbean genotypes VC-6379 (5.74 mg g^{-1}) and YBSM (5.85 mg g^{-1}) as compared to mean of 104 genotypes (8.26 mg g^{-1}) (Dhole and Reddy 2015). Low PA content was observed in the blackgram genotypes KUG-365 (1.7 mg g^{-1}), Shekhar-2 (3.7 mg g^{-1}), and KUG-230 (4.0 mg g^{-1}) (Suneja et al. 2011).

Genetics and Molecular Markers Linked to Phytic Acid

In mungbean, high phytic acid phosphorus was controlled by dominant alleles at two independent loci of major genes showing duplicate recessive epistasis (Sompong et al. 2010). Two major QTLs, viz., SDPAP4.1 and SDPAP11.1, were reported to be present on linkage group 4A and 11A in interval markers CEDG139-MBSSR179 and BM141-VR222, respectively, in F_2 population of mungbean (Sompong et al.2012). The high heritability (0.80–0.88) was reported for phytic acid content in mungbean (Sompong et al. 2010; Dhole and Reddy 2015). The high heritability values suggested that the trait may be controlled either by major genes or by polygenes with additive gene action. Hence, selection will be effective for this trait provided adequate genetic variation present in the population.

Priority Traits for Genetic Bio-fortification

Mungbean and blackgram are nutritious pulses with high protein, carbohydrate, minerals, and vitamins. Still there is a scope for genetic bio-fortification for these traits. Apart from that, lowering the antinutritional factors like PA, trypsin inhibitor, and oligosaccharides will improve the bioavailability of nutrients already present in seeds. Recently, there is an increasing interest in the development of crops with low PA (*lpa*) content to enhance the bioavailability of minerals and other nutrients. Earlier breeding efforts have identified several *lpa* mutants resulting in reduction of seed PAP from 50% to >95% in crops, such as barley, wheat, maize, soybean, and common bean (Raboy and Gerbasi 1996; Larson et al. 1998; Guttieri et al. 2004; Yuan et al. 2007; Campion et al. 2009). Such *lpa* mutants or genotypes were not reported in both mungbean and blackgram until now.

α-Galactosides and Raffinose Family of Oligosaccharides (RFOs)

α-Galactosides are considered as an important group of low molecular weight nonreducing sugars that are soluble in water and water-alcohol solutions (Martínez-Villaluenga et al. 2008). These oligosaccharides are ubiquitous in plant kingdom and ranks next to sucrose among soluble sugars (Frias et al. 1999). α-Galactosides get accumulated in higher amount in storage organs like seeds during later stages of development and maturation (Peterbauer et al. 2001). Pulse seeds contain some of the highest concentrations of oligosaccharides among all the crops. RFOs, in particular raffinose, stachyose and verbascose, are predominant in legumes; higher homologs such as ajugose are found in trace quantities in crops like blackgram (Peterbauer and Richter 2001; Girigowda et al. 2005; Girigowda et al. 2006).

Humans and monogastric animals do not possess the enzyme called α-galactosidase necessary for hydrolyzing the linkages present in these oligosaccharides, so that they cannot be digested when consumed. Intact oligosaccharides reach the lower intestine and undergo anaerobic fermentation by bacteria with gas expulsion (H_2, CO_2, and traces of CH_4), causing the flatus effect and sometimes diarrhea and abdominal pain (Reddy et al. 1980). Increase in fermentable carbohydrates in lower part of digestive tract may cause the disturbance in the existing microbial balance, causing diarrhea (Veldman et al. 1993). But removal of these compounds from beans could not reduce the flatulence problem completely, and hence involvement of indigestible polysaccharides was also associated with intestinal gas production (Reddy et al. 1984). The presence of RFO in diet can reduce the available dietary energy and interferes with the digestion of other nutrients (Martínez-Villaluenga et al. 2008). Legume scientists and growers consider flatulence to be the single most important factor that deters from eating more of them.

Chemical Structures of Alpha-Galactosides

Alpha-galactosides are considered as sucrosyl galactosides that consist of linear chains of galactopyranosyl residues attached to the C-6 of the glucose moiety of sucrose via an α–(1 → 6) galactopyranosidic linkage (Avigad and Dey 1997). The chemical structures of important α-galactosides are presented in Fig. 3. Alpha-galactosides can further be classified into two groups. Raffinose family of oligosaccharides (RFO) constitutes the first group, and the first member of this group, raffinose (*α*-D-galactopyranosyl-[1 → 6]-*α*-D-glucopyranosyl-[1 → 2]-*β*-D-fructofura-noside; degree of polymerization [DP] = 3), is the main RFO in most monocotyledon seeds, while its higher homologs, stachyose (*α*-D-galactopyranosyl-[1 → 6]-*α*-D-galactopyranosyl-[1 → 6]-*α*-D-glucopyranosyl-[1 → 2]-*β*-D-fructofuranoside; DP = 4), verbascose (*α*-D-galactopyranosyl-[1 → 6]-[*α*-D-galactopyranosyl-(1 → 6)]2-*α*-D-glucopyranosyl-[1 → 2]*β*-D-fructofuranoside; DP = 5), and ajugose (*α*-D-galactopyranosyl-[1 → 6]-[*α*-D-galactopyranosyl-(1 → 6)]3-*α*-D-glucopyranosyl-[1 → 2]-*β*-D-fructofuranoside; DP = 6), accumulate predominantly in seeds of dicotyledons (Sprenger and Keller 2000). Higher members like ajugose are generally found in trace quantities in seeds (Peterbauer and Richter 2001). The second group of α-galactosides includes galactosyl cyclitols (Lahuta et al. 2010). Ciceritol (*α*-D-galactopyranosyl-[1 → 6]-*α*-D-galactopyranosyl-[1 → 2]-4-O-methyl-*quiro*-inositol) is the most common galactosyl cyclitols and was first reported from chickpea (*Cicer arietinum*; Quemener and Brillouet 1983) followed by lentil (*Lens culinaris*; Bernabe et al. 1993; Martínez-Villaluenga et al. 2008).

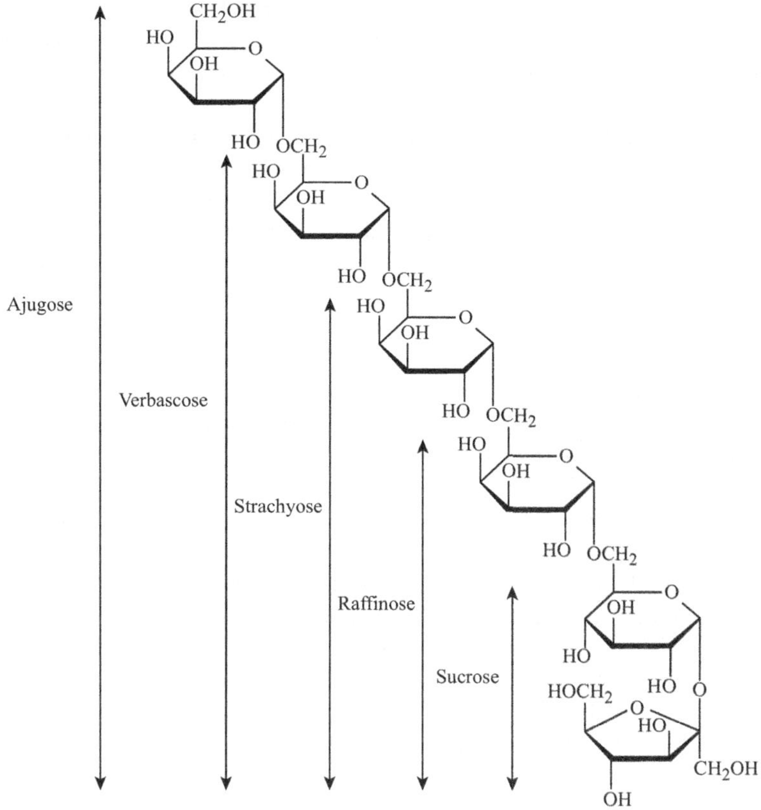

Fig. 3 Chemical structures of important α-galactosides

Biosynthesis of RFO

Sucrose is formed as a major output of photosynthesis in higher plants. The galactosyl group of RFOs is donated by galactinol (Gol; 1-O-α-D-galactopyranosyl-L-myo-inositol). Synthesis of Gol is a key and absolute requirement for entering into the pathway of RFO biosynthesis. The biosynthesis of *α*-D-galactosyl derivatives of sucrose is initiated by enzyme galactinol synthase (GS, UDP-*α*-D-galactose:1-L-myo-inositol-*O*-*α* galactopyranosyltransferase, EC 2.4.1.123). GS catalyzes the transfer of galactosyl unit from UDP-D-galactose (derived from the common nucleotide pathway via UDP-D-galactose 4-epimerase to *myo*-inositol generating galactinol) (Joersbo et al. 1999). The key enzyme galactinol synthase (GolS, EC 2.4.1.123) thus is the primary checkpoint in RFO flux, which synthesizes Gol in plants using UDP-galactose (UDP-Gal) and L-myo-inositol. GolS serves as a cross-link between central inositol (Ino) metabolism and RFO biosynthesis and also controls entry of Ino into the process (Fig. 2).

The biosynthesis of RFO proceeds by reversible transfer of the galactosyl residue from donor galactinol to sucrose that results in synthesis of raffinose (trisaccharide) and inositol is released. This reaction is catalyzed by raffinose synthase (RS; EC 2.4.1.82; Martínez-Villaluenga et al. 2008). Higher members of raffinose family of oligosaccharides can be synthesized either following galactinol-dependent pathway or galactinol-independent pathway. In galactinol-dependent pathway, raffinose serves as an acceptor for another galactosyl residue from galactinol, yielding tetrasaccharide stachyose in the presence of stachyose synthase (STS; EC 2.4.1.67) enzyme (Karner et al. 2004). Likewise, verbascose is synthesized from stachyose in the presence of enzyme verbascose synthase (VS) (Fig. 2). Activity of verbascose synthase was observed in purified stachyose synthase from seeds of pea (Peterbauer et al. 2002), while stachyose synthase from adzuki bean seeds was devoid of verbascose synthase activity (Peterbauer and Richter 1998). Therefore, a new galactinol-independent pathway has been proposed for the biosynthesis of higher members of raffinose family (Bachmann et al. 1994; Haab and Keller 2002). According to this, an already present RFO molecule transfers its terminal galactosyl residue to others yielding a higher member of raffinose family. Galactan:galactan galactosyltransferase (GGT) has been proposed to catalyze this reaction (Haab and Keller 2002; Tapernoux-Luthi et al. 2004). Galactosyl cyclitols are also described to support stachyose biosynthesis by acting as a galactosyl donor (Peterbauer et al. 2001). In brief, galactinol synthase (GS) and raffinose synthase (RS) catalyze the initial consecutive committed steps in RFO biosynthesis (Keller 1992), whereas STS, VS, and GGT are responsible for the synthesis of higher members of RFO. RFO biosynthetic genes and corresponding enzymes have been studied in various legume crops (Table 3).

Physiological Role of RFO in Plants

RFOs play an important role in plant growth and development. RFO participates in different physiological mechanisms including desiccation tolerance (Martínez-Villaluenga et al. 2008), seed storability (Horbowicz and Obendorf 1994), biotic

Table 3 RFO biosynthetic pathway enzymes and their genes reported in legume crops

Enzyme	Plant	Accession no./work done	References
Galactinol synthase	*Phaseolus vulgaris*	Purified protein	Liu et al. (1995)
	Medicago sativa	AY126615	Cunningham et al. (2003)
	Glycine max	AY126715	Obendorf et al. (2004)
	Medicago falcata	FJ607306	Zhuo et al. (2013)
Raffinose synthase	*Vicia faba*	Purified protein	Lehle and Tanner (1973)
	Pisum sativum	AJ426475	Peterbauer et al. (2002)
Stachyose synthase	*Vigna angularis*	Purified protein	Peterbauer and Richter (1998)
	Vigna angularis	Y19024	Peterbauer et al. (1999)
	Lens culinaris	Purified protein	Hoch et al. (1999)
	Pisum sativum	AJ311087	Peterbauer et al. (2002)

and abiotic stress tolerance (Nishizawa et al. 2008a), photoassimilate translocation (Dinant and Lemoine 2010), and seed germination (Blöchl et al. 2007).

Seed Development and Desiccation Tolerance

In seeds, major loss of water takes place during seed maturation which is termed as "desiccation" that may lead to membrane damage and death of embryo. Tolerance against desiccation can be achieved by the accumulation of certain nonreducing sugars like sucrose and RFO (Koster and Leopold 1988). Many reports suggested the role of RFO in desiccation tolerance (Blackman et al. 1992; Corbineau et al. 2000; Angelovici et al. 2010), and the first mechanism by which they provide protection is water replacement. The hydroxyl groups of RFO are capable of replacing water molecules and maintaining the hydrophilic interactions within the cell that is necessary for stabilizing native macromolecules (like protein) and membrane structure during dehydration process (Koster 1991). The second mechanism for RFO's role in desiccation tolerance is "vitrification" or formation of glass within the cell. This is the state of a cell solution having very high viscosity due to loss of water. At this state, cell solution has the properties of a plastic solid. It is responsible for ensuring stability (by preventing the reactions required diffusion), preventing cellular collapse (by filling the blank spaces within the biomolecules), and maintaining hydrogen bonding within the cell (Koster and Leopold 1988; Koster 1991; Martínez-Villaluenga et al. 2008; Angelovici et al. 2010). It has been reported that late embryogenesis abundant (LEA) proteins and small heat shock proteins (sHSP) along with RFO are responsible for the glassy state (Pukacka et al. 2009). Increased biosynthesis of these oligosaccharides restricts the synthesis of monosaccharides, resulting in decreased respiration rate (site of reactive oxygen species formation). Alpha-galactosides along with sucrose have also been associated with seed storability. Horbowicz and Obendorf (1994) found storability half-viability periods >10 years when sucrose-to-oligosaccharide ratio was <1.0, while this period was <10 years in case of ratio > 1.0.

Abiotic and Biotic Stress Tolerance

Both biotic and abiotic stresses accumulate reactive oxygen species (ROS) within the plant cell. These ROS in higher concentrations are capable of damaging proteins, lipids, nucleic acids, and other biomolecules irreversibly (Scandalios 2005). Carbohydrates including RFO and sugar alcohols also contribute to protecting cells from oxidative damage and maintaining redox homeostasis (Nishizawa et al. 2008b; Keunen et al. 2013). RFO might have the capability to scavenge ROS. During this detoxification process, RFOs are proposed to convert in their oxidized radical forms that are further regenerated by reacting with other antioxidants like ascorbic acid (ASC) or flavonoids (Van den Ende and Valluru 2009). There are reports of Gol switching on early pathogen-attack-related transcripts (such as PR1a, PR1b, and NtACS1) in tobacco (Kim et al. 2008), suggesting a role in biotic stress signaling. Gol induces the expression of the PR-1a gene, via a salicylic acid-dependent pathway (Couée et al. 2006). Both GolS and RafS contain W-box cis-elements in their promoters, regulated by ABA-inducible WRKY (Wang et al. 2009). This suggests a possible role of GolS and RafS downstream ABA signaling. A recent study demonstrated that starch hydrolysis results in hexose and Raf accumulation during the first 24 h after a cold shock treatment in *Arabidopsis*. Nishizawa et al. (2008a, 2008b) reported tolerance to oxidative stress in GolS and RafS overexpressing transgenic plants predicting a role of Gol and Raf as scavengers of ROS, thus playing a novel role in the protection of cellular metabolism.

Seed Germination

RFOs get accumulated in storage organs (like tubers, seeds) of most of the plants mainly in legumes (Peterbauer et al. 2002) by phloem loading and transport (Turgeon 1996; Sprenger and Keller 2000). RFOs accumulate late in seed development, starting at about the beginning of seed fill and continuing up to maturation drying. They are deposited in all parts of the seed (endosperm, embryo, and the seed coat), although the levels of individual α-galactosides may vary considerably in these tissues (Kuo et al. 1988; Horbowicz and Obendorf 1994; Frias et al. 1999). RFOs protect the embryo during the desiccation that occurs during seed maturation and thus play an important role in prolonged seed survival (Peterbauer et al. 2002). During early stages of seed germination, they are rapidly mobilized by α-galactosidases (α-D-galactoside galactohydrolase, E.C.3.2.1.22) and provide readily available energy and carbon (Zhao et al. 2006). Alpha-galactosidase cleaves the terminal nonreducing α(1 → 6)-linked galactose residues of α-galactosides (Anisha et al. 2011). The resulted increasing sucrose concentration during transition phase, from germination to plant growth, was attributed to induce the expression of alkaline α-galactosidase resulting in mobilization of remaining RFO.

Raffinose and stachyose also serve as the main transportable solute in the orders Lamiales, Cucurbitales, and Cornales and in one family of the Celastrales and are mechanistically linked with phloem loading. There are two RFO pools in its leaves: a storage pool associated with leaf mesophyll and a transport pool associated with the phloem loading sites (Bachmann et al. 1994) where Raf and especially Sta are produced and loaded in the phloem, according to the polymer trapping model (Turgeon 1991). RFOs are loaded into phloem suggestively in symplastic type II plants using polymer trapping model. Briefly, sucrose from source cells (mesophyll) moves into the intermediate cells via bundle sheath where the enzymes for RFO biosynthesis are localized. The RFOs (Raf/Sta) cannot diffuse back to the source because of their higher size, and that traps them in the intermediate cells. The only way to move is within the sieve cells, and due to the high osmotic pressure buildup, the sugars are thus loaded into the sieve cells. This model is highly species specific, and most of the experimental evidences come from the Cucurbitaceae.

Genetic Variability for Oligosaccharides in Mungbean and Blackgram

Considerable varietal differences were observed in the oligosaccharides of the four pulses (pigeon pea, chickpea, blackgram, and mungbean) studied, and it was observed that the capacity to induce flatus could be graded in the following order: Bengal gram > blackgram > red gram > green gram (Rao and Belvady 1978). Oligosaccharide content reported in mungbean and blackgram in comparison to other pulse crops is given in Table 4. The concentration and composition of α-galactosides depend on the type of crop, growing environment, and also the genotype (Reddy

Table 4 Oligosaccharide content (%) of mungbean and blackgram in comparison to other pulses

Crop	Sucrose	Raffinose	Stachyose	Verbascose	Total	References
Chickpea	0.7–2.9	0.7–2.4	2.1–2.6	0.4–4.5	3.5–9.0	
	1.89	1.45	2.56	0.19	–	Alajaji and El-Adawy (2006)
	4.3	1.0	2.8	Traces	–	Aman (1979)
	K: 3.8 (3.1–4.4) D: 2.0 (1.56–2.85)	K: 0.6 (0.48–0.73) D: 0.5 (0.46–0.77)	K: 2.2 (1.76–2.72) D: 1.6 (1.25–1.98)	–	–	Wang and Daun (2004)
	–	K: 5.2	K: 2.7	ND (Ciceritol: K: 6.7)	–	Han and Baik (2006)
	1.52	0.32	1.7	ND (Ciceritol: 2.8)	–	Aguilera et al. (2009)
Pigeon pea	2.7	1.0–1.1	2.7–3.0	4.0–4.1	3.5–10.2	Asif et al. (2013)
Mung-bean	0.3–0.2	0.3–2.6	1.2–2.8	1.7–2.8	3.9–7.2	Asif et al. (2013)
	–	0.41	1.49	–	1.9	Mubarak (2005)
					7.2	Kakati et al. (2010)
Black-gram	–	–	–	–	4.7–10.9 (mg/g)	Yadhu et al. (2011)
	–	–	–	–	7.4	Kakati et al. (2010)
	3.28 mg/g to 32.87 mg/g	0.18–8.08 mg/g	8.90–37.27 mg/g	13.95–31.02 mg/g	0.30–4.48 mg/g	Souframanien et al. (2014)
Lentil	1.8–2.5	0.4–1.0	1.9–2.7	1.0–3.1	4.2–6.1	Asif et al. (2013)
Peas	2.3–2.4	0.3–0.9	2.2–2.9	1.7–3.2	5.3–8.7	Asif et al. (2013)

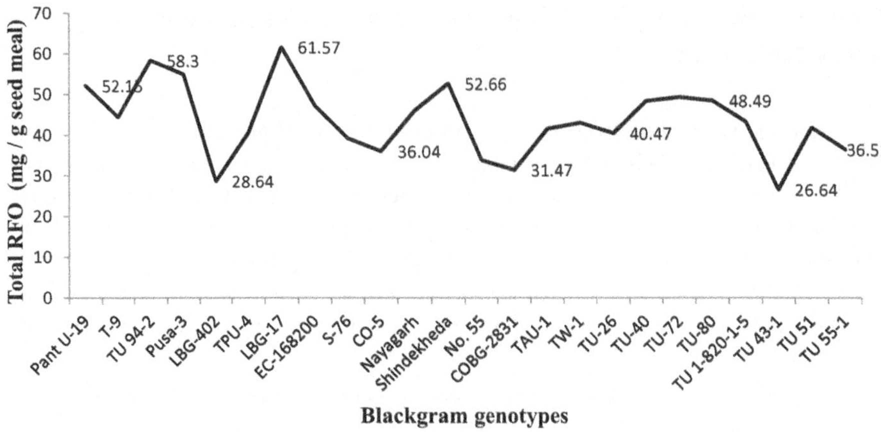

Fig. 4 Variation in total RFO content among blackgram genotypes

et al. 1980). Sosulski et al. (1982) studied the variation in α-galactoside concentration in 11 legumes and reported verbascose as the predominant α-galactoside in mungbean and faba bean. Verbascose was previously reported as a major RFO present in blackgram (Reddy et al. 1980; Girigowda et al. 2005) and in related *Vigna* species such as mungbean and cowpea (Aman 1979; Kuo et al. 1988). Similarly, in blackgram, Souframanien et al. (2014) identified verbascose as the major RFO, followed by stachyose among the blackgram genotypes studied. The total RFO concentration in blackgram ranged from 26.64 to 61.57 mg/g with an average of 43.6 mg/g (Fig. 4). The highest total RFO content was observed in the genotype LBG-17 (61.57 mg/g). Low content of total RFOs was observed in TU43–1 (26.64 mg/g). A wide variation was observed for individual RFOs such as raffinose, stachyose, verbascose, and ajugose, which ranged from 0.18 to 8.08 mg/g, 8.90 to 37.27 mg/g, 13.95 to 31.02 mg/g, and 0.30 to 4.48 mg/g, respectively. TU43–1 and LBG- 402 which recorded low content of total RFOs also showed lowest verbascose and stachyose content, respectively. TU1–820–1-5 recorded high verbascose (31.02 mg/g) with low stachyose (11.13 mg/g) and raffinose (0.18 mg/g) content (Souframanien et al. 2014). Ajugose was a minor sugar followed by raffinose in blackgram. Occurrence of higher oligosaccharide has been reported earlier in blackgram (Girigowda et al. 2006). Souframanien et al. (2014) reported the variation in ajugose content among back gram genotypes studied. The extent of ajugose content varied from 4.48 mg/g (LBG-17) to 0.30 mg/g (TU-72) in blackgram.

Breeding Methods

In mungbean and blackgram improvement program, most commonly applied methods are pureline selection in landraces and hybridization in diverse parents followed by pedigree selection, mutation breeding, and wide hybridization. But these methods are mostly employed to develop high-yielding and disease-resistant varieties. Quality improvement was never a major objective in these neglected pulse crops. Very few breeding efforts were made as bio-fortification of mungbean and blackgram is concerned. Recent advances in genomics and biotechnology offer new tools like transgenic and RNAi-mediated gene silencing for bio-fortification. The varieties developed through such technologies are in question due to strict legislative limits on the release of genetically engineered plants and consumers' acceptance especially to edible crops. This makes the role of conventional breeding more significant for genetic bio-fortification. Variability for phytic acid and oligosaccharides in these crops is not well studied and documented. Genetic variability is basic need for the any crop improvement program through conventional breeding methods. Lack of variability for these traits limits the use of pureline selection and pedigree method. Hence, mutation breeding method is a better option to address these issues. Mutants for *lpa* were successfully identified, isolated, and characterized in several crops like barley, maize, rice, wheat, soybean, common bean, and pea. There are three important steps in phytic acid biosynthetic pathway, i.e., supply pathway (from glucose 6-P to myo-inositol 3-phosphate), the end of the pathway (from myo-inositol 3-phosphate to InsP6), and tissue compartmentation of InsP6 and/or its transport and storage to the vacuole (MRP transporter). On the basis of affected step of the biosynthetic pathway due to mutation, mutants can be grouped into three classes. Mutants belonging to the first and the third classes are generally characterized by decreased InsP6 levels accompanied by a molar equivalent increase in inorganic Pi but not by accumulation of lower InsPs (inositol phosphates with up to five phosphate residues), a characteristic specific of the second class of mutants (Sparvoli and Cominelli 2015). Many *lpa* mutants were reported in various crops with their agronomic studies. Reduced germination and seed development and stunted vegetative growth have been observed in some of the *lpa* mutants. Thus, while breeding *lpa* mutant, breeders should not overlook negative effects on the plant health. Otherwise, *lpa* genotypes with poor agronomic traits may not be useful for cultivation and further breeding.

Genetically manipulating the level of RFO—by inhibiting galactinol synthase activity—has been patented (Kerr et al. 1998). This is the first committed reaction in the pathway and involves the synthesis of galactinol from UDP-Gal and myo-inositol. The individual members of the RFO are then synthesized by distinct galactosyltransferases (e.g., raffinose synthase and stachyose synthase). The physiological importance of the RFO during seed development and storage (see below) suggests that a better strategy would be based on the activation of α-galactosidase to degrade the RFO after harvesting or based on the transfer of α-galactosidase from a thermophilic bacterium (*Thermotoga neapolitana*) into grain legumes (Griga et al. 2001). This has a temperature optimum close to 100 °C and could be activated by, for example, canning. Frias et al. (1999) have suggested an alternative: reducing the level of the RFO while promoting the synthesis of related compounds such as the galactosyl cyclitols. This would maintain the protective nature of these compounds but decrease their flatus potential, because there is evidence that ciceritol is more slowly hydrolyzed by α-galactosidase than the RFO. Ciceritol is present in chickpea and lentil (*Lens culinaris*) but has not been detected in pea. The key to introducing galactinol cyclitols into pea with an accompanied reduction in the RFO content appears to lie with stachyose synthase, which has a central role in the synthesis of the galactinol cyclitols and in the synthesis of stachyose (Peterbauer and Richter 2001). It represents a link, therefore, between the RFO and galactinol cyclitol pathways. Plant breeders or molecular biologists wish to manipulate levels of these compounds to obtain optimal effect during cultivation while ensuring the quality of the harvested crop. In addition, the balance of adverse and beneficial properties should be borne in mind when breeding strategies are planned.

Upregulation and Downregulation of Key Biosynthetic Enzyme

Alpha-galactosidase is a well-known enzyme for RFO breakdown by hydrolyzing $\alpha(1 \rightarrow 6)$ linkage (Blöchl et al. 2008). Using this characteristic together with transformation approach, Polowick et al. (2009) developed transgenic pea lines overexpressing α-galactosidase from coffee (*Coffea arabica* L.). These transgenic lines showed up to 40% reduction in raffinose and stachyose concentration without affecting seed germination rate (96%). Galactinol synthase (GS) is considered as the first committed and key regulating step of RFO biosynthesis influencing carbon partitioning between sucrose and RFO (Peterbauer et al. 2001; Nishizawa et al. 2008a). Bock et al. (2009) downregulated the expression of galactinol synthase in canola (*Brassica napus* L.) using antisense approach. Consequently, they observed a decrease in galactinol and stachyose concentration in transgenic canola seeds.

G × E Influences RFO Concentration in Seeds

The effect of G×E on seed RFO concentration has been reported in some crops like peanut (*Arachis hypogaea* L.; Pattee et al. 2000), soybean (*Glycine max* L. Merr.; Cicek et al. 2006; Jaureguy et al. 2011), sugar beet (*Beta vulgaris* L.; Hoffmann et al. 2009), and lentil (*Lens culinaris* Medikus subsp. culinaris; Tahir et al. 2011). Most of the studies showed significant effect of G × E on seed RFO concentration as RFOs act as antioxidants during stress tolerance. Therefore, environmental conditions affect RFO level, i.e., more adverse conditions may result in higher RFO concentration.

Analytical Methods

Analytical method for determination of biochemicals like PA and oligosaccharide is very important to find out the desired segregants or mutant from large population. Screening method should be rapid, simple, and accurate. Many different methods of analysis had been employed to determine PA from seeds of different crops. Most widely used method for PA determination is colorimetric method with different modifications like simple colorimetric method (Latta and Eskin 1980) and modified colorimetric method (Gao et al. 2007). These methods are simpler and less expensive for assaying a large number of samples, allowing its effective application in breeding and genetic studies of *lpa*. In mungbean, modified colorimetric method was used effectively for a large number of samples (Dhole and Reddy 2015; Dhole and Reddy 2016). For more accuracy, other methods like enzymatic spectrophotometric method (March et al. 1995), LC-MS bioanalytical method (Tur et al. 2013), gas chromatography-mass spectroscopy (March et al. 2001), reversed-phase high-performance liquid chromatography (Dost and Tokul 2006), etc., were used for estimation of PA from different food samples. But these methods are costly and time-consuming. All the methods for estimation of PA are destructive. To breed the *lpa* genotypes, nondestructive, quick, simple, and cost-effective method is required to handle a large number of samples. Recently, Fourier transform near-infrared (FT-NIR) spectroscopy was validated for estimation of PA from mungbean seeds (Pande and Mishra 2015). One sample can be estimated in 1–2 min without destruction of seeds; hence, a large number of samples can be analyzed without any chemical use. But it required prior standardization for a given crop.

Determination of Sugar Concentration

Analytical estimation is the first step of all RFO-related studies. It is also helpful in selecting genotypes with high and low RFO concentration that can be utilized to understand RFO biosynthesis, identify key regulating biosynthetic step, and study natural variation together with impact of genotype, environment, and their interaction on RFO concentration. RFO represent a class of soluble and nonreducing oligosaccharide sugars. The analytical methods to determine sugars can be categorized into four main groups: (1) chemical, (2) physical, (3) enzymatic, and (4) chromatographic method (http://people.umass.edu/~mcclemen/581Carbohydrates.html). Chemical method reported follows the gravimetric principles in which reducing sugars are oxidized by heating with excess of copper sulfate and alkaline tartrate under carefully controlled conditions. The resulted copper oxide precipitate was determined by filtration, drying, and weighing. Chemical methods cannot estimate the composition of reducing sugars and direct concentration of nonreducing sugars.

Colorimetric approach includes phenol-sulfuric acid and anthrone-based methods that determine the concentration of total sugars in the sample. Sugars react with anthrone (with sulfuric acid) or phenol (with sulfuric acid) and produce blue-green or yellow-orange color having absorption maxima at 625 and 490 (480) nm, respectively (Brummer and Cui 2005). Phenol-sulfuric acid method is the most widely used approach to determine total sugars in aqueous solutions (Albalasmeh et al. 2013).

Physical methods to determine sugars utilize polarimetry, refractive index, potentiometry, etc., methodologies (http://people.umass.edu/~mcclemen/581Carbohydrates.html; Moresco et al. 2008). All the abovementioned methods are unable to predict the composition of either total or reducing/nonreducing sugars. Therefore, concentration and composition of RFO cannot be determined. To determine total RFO concentration, enzymatic method includes hydrolysis of RFO and sucrose into glucose by α-galactosidase and invertase. Thereafter, absorbance of glucose concentration can be measured using spectrophotometer. This approach was adopted by Megazyme (Megazyme International Ireland Ltd., Wicklow, Ireland), and a kit was developed to determine concentration of total RFO.

To perform compositional study for RFO and other soluble sugars, chromatographic techniques have been described as reliable and efficient approach. Among different chromatographic methods reported, high-performance liquid chromatography with refractive index detector (HPLC-RI) and high-performance anion exchange chromatography with pulsed amperometric detector (HPAEC-PAD) are widely used approaches. Jones et al. (1999) reported a TLC (thin-layer chromatography)-based method for qualitative estimation of individual RFO concentration. This method can be utilized to screen a large number of genotypes in a population, and selected genotypes can be used for further study. However, TLC is not capable of quantifying individual RFO concentration; hence, HPLC-RI or HPAEC-PAD methods should be utilized.

Sánchez-Mata et al. (1999) developed a HPLC with differential refractometer detector-based method using Waters μBondapak/carbohydrate column and acetonitrile-water (80:20. v/v) as mobile phase with a flow rate of 0.9 mL/min. Using this method, they reported the concentration of ribose, fructose, glucose, galactose, sucrose, maltose, raffinose, and stachyose in seeds of lentils (*Lens esculenta* L.), dry peas (*Pisum sativum* L.), white kidney beans (*Phaseolus vulgaris* L.), pinto beans (*Phaseolus vulgaris* L.), and chickpeas (*Cicer arietinum* L.).

Processing Methods to Reduce RFOs and PA

Different processing methods like de-hulling, cooking (boiling, autoclaving, and microwave cooking), soaking, germination, gamma irradiation, α-galactosidase treatment, ultrasound, hydrostatic pressure, and thermal dehydration have been reported to reduce RFO concentration significantly in seeds of crops like green gram (*Phaseolus aureus*; Rao and Vakil 1983), blackgram (*Vigna mungo* L.;

Girigowda et al. 2005), and mungbean (*Vigna radiata* L.; Anisha and Prema 2008; Tajoddin et al. 2010). However, such physical and mechanical treatments also reduce concentration of protein, B vitamins, minerals, and amino acid. The application of a single technique is frequently insufficient for effective treatment, and so combination is commonly employed. Low molecular compounds are leached out into the cooking water, which is often discarded. Overnight soaking of pulses results in sizable reduction of the concentration of oligosaccharides (Frias et al. 2000). Cooking was shown to reduce RFOs in horse gram and green gram (Mulimani and Devindra 1998). Germination pulse seeds reduce the contents of oligosaccharides and other N-containing ANFs served to improve the palatability. Germination is reported to reduce the RFO concentration in pulses, as complex sugars are converted into simple sugars (Martin-Cabrejas et al. 2008). During germination, phytate is degraded by native phytase. Fermentation is widely used in food detoxification process (Salih et al. 1991), and a variety of fermented foods are eaten around the world (Reddy and Salunkhe 1989).

Conclusion

The presence of phytic acid and oligosaccharides in mungbean and blackgram is the major concern from the nutritional point of view. However, they have predominant roles in plant transport, storage, and stress tolerance. Genetic manipulation of phytic acid and oligosaccharides for their reduced content should consider their biological consequences and economic constraints. Efforts in the past toward understanding the variation and isolation of low phytic acid and reduced oligosaccharide contents in the other crops offer a scope to improve them in mungbean and blackgram. Appropriate screening method coupled with screening of a large population can help in identifying the suitable genotypes/mutants with desired concentration of phytic acid and oligosaccharide contents in mungbean and blackgram.

References

Abernethy RH, Paulsen GM, Roscoe E Jr (1973) Relationship among phytic acid, phosphorus, and zinc during maturation of winter wheat. J Agric Food Chem 21:282–284

Aguilera Y, Martin-Cabrejas MA, Benitez V, Molla E, Lopez-Andreu FJ, Esteban RM (2009) Changes in carbohydrate fraction during dehydration process of common legumes. J Food Compos Anal 22:678–683

Alajaji SA, El-Adawy TA (2006) Nutritional composition of chickpea (*Cicer arietinum* L.) as affected by microwave cooking and other traditional cooking methods. J Food Compos Anal 19:806–812

Albalasmeh AA, Berhe AA, Ghezzehei TA (2013) A new method for rapid determination of carbohydrate and total carbon concentrations using UV spectrophotometry. Carbohydr Polym 97:253–261

Aman P (1979) Carbohydrates in raw and germinated seeds from mungbean and chickpea. J Sci Food Agric 30:869–875
Angelovici R, Galili G, Fernie AR, Fait A (2010) Seed desiccation: a bridge between maturation and germination. Trends Plant Sci 15:211–218
Anisha GS, Prema P (2008) Reduction of non-digestible oligosaccharides in horse gram and green gram flours using crude α-galactosidase from *Streptomyces griseoloalbus*. Food Chem 106:1175–1179
Anisha GS, John RP, Prema P (2011) Substrate specificities and mechanism of action of multiple α-galactosidases from *Streptomyces griseoloalbus*. Food Chem 124:349–353
Asif MI, Wani SA, Lone AA, Dar ZA, Nehvi FA (2013) Breeding for quality traits in grain legumes. In: Malik CP, Sanghera G, Wani SH (eds) Conventional and non-conventional interventions in crop improvement, pp 1–20
Avigad SM, Dey PM (1997) In Plant Biochemistry (Eds. Dey PM, Harbourne J) Academic Press: San Diego, pp 143–204
Bachmann M, Matile P, Keller F (1994) Metabolism of the raffinose family oligosaccharides in leaves of *Ajuga reptans* L. (Cold acclimation, translocation, and sink to source transition: discovery of chain elongation enzyme). Plant Physiol 105:1335–1345
Bardocz S, Gelencser E, Pusztai A (1996) Effects of antinutrients on the nutritional value of legume diets, vol 1. ESSE-EC-EAEC, Brussels
Bernabe M, Fenwick R, Frias J, Jimenez-Barbero J, Price K, Valverde S, Vidal-Valverde C (1993) Determination, by NMR spectroscopy, of the structure of ciceritol, a pseudotrisaccharide isolated from lentils. J Agric Food Chem 41:870–872
Bhati K, Aggarwal S, Sharma S, Mantri S, Singh S, Bhalla S, Kaur J, Tiwari S, Roy J, Tuli R (2014) Differential expression of structural genes for the late phase of phytic acid biosynthesis in developing seeds of wheat (*Triticum aestivum* L.). Plant Sci 224:74–85
Blackman SA, Obendorf RL, Leopold AC (1992) Maturation proteins and sugars in desiccation tolerance of developing soybean seeds. Plant Physiol 100:225–230
Blöchl A, Peterbauer T, Richter A (2007) Inhibition of raffinose oligosaccharide breakdown delays germination of pea seeds. J Plant Physiol 164:1093–1096
Blöchl A, Peterbauer T, Hofmann J, Richter A (2008) Enzymatic breakdown of raffinose oligosaccharides in pea seeds. Planta 228:99–110
Bock C, Ray H, Georges F (2009) Down-regulation of galactinol synthesis in oilseed *Brassica napus* leads to significant reduction of anti-nutritional oligosaccharides. Botany 87:597–603
Bregitzer P, Raboy V (2006) Effects of four independent low phytate mutations on barley agronomic performance. Crop Sci 46:1318–1322
Brown KH, Solomons NW (1991) Nutritional problems of developing countries. Infect Dis Clin N Am 5:297–317
Brummer Y, Cui SW (2005) Understanding Carbohydrate analysis. In: Food carbohydrates: Chemistry, physical properties and applications; Cui SW. Eds.; CRC Press, Taylor & Francis Group (USA) Boca Raton
Calderon-Villalobos LI, Tan X, Zheng N, Estelle M (2010) Auxin perception-structural insights. Cold Spring Harb Perspect Biol 2:a005546
Campion B, Sparvoli F, Doria E, Tagliabue G, Galasso-Fileppi IM, Bollini R, Nielsen E (2009) Isolation and characterization of an *lpa* (low phytic acid) mutant in common bean (Phaseolus vulgaris L.). Theor Appl Genet 118:1211–1221
Chappell A, Scaboo A, Wu X, Nguyen H, Pantalone V, Bilyeu K (2006) Characterization of the *MIPS* gene family in *Glycine max*. Plant Breed 125:493–500
Chen Q, Niu X, Chai M, Chen J, Liu Q, Wang X (2003) Isolation of an Arabidopsis gene encoding Ins (1,3,4) p-3 5/6-kinase-like protein and involved in plant response to abiotic stresses. Acta Bot Sin 45:211–218
Cheryan M (1980) Phytic acid interactions in food systems. CRC Crit Rev Food Sci Nutr 13:297–325
Chitra U, Vimala V, Singh U, Geervani P (1995) Variability in phytic acid content and protein digestibility of grain legumes. Plant Foods Hum Nutr 47(2):163–172

Cicek MS, Chen P, Saghai Maroof MA, Buss GR (2006) Interrelationships among agronomic and seed quality traits in an interspecific soybean recombinant inbred population. Crop Sci 46:1253–1259
Corbineau F, Picard MA, Fougereux J, Ladonne F, Côme D (2000) Effects of dehydration conditions on desiccation tolerance of developing pea seeds as related to oligosaccharide content and cell membrane properties. Seed Sci Res 10:329–339
Couée C, Sulmon G, Gouesbet A, Amrani E (2006) Involvement of soluble sugars in reactive oxygen species balance and responses to oxidative stress in plants. J Exp Bot 57:449–459
Cromwell GL, Coffey RD (1991) Phosphorous- a key essential nutrient, yet a possible major pollutant-its central role in animal nutrition. In: Lyons TP (ed) Biotechnology in the feed industry. Alltech Tech Publishers, Nicholaville, pp 133–145
Cunningham SM, Nadeau P, Castonguay Y, Laberge S, Volenec JJ (2003) Raffinose and stachyose accumulation, galactinol synthase expression, and winter injury of contrasting alfalfa germplasms. Crop Sci 43:562–570
Dahiya PK, Linnemann AR, Nout MJR, Boekel MAJS, Grewal RB (2013) Nutrient composition of selected newly bred and established mungbean varieties. LWT – Food Sci Tech 54:249–256
Das A, Raychaudhuri U, Chakraborty R (2012) Cereal based functional food of Indian subcontinent: a review. J Food Sci Tech 49(6):665–672
Debnath D, Sahu NP, Pal AK, Baruah K, Yengkokpam S, Mukherjee SC (2005) Present scenario and future prospects of phytase in aquafeed -review. Asian-Australas J Anim Sci 18:1800–1812
Desai M, Rangarajan P, Donahue J, Williams S, Land E, Mandal M, Phillippy B, Perera I, Raboy V, Gillaspy G (2014) Two inositol hexakisphosphate kinases drive inositol pyrophosphate synthesis in plants. Plant J 80:642–653
Dhole VJ, Reddy KS (2015) Genetic variation for phytic acid content in mungbean (*Vigna radiata* L. Wilczek). Crop J 3:157–162
Dhole VJ, Reddy KS (2016) Association of phytic acid content with biotic stress tolerance in mungbean (*Vigna radiata* L. Wilczek). Phytoparasitica 44(2):261–267
Dinant S, Lemoine R (2010) The phloem pathway: new issues and old debates. C R Biol 333:307–319
Donahue J, Alford S, Torabinejad J, Kerwin R, Nourbakhsh A, Ray W, Hernick M, Huang X, Lyons B, Hein P (2010) The *Arabidopsis thaliana myo*-inositol 1-phosphate synthase1 gene is required for *myo*-inositol synthesis and suppression of cell death. Plant Cell 22:888–903
Dost K, Tokul O (2006) Determination of phytic acid in wheat and wheat products by reverse phase high performance liquid chromatography. Anal Chim Acta 558:22–27
Duhan A, Chauhan BM, Kapoor AC (1989) Phytic acid content of chickpea (*Cicer arietinum*) and black gram (*Vigna mungo*): varietal differences and effect of domestic processing and cooking methods. J Sci Food Agric 49:449–455
Enneking D, Wink M (2000) Towards elimination of anti-nutritional factors in grain legumes. In: Knight R (ed) Linking research and marketing opportunities for pulses in the 21st century. Kluwer Academic Publishers, Dordrecht, pp 671–683
Erdman JW (1981) Bioavailability of trace minerals from cereals and legumes. Cereal Chem 58:21–26
Ertl DS, Young KA, Raboy V (1998) Plant genetic approaches to phosphorus management in agricultural production. J Environ Qual 27:299–304
Fassetti F, Leone O, Palopoli L, Rombo SE, Saiardi A (2011) IP6K gene identification in plant genomes by tag searching. BMC Proc 5. https://doi.org/10.1186/1753-6561-5-S2-S1
Fileppi M, Galasso I, Tagliabue G, Daminati M, Campion B, Doria E, Sparvoli F (2010) Characterisation of structural genes involved in phytic acid biosynthesis in common bean (*Phaseolus vulgaris* L.). Mol Breed 25:453–470
Frias J, Bakhsh A, Jones DA, Arthur AE, Vidal-Valverde C, Rhodes MJC et al (1999) Genetic analysis of the raffinose oligosaccharide pathway in lentil seeds. J Exp Bot 50:469–476
Frias J, Vidal-Valverde C, Sotomayor C, Diaz-Pollan C, Urbano G (2000) Influence of processing on available carbohydrate content and antinutritional factors of chickpeas. Eur Food Res Technol 210:340–345

Fu J, Peterson K, Guttieri M, Souza E, Raboy V (2008) Barley (*Hordeum vulgare* L.) inositol monophosphatase: gene structure and enzyme characteristics. Plant Mol Biol 67:629–642

Gao Y, Shang C, Saghai Maroof MA, Biyashev RM, Grabau EA, Kwanyuen P, Burton JW, Buss GR (2007) A modified colorimetric method for phytic acid analysis in soybean. Crop Sci 47:1797–1803

Gillaspy GE, Keddie JS, Oda K, Gruissem W (1995) Plant inositol monophosphatase is a lithium-sensitive enzyme encoded by a multigene family. Plant Cell 7:2175–2185

Gillman J, Pantalone V, Bilyeu K (2009) The low phytic acid phenotype in soybean line CX1834 is due to mutations in two homologs of the maize low phytic acid gene. Plant Genome 2:179–190

Girigowda K, Prashanth SJ, Mulimani VH (2005) Oligosaccharides of black gram (*Vigna mungo* L.) as affected by processing methods. Plant Foods Hum Nutr 60:173–180

Girigowda K, Peterbauer T, Mulimani VH (2006) Isolation and structural analysis of ajugose from *Vigna mungo* L. Carbohydr Res 341:2156–2160

Griga M, Kosturkova G, Kuchuk N, Ilieva-Stoilova M (2001) Biotechnology. In: Hedley CL (ed) Carbohydrates in grain legume seeds. Improving nutritional quality and agronomic characteristics. CAB International, Wallingford, pp 145–207

Guttieri M, Bowen D, Dorsch JA, Raboy V, Souza E (2004) Identification and characterization of low phytic acid wheat. Crop Sci 44:418–424

Haab CI, Keller F (2002) Purification and characterization of the raffinose oligosaccharide chain elongation enzyme, galactan:galactan galactosyltransferase (GGT), from *Ajuga reptans* leaves. Physiol Plant 114:361–371

Han IH, Baik B (2006) Oligosaccharide content and composition of legumes and their reduction by soaking, cooking, ultrasound, and high hydrostatic pressure. Cereal Chem 83:428–433

Hanakahi LA, Bartlet-Jones M, Chappell C, Pappin D, West SC (2000) Binding of inositol phosphate to DNA-PK and stimulation of double-strand break repair. Cell 102:721–729

Hande PA, Mondal S, Badigannavar AM, D'Souza SF (2013) Genetic variability of phytic acid phosphorus and inorganic phosphorus in cultivated groundnut (*Arachis hypogaea* L.). Plant Genet Resour 11(3):190–195

Hoch G, Peterbauer T, Richter A(1999) Purification and characterization of stachyose synthase from lentil (Lens culinaris) seeds: galactopinitol and stachyose synthesis. Arch. Biochem. Biophys. 366: 75–81.

Hoffmann CM, Huijbregts T, van Swaaij N, Jansen R (2009) Impact of different environments in Europe on yield and quality of sugar beet genotypes. Eur J Agron 30:17–26

Horbowicz M, Obendorf RL (1994) Seed desiccation tolerance and storability: dependence on flatulence-producing oligosaccharides and cyclitols-review and survey. Seed Sci Res 4:385

Hoy M, Efanov AM, Bertorello AM, Zaitsev SV, Olsen HL, Bokvist K, Leibiger B, Leibiger IB, Zwiller J, Berggren PO (2002) Inositol hexakisphosphate promotes dynamin I-mediated endocytosis. Proc Natl Acad Sci:6773–6777

Institute of Medicine, Food and Nutrition Board (2001) Dietary reference intakes for vitamin a, vitamin K, arsenic, boron, chromium, copper, iodine, Iron, manganese, molybdenum, nickel, silicon, vanadium, and zinc. National Academy Press, Washington, DC. http://ods.od.nih.gov/factsheets/zinc-HealthProfessional/

Jaureguy LM, Chen P, Scaboo AM (2011) Heritability and correlations among food-grade traits in soybean. Plant Breed 13:647–652

Joersbo M, Guldager Pedersen S, Nielsen JE, Marcussen J, Brunstedt J (1999) Isolation and expression of two cDNA clones encoding UDP-galactose epimerase expressed in developing seeds of the endospermous legume guar. Plant Sci 142:147–154

Jones GP (1997) Minerals. In: Wahlquist ML (ed) Food and nutrition. Allen & Unwin, Sydney, pp 249–254

Jones DA, DuPont MS, Ambrose MJ, Frias J, Hedley CL (1999) The discovery of compositional variation for the raffinose family of oligosaccharides in pea seeds. Seed Sci Res 9:305–310

Joyce C, Deneau A, Peterson K, Ockenden I, Raboy V, Lott JNA (2005) The concentrations and distributions of phytic acid phosphorus and other mineral nutrients in wild-type and low phytic acid Js-12-LPA wheat (*Triticum aestivum*) grain parts. Can J Bot 83(12):1599–1607

Kakati P, Deka SC, Kotoki D, Saikia S (2010) Effect of traditional methods of processing on the nutrient contents and some antinutritional factors in newly developed cultivars of green gram [*Vigna radiata* (L.) Wilezek] and black gram [*Vigna mungo* (L.) Hepper] of Assam, India. Int Food Res J 17:377–384

Kang J, Park J, Choi H, Burla B, Kretzschmar T, Lee Y, Martinoia E (2011) Plant ABC transporters. Arab Book 9:e0153

Karner U, Peterbauer T, Raboy V, Jones DA, Hedley CL, Richter A (2004) *Myo*-inositol and sucrose concentrations affect the accumulation of raffinose family oligosaccharides in seeds. J Exp Bot 55:1981–1987

Kataria A, Chauhan BM, Punia D (1989) Anti-nutrients in amphidiploids (Black gram x Mungbean): varietal differences and effect of domestic processing and cooking. Plant Foods Hum Nutr 39:257–266

Keller F (1992) Galactinol synthase is an extravacuolar enzyme in tubers of japanese artichoke (*Stachys sieboldii*). Plant Physiol 99:1251–1253

Kerr PS, Pearlstein RW, Schweiger BJ, Becker-Manley MF, Pierce JW, inventors (1998) Nucleotide sequences of galactinol synthase from zucchini and soybean. U.S. Patent Application No. 5773699. June 30, 1998

Keunen E, Peshev D, Vangronsveld J, Van den Ende W, Cuypers A (2013) Plant sugars are crucial players in the oxidative challenge during abiotic stress: extending the traditional concept. Plant Cell Environ 36:1242–1255

Kim S, Tai T (2011) Identification of genes necessary for wild-type levels of seed phytic acid in Arabidopsis thaliana using a reverse genetics approach. Mol Genet Genomics 286:119–133

Kim S, Tai T (2014) Identification of novel rice low phytic acid mutations via TILLING by sequencing. Mol Breed 34:1717–1729

Kim S, Andaya C, Goyal S, Tai T (2008) The rice OsLpa1 gene encodes a novel protein involved in phytic acid metabolism. Theor Appl Genet 117:769–779

Klein M, Burla B, Martinoia E (2006) The multidrug resistance-associated protein (MRP/ABCC) subfamily of ATP-binding cassette transporters in plants. FEBS Lett 580:1112–1122

Koster KL (1991) Glass formation and desiccation tolerance in seeds. Plant Physiol 96:302–304

Koster KL, Leopold AC (1988) Sugars and desiccation tolerance in seeds. Plant Physiol 88:829–832

Kumar V, Sinha AK, Makkar HPS, Beckeret K (2010) Dietary roles of phytate and phytase in human nutrition: a review. Food Chem 120:945–959

Kuo TM, Van Middlesworth JF, Wolf WJ (1988) Content of raffinose oligosaccharides and sucrose in various plant seeds. J Agric Food Chem 36:32–36

Laha D, Johnen P, Azevedo C, Dynowski M, Weiss M, Capolicchio S, Mao H, Iven T, Steenbergen M, Freyer M (2015) VIH2 regulates the synthesis of inositol pyrophosphate InsP8 and jasmonate-dependent defenses in *Arabidopsis*. Plant Cell 27:1082–1097

Lahuta LB, Goszczyńska J, Horbowicz M, Hołdyński C, Górecki RJ (2010) Cyclitols affect accumulation of α-d-galactosides in developing Vicia seeds. Acta Physiol Plant 32:933–944

Larson SR, Young KA, Cook A, Blake TK, Raboy V (1998) Linkage mapping of two mutations that reduce phytic acid content of barley grain. Theor Appl Genet 97:141–146

Latta M, Eskin M (1980) A simple and rapid colorimetric method for phytate determination. J Agric Food Chem 28:1313–1315

Lattanzio V, Terzano R, Cicco N, Cardinali A, Di-Venere D, Linsalata V (2005) Seed coat tannins and bruchid resistance in stored cowpea seeds. J Sci Food Agric 85:839–846

Leeson S (1993) Recent advances in fat utilisation by poultry. In: Recent advances in animal nutrition in Australia. The University of New England, Armidale, pp 170–1981

Lehle L, Tanner W (1973) The function of myo-inositol in the biosynthesis of raffinose. Eur J Biochem 38:103–110

Liu JJ, Odegard W, de Lumen BO (1995) Galactinol synthase from kidney bean cotyledon and zucchini leaf (purification and n-terminal sequences). Plant Physiol 109:505–511

Loewus F (2002) Biosynthesis of phytate in food grains and seeds. In: Reddy NR, Sathe SK (eds) Food phytates. CRC Press, Boca Raton, pp 53–61

Loewus FA, Murthy PPN (2000) myo-Inositol metabolism in plants. Plant Sci 150:1–19
Lott JNA (1980) Protein Bodies. In: Tolbert NE (ed) The biochemistry of plants. Academic Press, New York, pp 589–623
Luo Y, Qin G, Zhang J, Liang Y, Song Y, Zhao M, Tsuge T, Aoyama T, Liu J, Gu H (2011) D-*myo*-inositol-3-phosphate affects phosphatidylinositol-mediated endomembrane function in *Arabidopsis* and is essential for auxin-regulated embryogenesis. Plant Cell 23:1352–1372
Macbeth MR, Schubert HL, Van Demark AP, Lingam AT, Hill CP, Bass BL (2005) Inositol hexakisphosphate is bound in the ADAR2 core and required for RNA editing. Science 309:1534–1539
Mali G, Sala M, Arcon I, Kaucic V, Kolar J (2006) Insight into the short-range structure of amorphous iron inositol hexaphosphate as provided by P-31 NMR and Fe X-ray absorption spectroscopy. J Phys Chem B 110(46):23060–23067
March JG, Villacampa AI, Grases F (1995) Enzymatic-spectrophotometric determination of phytic acid with phytase from Aspergillus ficuum. Anal Chim Acta 300:269–272
March JG, Simonet BM, Grases F (2001) Determination of phytic acid by gas chromatography–mass spectroscopy: application to biological samples. J Chromatogr B 757:247–255
Martínez-Villaluenga C, Frias J, Vidal-Valverde C (2008) Alpha-galactosides: antinutritional factors or functional ingredients? Crit Rev Food Sci Nutr 48:301–316
Matyka S, Korol W, Bogusz G (1990) The retention of phytin phosphorus from diets with fat supplements in broiler chickens. Anim Feed Sci Technol 31:223–230
Moresco H, Sansón P, Seoane G (2008) Simple potentiometric determination of reducing sugars. J Chem Edu 85:1091–1093
Mubarak AE (2005) Nutritional composition and anti-nutritional factors of mungbean seeds 40 (*Phaseolus aureus*) as affected by some home traditional processes. Food Chem 89:489–495
Mulimani VH, Devindra S (1998) Effect of soaking, cooking and crude α-galactosidase treatment on the oligosaccharide content of redgram flour. Food Chem 61:475–479
Murphy AM, Otto B, Brearley CA, Carr JP, Hanke DE (2008) A role for inositol hexakisphosphate in the maintenance of basal resistance to plant pathogens. Plant J 56:638–652
Nishizawa A, Yabuta Y, Shigeoka S (2008a) The contribution of carbohydrates including raffinose family oligosaccharides and sugar alcohols to protection of plant cells from oxidative damage. Plant Signal Behav 3:1016–1018
Nishizawa A, Yabuta Y, Shigeoka S (2008b) Galactinol and raffinose constitute a novel function to protect plants from oxidative damage. Plant Physiol 147:1251–1263. https://doi.org/10.1104/pp.108.122465
Obendorf RL, Odorcic S, Ueda T, Coseo MP, Vassallo E(2004) Soybean galactinol synthase forms fagopyritol B1 but not galactopinitols: substrate feeding of isolated embryos and heterologous expression. Seed Sci. Res. 14: 321–333.
O'Dell BL, de Boland A (1976) Complexation of phytate with proteins and cations in corn rain and oilseed meals. J Agric Food Chem 24:804–808
O'Dell BL, de Boland AR, Koirtyohann SR (1972) Distribution of phytate and nutritionally important elements among morphological components of cereal grains. J Agric Food Chem 20(3):718–721
Ockenden I, Dorsch JA, Reid MM, Lin L, Grant LK, Raboy V, Lott JNA (2004) Characterization of the storage of phosphorus, inositol phosphate and cations in grain issues of four barley (*Hordeum vulgare* L.) low phytic acid genotypes. Plant Sci 167:1131–1142
Odom A, Stahlberg A, Wente S, York J (2000) A role for nuclear inositol 1,4,5-trisphosphate kinase in transcriptional control. Science 287:2026–2029
Pande R, Mishra HN (2015) Fourier transform near-infrared spectroscopy for rapid and simple determination of phytic acid content in green gram seeds (*Vigna radiata*). Food Chem 172:880–884
Panzeri D, Cassani E, Doria E, Tagliabue G, Forti L, Campion B, Bollini R, Brearley CA, Pilu R, Nielsen E et al (2011) A defective ABC transporter of the MRP family, responsible for the bean lpa1 mutation, affects the regulation of the phytic acid pathway, reduces seed *myo*-inositol and alters aba sensitivity. New Phytol 191:70–83

Pattee HE, Isleib TG, Giesbrecht FG, McFeeters RF (2000) Investigations into genotypic variations of peanut carbohydrates. J Agric Food Chem 48:750–756
Peterbauer T, Richter A (1998) Galactosylononitol and stachyose synthesis in seeds of adzuki bean: purification and characterization of stachyose synthase. Plant Physiol 117:165–172
Peterbauer T, Richter A (2001) Biochemistry and physiology of raffinose family oligosaccharides and galactosyl cyclitols in seeds. Seed Sci Res 11:85–197
Peterbauer T, Mucha J, Mayer U, Popp M, Glossl J, Richter A (1999) Stachyose synthesis in seeds of adzuki bean (*Vigna angularis*): molecular cloning and functional expression of stachyose synthase. Plant J 20:509–518
Peterbauer T, Lahuta LB, Blochl A, Mucha J, Jones DA, Hedley CL, Gorecki RJ, Richter A (2001) Analysis of the raffinose family oligosaccharide pathway in pea seeds with contrasting carbohydrate composition. Plant Physiol 127:1764–1772
Peterbauer T, Mucha J, Mach L, Richter A (2002) Chain elongation of raffinose in pea seeds. Isolation, characterization, and molecular cloning of a multifunctional enzyme catalyzing the synthesis of stachyose and verbascose. J Biol Chem 277:194–200
Polowick PL, Baliski DS, Bock C, Ray H, Georges F (2009) Over-expression of α-galactosidase in pea seeds to reduce raffinose oligosaccharide content. Botany 87:526–532
Pukacka S, Ratajczak E, Kalemba E (2009) Non-reducing sugar levels in beech (*Fagus sylvatica*) seeds as related to withstanding desiccation and storage. J Plant Physiol 166:1381–1390
Quemener B, Brillouet J (1983) Ciceritol, a pinitol digalactoside form seeds of chickpea, lentil and white lupin. Phytochemistry 22:1745–1751
Raboy V (1997) Accumulation and storage of phosphate and minerals. In: Larkins BA, Vasil IK (eds) Cellular and molecular biology of plant seed development. Kluwer Academic Publishers, Dordrecht, pp 441–477
Raboy V (2002) Progress in breeding low phytate crops. J Nutr 132:503S–505S
Raboy V, Gerbasi P (1996) Genetics of myo-inositol phosphate synthesis and accumulation. In: Biswas B, Biswas S (eds) Myo-inositol phosphates, phosphoinositides and signal transduction. Plenum Press, New York, pp 257–285
Raboy V, Gerbasi P, Young K, Stoneberg S, Pickett S, Bauman A, Murthy P, Sheridan W, Ertl D (2000) Origin and seed phenotype of maize low phytic acid 1-1 and low phytic acid 2-1. Plant Physiol 124:355–368
Rao PU, Belvady B (1978) Oligosaccharides in pulses: varietal differences and effects of cooking and germination. J Agric Food Chem 26:316–319
Rao VS, Vakil UK (1983) Effects of gamma-irradiation on flatulence-causing oligosaccharides in green gram (*Phaseolus aureus*). J Food Sci 48:1791–1795
Ravindran V, Bryden WL, Kornegay ET (1995) Phytates: occurrence, bioavailability, and implications in poultry nutrition. Avian Poult Biol Rev 6:125–143
Reddy NR (2002) Occurrence, distribution, content, and dietary intake of phytate. In: Reddy NR, Sathe SK (eds) Food phytates. CRC Press, Boca Raton, pp 25–51
Reddy NR, Salunkhe DK (1989) Fermentation. In: Salunkhe DK, Kadam SS (eds) CRC handbook of world food legumes: Nutritional chemistry, processing technology, and utilisation, vol 3. CRC Press, Boca Raton, pp 177–1217
Reddy NR, Salunkhe DK, Sharma RP (1980) Flatulence in rats following ingestion of cooked and germinated black gram and a fermented product of black gram and rice blend. J Food Sci 45:1161–1164
Reddy NR, Pierson MD, Sathe SK, Salunke DK (1984) Chemical, nutritional and physiological aspects of dry bean carbohydrates; a review. Food Chem 13:25–68
Rickard SE, Thompson LU (1997) Interactions and biological effects of phytic acid. In: Shaidi F (ed) Antinutrients and phytochemicals in food. American Chemical Society, Washington, DC, pp 294–312
Salih OM, Nour AM, Harper DB (1991) Nutritional quality of uncultivated cereal grains utilised as famine foods in western Sudan as measured by chemical analysis. J Sci Food Agric 57:367–377

Sánchez-Mata MC, Cámara Hurtado M, Díez-Marqués C (1999) Effect of domestic processes and water hardness on soluble sugars content of chickpeas (*Cicer arietinum* L.). Food Chem 65:331–338

Scandalios JG (2005) Oxidative stress: molecular perception and transduction of signals triggering antioxidant gene defenses. Braz J Med Biol Res 38:995–1014

Selle PH, Ravindran V, Caldwell RA, Bryden WL (2000) Phytate and phytase: consequences for protein utilisation. Nutr Res Rev 13:255–278

Shears SB (2015) Inositol pyrophosphates: why so many phosphates? Adv Biol Regul 57:203–216

Shi J, Wang H, Wu Y, Hazebroek J, Meeley R, Ertl D (2003) The maize low-phytic acid mutant lpa2 is caused by mutation in an inositol phosphate kinase gene. Plant Physiol 131:507–515

Shi J, Wang H, Hazebroek J, Ertl D, Harp T (2005) The maize low-phytic acid 3 encodes a myo-inositol kinase that plays a role in phytic acid biosynthesis in developing seeds. Plant J 42:708–719

Shi J, Wang H, Schellin K, Li B, Faller M, Stoop J, Meeley R, Ertl D, Ranch J, Glassman K (2007) Embryo-specific silencing of a transporter reduces phytic acid content of maize and soybean seeds. Nat Biotech 25:930–937

Shitre AS, Gadekar DA, Ramachandran V, Vikas R, Bakshi S, Kumar V, Vishwakarma G, Das BK (2015) Genotypic variation for phytic acid, inorganic phosphate and mineral contents in advanced breeding lines of wheat (*Triticum aestivum* L.). Electron J Plant Breed 6(2):395–402

Singh M, Krikorian AD (1982) Inhibition of trypsin activity *in vitro* by phytate. J Agric Food Chem 30:799–800

Sompong U, Kaewprasit C, Nakasathienl S, Srinives P (2010) Inheritance of seed phytate in mungbean (*Vigna radiata* L. Wilczek). Euphytica 171:389–396

Sompong U, Somta P, Raboy V, Srinives P (2012) Mapping of quantitative trait loci for phytic acid and phosphorus contents in seed and seedling of mungbean (*Vigna radiata* (L.) Wilczek). Breed Sci 62:87–92

Sosulski FW, Elkowicz L, Reichert RD (1982) Oligosaccharides in eleven legumes and their air-classified protein and starch fractions. J Food Sci 47:498–502

Souframanien J, Roja G, Gopalakrishna T (2014) Genetic variation in raffinose family oligosaccharides and sucrose content in black gram [*Vigna mungo* L. (Hepper)]. J Food Legumes 27:37–41

Sparvoli F, Cominelli E (2015) Seed biofortification and phytic acid reduction: a conflict of interest for the plant? Plan Theory 4:728–755

Sprenger N, Keller F (2000) Allocation of raffinose family oligosaccharides to transport and storage pools in *Ajuga reptans*: the roles of two distinct galactinol synthases. Plant J 21:249–258

Steger DJ, Haswell ES, Miller AL, Wente SR, O'Shea EK (2003) Regulation of chromatin remodeling by inositol polyphosphates. Science 299:114–116

Stiles A, Qian X, Shears S, Grabau E (2008) Metabolic and signaling properties of an ITPK gene family in *Glycine max*. FEBS Lett 582:1853–1858

Suneja Y, Kaur S, Gupta AK, Kaur N (2011) Levels of nutritional constituents and antinutritional factors in black gram (*Vigna mungo* L. Hepper). Food Res Int 44:621–628

Suzuki M, Tanaka K, Kuwano M, Yoshida KT (2007) Expression pattern of inositol phosphate related enzymes in rice (*Oryza sativa* L.): implications for the phytic acid biosynthetic pathway. Gene 405:55–64

Sweetman D, Stavridou I, Johnson S, Green P, Caddick S, Brearley C (2007) Arabidopsis thaliana inositol 1,3,4-trisphosphate 5/6-kinase 4 (AtITPK4) is an outlier to a family of ATP-grasp fold proteins from Arabidopsis. FEBS Lett 581:4165–4171

Tahir M, Vandenberg A, Chibbar RN (2011) Influence of environment on seed soluble carbohydrates in selected lentil cultivars. J Food Compos Anal 24:596–602

Tajoddin M, Shinde M, Junna L (2010) Raffinose, stachyose and sucrose contents of mung bean cultivars differing in seed coat color from Hyderabad-Karnataka region of India: effect of soaking and germination. Bioscan 5:343–346

Tajoddin MD, Shinde M, Lalitha J (2011) In vivo reduction the phytic acid content of mungbean (Phaseolus aureus L.) cultivars during germination. Am-Euras J Agric Environ Sci 10:127–132
Tan X, Calderon-Villalobos LI, Sharon AM, Zheng C, Robinson CV, Estelle M, Zheng N (2007) Mechanism of auxin perception by the TIR1 ubiquitin ligase. Nature 446:640–645
Tapernoux-Luthi E, Bohm A, Keller F (2004) Cloning, functional expression, and characterization of the raffinose oligosaccharide chain elongation enzyme, galactan:galactan galactosyltransferase, from common bugle leaves. Plant Physiol 134:1377–1387
Taunk J, Yadav NR, Yadav RC, Kumar R (2012) Genetic diversity among green gram [*Vigna radiata* (L.) Wilczek] genotypes varying in micronutrients (Fe and Zn) content using RAPD markers. Indian J Biotechnol 11:48–53
Thompson DB, Erdman JW (1982) Structural model for ferric phytate-implications for phytic acid analysis. Cereal Chem 59(6):525–528
Torabinejad J, Donahue J, Gunesekera B, Allen-Daniels M, Gillaspy G (2009) VTC4 is a bifunctional enzyme that affects myoinositol and ascorbate biosynthesis in plants. Plant Physiol 150:951–961
Tur F, Tur E, Lentheric I, Mendoza P, Encabo M, Isern B, Grases F, Maraschiello C, Perelló J (2013) Validation of an LC–MS bioanalytical method for quantification of phytate levels in rat, dog and human plasma. J Chromatogr B 928:146–154
Turgeon R (1991) Symplastic phloem loading and the sink-source transition in leaves: a model. In: Bonnemain VL, Delrot S, Dainty J, Lucas WJ (eds) Recent advances phloem transport and assimilate compartmentation. Ouest Editions, Nantes, pp 18–22
Turgeon R (1996) Phloem loading and plasmodesmata. Trends Plant Sci 1:418–423
Van den Ende W, Valluru R (2009) Sucrose, sucrosyl oligosaccharides, and oxidative stress: scavenging and salvaging? J Exp Bot 60:9–18
Veldman A, Veen WAG, Barug D, Van Paridon PA (1993) Effect of α - galactosides and α-galactosidase in feed on ileal piglet digestive physiology. J Anim Physiol Anim Nutr 69:57–65
Vohra A, Satyanarayana T (2003) Phytases: microbial sources, production, purification, and potential biotechnological applications. Crit Rev Biotechnol 23:29–60
Wang N, Daun J K (2004) The chemical composition and nutritive value of Canadian pulses. In: Canadian Grain Commission Report. pp. 19–29
Wang Z, Zhu Y, Wang L, Liu X, Liu Y, Phillips J et al (2009) A WRKY transcription factor participates in dehydration tolerance in Boea hygrometrica by binding to the W-box elements of the galactinol synthase (BhGolS1) promoter. Planta 230:1155–1166
Wanke D, Kolukisaoglu H (2010) An update on the ABCC transporter family in plants: many genes, many proteins, but how many functions? Plant Biol 12:15–25
Williams M, Torabinejad J, Cohick E, Parker K, Drake E, Thompson J, Hortter M, DeWald D (2005) Mutations in the *Arabidopsis* phosphoinositide phosphatase gene SAC9 lead to overaccumulation of Ptdins(4,5)P2 and constitutive expression of the stress-response pathway. Plant Physiol 138:686–700
Wilson M, Majerus P (1997) Characterization of a cDNA encoding *Arabidopsis thaliana* inositol 1,3,4-trisphosphate 5/6-kinase. Biochem Biophys Res Commun 232:678–681
Wongkaew A, Nakasathien S, Srinives P (2010) Isolation and characterization of d -myo-inositol-3-phosphate synthase from mungbean (Vigna radiata). Plant Mol Biol Rep 28:122–127
World Health Organization (WHO) (2004) Iron. In: Vitamin and mineral requirements in human nutrition, 2nd edn. World Health Organization, Geneva, pp 246–272
Xu J, Brearley CA, Lin WH, Wang Y, Ye R, Mueller-Roeber B, Xu ZH, Xue HW (2005) A role of Arabidopsis inositol polyphosphate kinase, AtIPK2α, in pollen germination and root growth. Plant Physiol 137:94–103
Xu X, Zhao H, Liu Q, Frank T, Engel K, An G, Shu Q (2009) Mutations of the multi-drug resistance-associated protein ABC transporter gene 5 result in reduction of phytic acid in rice seeds. Theor Appl Genet 119:75–83
Yadhu S, Satvir K, Anil KG, Narinder K (2011) Levels of nutritional constituents and antinutritional factors in black gram (*Vigna mungo* L. Hepper). Food Res Int 44:621–628

York JD, Odom AR, Murphy R, Ives EB, Wente SR (1999) A phospholipase C-dependent inositol polyphosphate kinase pathway required for efficient messenger RNA export. Science 285:96–100

Yuan S, Zhao FJ, Ren HJ, Zhu XL, Fu SL, Shu XJ (2007) Generation and characterization of two novel low phytate mutations in soybean (*Glycine max* L. Merr.). Theor Appl Genet 115:945–957

Yuan F, Zhu D, Tan Y, Dong D, Fu X, Zhu S, Li B, Shu Q (2012) Identification and characterization of the soybean IPK1 ortholog of a low phytic acid mutant reveals an exon-excluding splice-site mutation. Theor Appl Genet 125:1413–1423

Zhai H, Wang F, Si Z, Huo J, Xing L, An Y, He S, Liu Q (2015) A *myo*-inositol-1-phosphate synthase gene, *IbMIPS1*, enhances salt and drought tolerance and stem nematode resistance in transgenic sweet potato. Plant Biotechnol J. https://doi.org/10.1111/pbi.12402

Zhang ZB, Yang G, Arana F, Chen Z, Li Y, Xia HJ (2007) *Arabidopsis* inositol polyphosphate 6–/3-kinase (AtIPK2β) is involved in axillary shoot branching via auxin signalling. Plant Physiol 144:942–951

Zhao T, Corum JW III, Mullen J, Meeley RB, Helentjaris T, Martin D, Downie B (2006) An alkaline α-galactosidase transcript is present in maize seeds and cultured embryo cells, and accumulates during stress. Seed Sci Res 16:107–121

Zhuo C, Wang T, Lu S, Zhao Y, Li X, Guo Z (2013) Cold responsive galactinol synthase gene from Medicago falcata (MfGolS1) is induced by myo-inositol and confers multiple tolerances to abiotic stresses. Physiol Plant 149:67–78

Breeding for Better Grain Quality in *Lathyrus*

Debjyoti Sen Gupta, Surendra Barpete, Jitendra Kumar, and Shiv Kumar

Abstract Grass pea (*Lathyrus sativus* L.) is one of the primitive plant species domesticated for human food and animal feed. Like any other food legumes, it is rich in protein, healthy fats, vitamins and micronutrients. However, among pulses, this crop has a distinction of having homoarginine and β-ODAP (β-N-oxalyl-diamino-L-propionic acid). While the former makes it a functional healthy food, the later a toxic food. The presence of ODAP has caused much damage to its cultivation and consumption among growers and consumers as it is known to cause irreversible spastic paraparesis (paralysis) of lower limbs, if overconsumed continuously for a longer period as survival food. Many countries have put forth ban of its trade, leading to serious marketing issue. In this chapter, priority traits for genetic biofortification of grass pea and the genetic variability reported in the existing germplasm have been reviewed. Suitable analytical methods for estimating ODAP, protein and homoarginine concentration have been outlined. Further genetic variability for the target traits has been discussed along with currently available breeding methods and tools for mainstreaming biofortification efforts in grass pea. For low or no ODAP concentration in future varieties, recent advances in biotechnology and genomics-assisted breeding approaches are pertinent to deploy in this crop species.

Keywords *Lathyrus sativum* · Micronutrients · β-ODAP · Neurolathyrism
Homoarginine · ICARDA · Low ODAP · Toxin

D. Sen Gupta (✉) · J. Kumar
ICAR-Indian Institute of Pulses Research, Kanpur, Uttar Pradesh, India

S. Barpete
International Center for Agricultural Research in the Dry Areas (ICARDA), Food Legume Research Platform (FLRP), Sehore, Madhya Pradesh, India

S. Kumar (✉)
International Center for Agricultural Research in the Dry Areas (ICARDA), Rabat, Morocco
e-mail: sk.agrawal@cgiar.org

D. S. Gupta et al. (eds.), *Breeding for Enhanced Nutrition and Bio-Active Compounds in Food Legumes*, https://doi.org/10.1007/978-3-030-59215-8_6

Introduction

Grass pea or chickling vetch (*Lathyrus sativus* L.) is one of the primitive plant species domesticated for human consumption and animal feed in the Mediterranean region that later spread to other environments especially in the semiarid environments of South Asia and highlands of East Africa. It is known by various names in different countries like *khesari* in India, Nepal and Bangladesh, *guaya* in Ethiopia, *san li dow* in China and *pois carré* in France (Campbell 1997). Taxonomically, *Lathyrus* along with *Pisum*, *Lens* and *Vicia* belongs to the family Fabaceae and tribe Vicieae. Genus *Lathyrus* has about 187 species (Allkin et al. 1983, 1986). Based on morphological traits, the genus is subdivided into 13 sections (Kupicha 1983), but the phylogenetic relationships among sections and species require further detailed investigation involving morphological, biochemical, cytogenetic and molecular markers. There is large species diversity distributed from temperate to tropical regions of the world (Narayan 1991). The word grass pea generally refers to the two cultivated species, *Lathyrus sativus* and *Lathyrus cicera*. Some species are used as fodder (*L. hirsutus* and *L. palustris*). Many species of *Lathyrus* are grown as green manure crop. *L. odoratus* is grown as ornamental species. *Lathyrus sativus* and *L. cicera* are diploid (2n = 14); however, autopolyploid (Sybenga 1995; Khawaja et al. 1997; Khawaja 1988) or allopolyploid (Gutiérrez et al. 1994) species also exist.

Grass pea is a very hardy pulse crop capable of growing in extreme moisture stress conditions, thereby making it a suitable crop of cultivation in vast drought-prone areas of Asia and Africa. There is a lack of global production statistics of grass pea; however, as reported by Campbell (1997), grass pea is grown in India, Bangladesh, Pakistan, Nepal and Ethiopia. It is also cultivated in Central, South and Eastern Europe and in West Asia and North Africa (Syria, Lebanon, Palestine, Egypt, Iraq, Afghanistan, Morocco and Algeria). Worldwide, 1.2 million tonnes is produced from ~1.5 million ha area (Kumar et al. 2011b). India is the largest producer with 384,800 tonnes followed by Bangladesh (232,500 tonnes) and Ethiopia (202,126 tonnes) (Gupta et al. 2018; Kumar et al. 2011b). It is largely grown in very low fertility soils, and due to its robust root system, the crop is able to survive as it can extract water and nutrients from deeper soil regimes. This pulse crop is a part of subsistence agriculture in India and other neighbouring countries like Bangladesh and Nepal where it is grown as broadcast crop in rice fields before the rice harvest. Due to its capacity to withstand extra soil moisture, seeds of grass pea show a stable germination and growth. This hardiness to both water stress either in excess or in shortage and almost nil input cost makes this crop an ideal protein source for millions of people residing in poverty-ridden regions.

Despite having the status of poor man's food, it has helped thousands of people to overcome famines in many regions in the past. Grass pea is high in protein content (Almeida 1980) and contains free L-homoarginine, a precursor of lysine (Quereshi et al. 1977). Grass pea has lost its charm and popularity due to the presence of β-ODAP (β-N-oxalyl-diamino-L-propionic acid), a neurotoxic non-protein amino acid that can cause irreversible spastic paraparesis (paralysis) of lower limbs (Adiga et al. 1962, 1963; Kuo et al. 1998; Murti et al. 1964; Nunn et al. 1987; Padmanaban 1980; Rao et al. 1964; Ross et al. 1989; Roy et al. 1963; Spencer et al. 1986; Wang et al. 2000; Zhao et al. 1999a, b; Kumar et al. 2011b). There were many reports where consumption of grass pea as a major portion of diet (>25%) by malnourished people for several months had resulted in paralysis among consumers during famine period. Malnutrition and oxidative stress are reported as triggering factors in neurolathyrism (Getahun et al. 2003, 2005). The β-ODAP chemically excites the neurons causing paralytic neurolathyrism. Neurolathyrism-affected children developed arrested growth with underdeveloped nervous system (Grela et al. 2001). Naturally occurring ODAP exists in two isomeric forms, α and β (Bell and O'Donovan 1966), with α-isomer being less toxic (De Bruyn et al. 1994; Harrison et al. 1977). The β form predominates 95% of the total ODAP. β-ODAP may convert to α-ODAP by temperature and pH (Padmajaprasad et al. 1997; Long et al. 1996). The equilibrium concentration of α- and β-ODAP varied with temperature (Zhao et al. 1999b). The conversion of β-form to α-form is known to follow zero order reaction kinetics, i.e. the rate of conversion is independent of the concentration of the reactants (Belay et al. 1997). It is necessary to reduce the existing ODAP concentration in grass pea cultivars. Despite concerted efforts, zero ODAP cultivars could not become a reality to date (Jackson and Yunus 1984; Smartt 1984; Datta and Varshney 2009; Kumar et al. 2011a). Despite having anti-nutritional properties, β-ODAP has positive roles as well. It is used as hemostatic agent for therapeutic use due to its haemorrhage-stopping property and thrombopoiesis treatment (Ding et al. 2018; Lambein et al. 2019). Tamburino et al. (2012) estimated nutritional factors among Italy-grown grass pea varieties and total lipid amount (1.67 ± 0.18 g/100 g); among these unsaturated fatty acids, α-linolenic, linoleic and γ-linolenic acids were most abundant. Ascorbic acid (13.50 ± 0.30 mg/100 g) and glutathione (15.90 ± 0.10 mg/100 g), the folic acid content (206.70 ± 8.30 g/100 g) and total phenolic content (174.91 ± 8.39/100 g) were also high. Total saturated fatty acids, total monounsaturated fatty acids, total polyunsaturated fatty acids and total fatty acids ranged from 295.72 to 436.94, 113.19 to 170.78, 127.39 to 179.39 and 538.04 to 778.98 mg 100 g−1, respectively. In addition, unsaturated fatty acids, oleic acid, linoleic acid, γ-linolenic acid and α-linolenic acid which are the main components of fatty acids ranged from 109.22 to 163.95, 59.57 to 82.98, 16.18 to 30.38 and 45.56 to 71.59 mg 100 g−1, respectively (Arslan et al. 2017). While testing 22 grass pea genotypes in Turkey, it was found that retinol, β-carotene, thiamine, riboflavin, niacin, pantethine, pyridoxine, folic acid and ascorbic acid ranged from 25.6 to 44.1 μg/kg, 240.8 to 410.1 μg/kg, 3.74 to 5.44, 1.86 to 2.76, 12.37 to 20.25, 14.43 to 22.41, 4.92 to 6.62, 4.04 to 6.77 and 33.4 to 58.2 mg/kg, respectively, in seeds. This significant variation observed among these genotypes with low β-ODAP content is important for the grain quality improvement in grass pea (Arslan et al. 2017). Among the free amino acids, arginine was found to be the most fluctuating one (0.10–506.85 mg/g) in grass pea seeds (Arslan et al. 2017).

Priority Traits for Quality Improvement in Grass Pea

ODAP Concentration

The non-protein amino acid which causes neurolathyrism is a glutamate analogue, β-oxalyl-L-α, β-diaminopropionic acid or ODAP or BOAA (Lambein et al. 1993; Yan et al. 2006; Dunlop et al. 2015). Mechanism of toxicity of ODAP is due to neuroexcitation leading to dysregulation of calcium homeostasis and oxidative stress resulting in neuronal death (Van Moorhem et al. 2010). β-ODAP is derived from precursor, β-(isoxazolin-5-on-2-yl)-alanine, in pods, pericarp and immature seed (Kuo et al. 1993). As the maturity approaches, there was a decrease in ODAP content in pods and pericarp and increase in matured seed (Kuo et al. 1993). The levels of ODAP synthesized by the plant in seeds, leaves and other plant parts vary during plant growth stages and soil mineral deficiency or excess, and the presence of heavy metals also increases ODAP concentration (Jiao et al. 2011). Abiotic stresses like osmotic stress, waterlogging, flooding and drought all impact ODAP concentration. Under harsh growing conditions, ODAP concentration is higher than under normal conditions (Dunlop et al. 2015). Berger et al. (1999) reported that in *Lathyrus* species, the extent of variation of ODAP in the seed was species related; *L. ochrus* was having higher concentration (0.69 ± 0.02%, range 0.40–1.25%), followed by *L. sativus* (0.40 ± 0.01%, range 0.26–0.78%) and *L. cicera* (0.18 ± 0.01%, range 0.12–0.30%) (Table 1). ODAP concentrations in *L. ochrus* and *L. sativus* seeds were positively correlated with soil phosphorus and negatively correlated with clay content and salinity (Berger et al. 1999). Lambein et al. (1994) reported that soil depleted in Zn and rich in Fe may lead to higher accumulation of ODAP in matured seed of *Lathyrus sativus*. This happens particularly in monsoon rain-depleted soils in Indian subcontinent and zinc-deficient soil in Ethiopia (Lambein et al. 1994). It is hypothesized that ODAP acts as a carrier molecule for Zn ions (Lambein et al. 1994). In a separate study, ten grass pea genotypes were grown at five diverse locations in Ethiopia (Fikre et al. 2011). Path analysis revealed that Zn^{2+}/P, days to maturity, grain yield and K^+ were variables affecting the ODAP concentration in seeds. Linear correlation analysis showed that ODAP level was positively correlated ($r > 0.70$) with K^+ and sunshine hours and negatively correlated ($r < -0.70$) with soil pH, days to maturity and grain yield. The correlation of sunshine hour with ODAP level in seeds was highest during crop maturity phase. It was further suggested that ODAP biosynthesis and its response to environmental stress are maximized during the post-anthesis stage (Fikre et al. 2011). It is generally found that *Lathyrus* genotypes with white flowers have less ODAP and other antinutritional factors compared to the genotypes having coloured flowers (Ochatt et al. 2007). Grass pea genotypes of Mediterranean origin have white flowers and consequently less ODAP content; in contrary, grass pea genotypes from Indian subcontinent have coloured flowers and are generally high in ODAP content (Ochatt et al. 2007).

Table 1 Range of protein, iron, zinc and ODAP content in *Lathyrus* species

Trait	Species	Ranges	References
Protein	*L. sativus L.*	30%	Aletor et al. (1994)
		25.60%	Tamburino et al. (2012)
		23.93–31.94%	Tadesse and Bekele (2003)
		23–29.9%	Bisignano et al. (2002)
		27.29–31.98%	Urga et al. (2005)
		18.2–34.6%	Girma and Korbu (2012)
		22.4–28.2%	Barpete et al. (2012)
		8.6–32.2%.	Kumari et al. (2018)
	Lathyrus cicera L	21.50%	Cavada et al. (2011)
		23.8–27.4%	Grela et al. (2012)
	Lathyrus maritimus L	10.7–28.0%,	Dnyanu (1998)
	L. tingitanus	25.60%	Cavada et al. (2011)
	L. amphicarpos	19.90%	Cavada et al. (2011)
	L. angulatus	23.80%	Cavada et al. (2011)
	L. annuus	21.00%	Cavada et al. (2011)
	L. aphaca	20.80%	Cavada et al. (2011)
	L. clymenum	21.60%	Cavada et al. (2011)
	L. filiformis	21.20%	Cavada et al. (2011)
	L. hirsutus	25.10%	Cavada et al. (2011)
	L. latifolius	24.40%	Cavada et al. (2011)
	L. ochrus	20.20%	Cavada et al. (2011)
	L. pratensis	23.20%	Cavada et al. (2011)
	L. sphaericus	23.50%	Cavada et al. (2011)
	L. setifolius	22.70%	Cavada et al. (2011)
Zinc	*L. sativus L.*	2.74–4.52 mg/100 g	Urga et al. (2005)
		2.46–36.7 mg/100 g	Grela et al. (2012)
	Lathyrus maritimus L	3.0 mg/100 g	Dnyanu (1998)
	Lathyrus cicera L	1.96–2. 77 mg/100 g	Grela et al. (2012)
Iron	*L. sativus L.*	4.64–8.74 mg/100 g	Urga et al. (2005)
		4.11–5.48 mg/100 g	Grela et al. (2012)
		7.3 mg/100 g	Duke (1981)
	Lathyrus maritimus L	9.4 mg/100 g	Dnyanu (1998)
	Lathyrus cicera L	4.11–5.28 mg/100 g	Grela et al. (2012)
ODAP	*L. sativus L.*	0.02–2.40%	Abd El Moneim et al. (2001)
		0.14–0.91%	Tadesse and Bekele (2003)
		0.50–2.50%	Xiong et al. (2015)
		0.02–0.54%	Fikre et al. (2008)
		0.15–0.95%	Kumar et al. (2011b, 2013)
		0.13–0.44%	Girma and Korbu (2012)
		0.21–0.55%	Shiferaw and Porceddu (2018)
	Lathyrus cicera L	0.09–0.13%	Grela et al. (2012)
		0.03–0.22%	Abd El Moneim et al. (2001)
	Lathyrus ochrus	0.46–2.5%	Abd El Moneim et al. (2001)
		1.01%	Aletor et al. (1994)

Xiong et al. (2015) used seven grass pea genotypes differing in seed β-ODAP concentration at three different levels of water availability to find out changes in the β-ODAP in leaves, pods and seeds. The concentration and amount of β-ODAP decreased in leaves in early reproductive development and in pods as they matured, while water stress increased β-ODAP concentration in leaves and pods. The net amount of β-ODAP in leaves and pods at early reproductive stage was positively associated with seed β-ODAP concentration at maturity. It was concluded that variation among grass pea genotypes in seed β-ODAP concentration results from variation in net accumulation of β-ODAP in leaves and pods during vegetative and early reproductive development.

Protein Content

The total seed protein in grass pea varies between 18.2% and 34.6% (Aletor et al. 1994, Tamburino et al. 2012, Tadesse and Bekele 2003, Bisignano et al. 2002, Urga et al. 2005, Girma and Korbu 2012, Barpete et al. 2012, Kumari et al. 2018) (Table 1). The grass pea seed protein is composed of globulins (66%), albumins (14%), glutelins (15%) and prolamins (5%) (Chandna and Matta 1994). The large differences in total seed protein content between *Lathyrus* species including *L. sativus, L. cicera, L. maritimus, L. tingitanus, L. amphicarpos, L. angulatus, L. annuus, L. aphaca, L. clymenum, L. filiformis, L. hirsutus, L. latifolius, L. ochrus, L. pratensis, L. setifolius and L. sphaericus* have been observed (Cavada et al. 2011; Grela et al. 2012; Kumari et al. 2018). The total seed protein content ranged from 10.7% to 28.0% in *L. maritimus* (Dnyanu 1998) and 21.5% to 27.4% in *L. Cicera* (Cavada et al. 2011; Grela et al. 2012) (Table 1). The matured grass pea leaves also contain 17% protein content (Rizvi et al. 2016). Grass pea contains higher protein content in seeds than other cool-season food legumes like field pea (23%), faba bean (24%), chickpea (22%) and lentil (27%) (Petterson et al. 1997; Ravindran and Blair 1992).

Like other grain legumes, it has two types of storage proteins and several minor proteins like protease and amylase inhibitors, lectins, lipoxygenase, etc. (Duranti 2006). The most abundant class of storage protein is the globulins (Duranti 2006; Duranti and Gius 1997). Some of the globulins (7S) also exhibited bioactive properties. 7S globulin showed activities related to upregulation of LDL receptors, plasma cholesterol, triglyceride reduction and anti-atheromatous activities (Duranti and Gius 1997; Duranti et al. 2004; Fukui et al. 2004; Desroches et al. 2004; Castiglioni et al. 2003; Adams et al. 2002). Chavan et al. (2001) extracted protein isolates from *L. maritimus* using sodium hydroxide (NaOH) and sodium hexametaphosphate (SHMH). NaOH extract was having more protein content than SHMH extract. Protein extracts were also having high predicted biological value and protein efficiency ratio. NaOH- and SHMP-extracted protein showed protein digestibility of 81–83% for pepsin-trypsin and 78.6–79.2% for pepsin-pancreatin (Chavan et al. 2001). Similarly, protein isolates from *L. clymenum* and *L. annuus* were analysed

(Cavada et al. 2010). These isolates were derived by alkaline extraction and acid precipitation of proteins at their isoelectric point (pH 4.5). The per cent protein recovery was around 60% for both these *Lathyrus* species. *L. annuus* and *L. clymenum* protein isolates contained 81.07% and 82.4% of proteins, respectively. The in vitro protein digestibility increased to 93% and 95% in the protein isolates of *L. annuus* and *L. clymenum* (Cavada et al. 2010).

In grass pea, protein constitutes 20% of the seed dry weight, >60% of which is composed by globulins and 30% by albumins (Rosa et al. 2000). A single, 24 kDa polypeptide comprises more than half of the protein present in the albumin fraction. The globulins may be fractionated into three main components, which were named: α-lathyrin (the major globulin), β-lathyrin and γ-lathyrin. R-Lathyrin with a sedimentation coefficient of 18S is composed of three main types of unglycosylated subunits (50–66 kDa), each of which produces, upon reduction, a heavy and a light polypeptide chain, by analogy with 11S subunit (Rosa et al. 2000).

Proteins have been used by many workers to delineate intraspecific variation and interspecific relationships in many crop species (Ayaz et al. 1999; Przybylska et al. 2000; Emre et al. 2006, 2007). Emre (2009) used dry seeds of seven *Lathyrus* species and found that *L. annuus* (2.237 lg/ml) and *L. cicera* (2.158 lg/ml) have highest albumin content, while *L. phaselitanus* (6.972 lg/ml) and *L. cicera* (6.881 lg/ml) have higher globulin A content. However, *L. chloranthus* and *L. stenophyllus* have the highest globulin B (6.213 and 6.118 lg/ml) and glutelin contents (4.306 and 4.293 lg/ml), and *L. hirsutus* has the highest prolamin amount (0.458 lg/ml). Based on protein profile, it was found that *L. stenophyllus* and *L. sativus* were close (51.2% similarity), and also *L. cicera* and *L. annuus* have close relationship (44.1% similarity). In addition, it was found that *L. hirsutus* has less homology with other tested species. *L. chloranthus* and *L. phaselitanus* (53.4% similarity) were found to be closely related.

Like most grain legumes, grass pea is deficient in the essential sulphur-containing amino acids, methionine and cysteine, but it is rich in lysine that is low in cereals (Gatel 1994; Ravindran and Blair 1992; Mahler-Slasky and Kislev 2010). The amino acid profiles of *grass pea* are like those reported for many grain legumes (Hanbury et al. 2000). The deficiency of essential sulphur-containing amino acids, such as methionine that plays a vital role in the central nervous system (Amara et al. 1995), may be overcome with balanced diet-containing cereals (Lambein and Kuo 2004).

Micronutrients

Micronutrients are chemical compounds important to human health. At least 30 essential micronutrients exist that cannot be synthesized within human body and must be supplied through food, either of plant or animal origin (Shergill-Bonner 2013). Most of the countries have their own standard recommendation for daily intake of micronutrients and vitamins. Micronutrients are required in trace quantities (ppm level), and recommended daily allowances are measured in milligrams per day, and they act as cofactors in metabolic pathways and biochemical reactions. For example, zinc is a cofactor in hundreds of enzymes and plays an important role in boosting immunity (Shergill-Bonner 2013).

Data of micronutrient profile of *Lathyrus* species are limited (Table 1). Grass pea has a good quantity of iron in its seeds that ranged from 4.11 mg to 8.74 mg/100 g (Duke 1981; Urga et al. 2005; Grela et al. 2012), whereas iron content in *L. cicera* and *L. maritimus* contained 4.11–5.28 mg/100 g (Grela et al. 2012) and 9.4 mg/100 g (Dnyanu 1998), respectively. The zinc content of *Lathyrus* genus in seeds ranged from 2.46 to 4.52 mg/100 g in *Lathyrus sativus* (Urga et al. 2005; Grela et al. 2012), 1.96 to 2.77 mg/100 g in *Lathyrus cicera* (Grela et al. 2012), and 3.0 mg/100 g in *Lathyrus maritimus* (Dnyanu 1998). Recently, ICARDA breeding programme has analysed 485 germplasm accessions representing many species (personal communication, under publication) and found large genetic variability for micronutrient concentration in grass pea (Table 2) which can be used for mainstreaming biofortification in grass pea. More such efforts are needed to phenotype *Lathyrus* germplasm for Fe and Zn concentration to find out spectrum of genetic diversity for various micronutrients existing within cultivated species.

Homoarginine Content

β-ODAP and homoarginine are the major free non-protein amino acids present in grass pea seeds. Together they make up about 90% of ninhydrin-reacting compounds in the 70% ethanol extracts (Zhao et al. 1999a, b). Grass pea, like other orphan legumes, is still an untouched treasure for compounds that can contribute to human health. For instance, it is the only known dietary source of L-homoarginine. Therefore, as nutraceutical, grass pea is an excellent example of a potential "functional food" (Singh and Rao 2013; Llorent-Martínez et al. 2017). The amino acid L-homoarginine provides benefits in cardiovascular disease treatments (Rao 2011; Singh and Rao 2013; van Wyk et al. 2016) and in overcoming the consequences of

Table 2 Genetic variability for micronutrient and macronutrient concentration in 485 grass pea accessions

Nutrient	Range	Minimum (ppm)	Maximum (ppm)	Mean (ppm)	Standard deviation
P	2830	953.74	3784.12	2249.84	721
K	2513	1478.90	3992.29	2901.25	543
Zn	31	20.39	51.05	32.81	6
Ca	709	688.45	1397.54	1028.39	182
Mg	1125	554.00	1679.12	986.03	236
Mn	38	0.12	38.29	6.82	8
Fe	46	12.98	59.35	29.68	9
Se	0.58	0.01	0.59	0.18	0.14

hypoxia, i.e. the inadequate oxygen supply at the tissue level, associated with cancer tumour development (Ke and Costa 2006; Jammulamadaka et al. 2011). Thus, a daily dietary intake of L-homoarginine through small quantities of grass pea may be valuable for human health and deserves to be studied further (Rao 2011). There is threefold variation (6.26 vs. 20.97 g kg^{-1}) for homoarginine amount in grass pea (Piergiovanni and Damascelli 2011). Fikre et al. (2008) analysed grass pea genotypes with different origin and concluded that the variation of homoarginine in grass pea varied between 0.68% and 0.86%. Other episodic studies available in the literature showed narrow variation for homoarginine in grass pea from 5.3 to 6.7 mg g^{-1} (Yan et al. 2005) and 3.2 to 10.6 mg g^{-1} (Zhao et al. 2011). However, year-to-year variation of grain yield did not affect the homoarginine because correlation analysis did not evidence a significant relationship between these traits (Piergiovanni et al. 2011). There is however a need to study the relative importance of soil composition, sowing and harvesting date, environmental conditions, genotype x environment interaction, etc., on the homoarginine content in grass pea.

Analytical Methods

A major priority for any breeding initiative is to have in place effective tools for assessing the genetic variation of the trait of interest. Characterization and evaluation of germplasm collections greatly assist in the identification of genetic materials that could be utilized in crop improvement. Diversity assessment can be carried out based on various types of data that emanate from morphological, biochemical, nutritional and DNA-based differences. Seed storage protein fractions are mixtures of components which show polymorphism both within and among genotypes of the same species. Ideally, these technologies should be relatively low in cost and also rapid in their analysis to allow for high throughput. For each nutritional trait, various analytical methods are developed and practised; some of them are described:

Protein Analysis

There are many methods in use for protein estimation, namely, the Lowry method (Lowry et al. 1951), Bradford method (Bradford 1976), Biuret method (Gornall et al. 1949), Kjeldahl method (Kjeldahl 1883), Dumas method (Dumas 1831), AACC International method 46-30.01 (n.d.), nondestructive NIR method and UV-visible spectroscopy. Few of the recent methods are described as follows:

Nondestructive NIRS Method Near-infrared reflectance spectroscopy (NIRS) is an indirect and efficient method of measuring the chemical composition of feedstuffs based on the unique near-infrared absorption properties of the major chemical components of a sample. The NIR method has advantages such as rapid determination, minimal preparation of samples, nonconsumptive analyses, multiplicity of sample preparation in one operation, no consumption of reagents and ultimately low marginal costs of analyses. The NIRS procedure includes scanning of each sample in a closed 3.5 cm diameter ring cup, using a Foss model DS2500 scanning monochromator device. Spectral absorbance values are recorded from 1,100 to 2,500 nm, every 0.5 nm, as log 1/R, where R represents the per cent of energy reflected, which must then be related to the amount of the component as determined by reference or standard method. The spectra are exported to the WinISI software version 4.4 and were combined with the chemical reference data. The relationship between the log (1/R) values and the reference method values is expressed as an approximation and always involves some form of regression equation.

UV-Visible Spectroscopy Protein concentrations can be determined directly by ultraviolet spectroscopy because of the presence of tyrosine and tryptophan which absorb at 280 nm. Because the levels of these two amino acids vary greatly from protein to protein, the UV absorbance per milligram protein is highly variable. The extinction coefficient (usually expressed as E1%, i.e. the absorbance at 280 nm of a 1% solution [10 mg/ml]) will generally fall between 4.0 and 15; however, examples of proteins at either extremes have been observed, e.g. parvalbumin (0.0), serum albumin (5.8), trypsin (14.3) and lysozyme (26.5). Thus, the absorbance at 280 nm will only give an estimation of the protein concentration unless the extinction coefficient for a pure protein has been accurately determined (by dry weight or by amino acid analysis). Alternatively, the absorbance in the far-ultraviolet region (190–220 nm) can be used. This method is much more sensitive and is less dependent on the amino acid composition because the absorbance is dominated by the peptide bond transition. The major advantages of this method include its high sensitivity, ease of performance and the fact that the method is nondestructive so valuable protein samples can be recovered. Major disadvantages include the requirement of UV spectrophotometers and quartz cuvettes and the fact that virtually everything including commonly used buffers absorbs in the UV regions. Nucleic acids also absorb strongly in the UV region (260 nm). A ratio of absorbance (280/260) can be used to correct for the presence of nucleic acids.

To determine the amino acid composition of proteins, a protein sample is first hydrolysed (strong acid) to release the amino acids, which are then separated using chromatography, e.g. cation exchange chromatography (Horn et al. 1946), affinity or absorption chromatography and HPLC-MS/MS system (Arslan et al. 2017).

β-ODAP Analysis

There are different methods to assess the ODAP content in grass pea seeds, namely, UV-spectrophotometer method (Rao 1978), capillary zone electrophoresis (Zhao et al. 1999a, b), high-performance liquid chromatography (Wang et al. 2000), nondestructive NIR method for ODAP analysis (El Haramein et al. 1998), HPLC-MS (mass spectrometry) method (Silva et al. 2019), liquid chromatography-mass spectrometry (LC-MS) (Emmrich et al. 2019) and thin layer chromatography (Ghosh et al. 2015). Various analytical methods are developed and practised; some of them are described below:

UV-spectrophotometer Method (Rao 1978) ODAP concentration is determined spectrophotometrically using an ortho-phthalaldehyde fluorescent dye. A quantitative relationship between the concentration of diaminopropionic acid and intensity of yellow colour produced when the ortho-phthalaldehyde reagent is added to the solution is the principle of this assay. In this method, 100 mg of the grass pea flour is extracted for 5 h with 10 mL ethanol 60% (v/v). The suspension is then centrifuged, and 75 μL of the supernatant is added to 92 μL of distilled water and to 0.33 mL of KOH 3N. The sample is kept in a boiling water bath for 30 min (alkaline hydrolysis) to convert from ODAP to DAP (diamino propionic acid) which can be determined and then bring to 1 mL with water. OPT reagent (2 mL) is added to the sample, and absorbance of resulting yellow solution is measured after 30 min using a spectrophotometer set at 420 nm.

Capillary Zone Electrophoresis (Zhao et al. 1999a, b) α-ODAP is determined using capillary zone electrophoresis (CZE) (Zhao et al. 1999a, b) with some modifications. CZE is carried out using Agilent HP3D machine and UV detection at 195 nm. The analyses are performed at a constant voltage of 20 kV at 20 °C in an electrolyte of 75 mM (H_3BO_3) buffer with pH 7.5. 0.5 g powder of *Lathyrus sativus* seeds used as sample is dissolved in 50 ml ethanol-water (30:70, v/v) solution and shaken for 2 h (in ice). After centrifugation (3500 rpm for 15 min), the upper clear solution is filtered with 0.45 μm filter paper. Then clear solution is diluted with ultra-distilled water (1:1) and is injected directly into the CZE system for 40 s at 50 mbar.

High-performance Liquid Chromatography (Wang et al. 2000) The HPLC method provides a simple accurate alternative to existing methods for plant screening purposes. The grass pea grinded sample is accurately weighed and added to ethanol-water (3:7, v/v), shaken briefly and sonicated for 30 min and then agitated with a magnetic stirrer for 2 h. The solution is separated after centrifugation (15 min at 15,000 g) and subsequently filtered. The HPLC system consists of a Waters Model 600E pump, an AccQ-Tag C_{18} (4 μm) column (15 × 0.39 cm), a column heater and a Model 2487 dual-wavelength absorbance detector set at 254 nm. The α- and β-ODAP are eluted at 17.16 min and 13.83 min, respectively, and should not be interfered with any of the compounds used. The HPLC detection limit for both isomers is 1.8 ng (signal/noise ratio = 2:1) which, when taking the pre-purification procedure into account, gives an apparent detection limit of 0.15 μg/g in the *L. sativus* samples. A positive correlation between the colorimetric and capillary electrophoresis was found (r = 0.83), but the colorimetric values showed, on average, 14% lower ODAP values (Tavoletti et al. 2005).

L-Homoarginine Analysis

Capillary zone electrophoresis (CZE) (Piergiovanni and Damascelli 2011) and ultrahigh-performance liquid chromatography tandem mass spectrometry (UHPLC-MS/MS) (Arslan et al. 2017) are used for the estimation of homoarginine content in *Lathyrus* seeds.

Mineral Analysis

Among several methods of mineral analysis, atomic absorption method (AAS), inductively coupled plasma optical emission spectrometry (ICP-OES), colorimetric techniques such as dithizone (for Zn) and Perls Prussian blue (for Fe) have been developed for high-throughput screening and are currently in use within some breeding programmes. Newer technologies are also being explored, and they include NIRS and both handheld and benchtop XRF. Results are promising and research in this area is continuing. Some of the methods used are described below:

Inductive Coupled Plasma Emission Spectrometry (ICPE) This is a rapid analysis for different elements within short time period. The total minerals (selenium, iron, zinc, calcium, magnesium, potassium, copper, etc.) are extracted using the modified HNO_3-H_2O_2 method described by Thavarajah et al. (2007, 2008, 2009). Approximately 500 mg of finely ground seeds is weighed into digestion tubes. The digestion is conducted at 90 °C using 6 ml of concentrated (70%) nitric acid (HNO_3) for 1 h, 3 ml of hydrogen peroxide (H_2O_2) for 15 min and 3 ml of 6 M hydrochloric acid (HCl) for 5 min. Measurement of total mineral concentration is validated using the National Institute of Standards and Technology (NIST) standard reference (1567a wheat flour) and a laboratory reference sample. The total mineral concentration is determined using inductively coupled plasma emission spectrometry (ICP-ES; ICP-6500 Duo, Thermo Fisher Scientific, PA, USA) with a detection limit of 5 μg/L.

Atomic Absorption Method (AAS) A Tecator digestion system (Tecator AB, Höganäs, Sweden), DS 40 aluminium block, autostep controller, 75 mL digestion tubes and Agilent atomic absorption spectrometer (AA-1475) are used in this method. Up to 5 g wet weight or 1.5 g dry weight of biological material, containing a maximum of 1 g fat, or from 100 to 1000 ng of selenium standard, is placed in digestion tubes, and 17 ml of a 7 + 3 mixture of nitric and perchloric acid is added together with a few alundum granules to prevent bumping. With the temperature program employed, digestion is complete within 20 h. When the maximum temperature in the aluminium block is 225 °C, giving a temperature in the digest near the boiling point of perchloric acid, maximum oxidation power is reached. After cooling, the volume of perchloric acid is adjusted by visual estimate to 4 ml. After that, 10 mL of 2.4 M HCl is added to each of the tubes, which are then warm in the aluminium block at 120 °C for 30 min. After cooling, the solution is diluted to 25 ml. The autosampler, hydride generator and the AA spectrometer are operated under the specific conditions. Sodium borohydride is dissolved in sodium hydroxide to give a 0.6% w/v NaBH4 in 0.5% w/v NaOH. The concentration of hydrochloric acid is 10 M, as recommended by the manufacturer.

Genetic Variability of ODAP and Protein Content

Past explorations have led to large ex situ collections of *Lathyrus* germplasm in different national and international gene banks. The *Lathyrus* database produced as a result of the *Lathyrus* global conservation strategy contains around 23,000 accessions with main collections held by University of Pau in France, ICARDA, NBPGR, India, and Genetic Resources Center in Bangladesh. Global collection at ICARDA represents 45 species from 45 countries (Kumar et al. 2013). This collection is unique because 45% and 54% of the accessions are wild relatives and landraces, respectively, mainly of *L. sativus*, followed by *L. cicera* and *L. ochrus* (Kumar et al. 2013). Furthermore, it is necessary to study the genetic diversity of the available collections in order to understand their full utilization potential and possible gaps. ICARDA has characterized more than 60% accessions for main descriptors (Robertson and Abd-El-Moneim 1997).

The evaluation of *Lathyrus* germplasm has been undertaken sporadically for different traits to identify useful donors for nutritional parameters including protein, micronutrients, homoarginine and ODAP content (Campbell et al. 1994; Grela et al. 2010; Hanbury et al. 1995; Kumar et al. 2013). For ODAP content, studies have shown a wide range of variation within the existing germplasm, ranging from 0.02% to 2.59%. Hanbury et al. (1999) reported a range of 0.04–0.76% for ODAP content in a set of 503 accessions procured from ICARDA. Pandey et al. (1997) reported a range of 0.128–0.872% for ODAP content among 1187 accessions. A detailed catalogue of grass pea germplasm comprising characterization and evaluation information on 63 traits for 1963 accessions has recently been published in India (Pandey et al. 2008). A wide range of variability was observed for all the traits of interest including ODAP content (0.067–0.712%). Some of the accessions having <0.1% ODAP are IPLY9, Prateek, AKL 19, BioL202, BioL203, Ratan, No. 2203 and No. 2208. Kumar et al. (2011b) also screened 1128 accessions of *L. sativus* and found a wide range (0.150–0.952%) for ODAP content. Only two accessions, IG118563 (0.150%) and IG64888 (0.198%), had low ODAP content. Multilocation evaluation of grass pea germplasm at ICARDA between 1999 and 2006 indicated the maximum variability for ODAP content in Ethiopian germplasm. Grass pea germplasm from Ethiopia and the Indian subcontinent is generally high in ODAP (0.7–2.4%) as compared to 0.02–1.2% in germplasm from the Near East (Abd-El-Moneim et al. 2000).

Wild crop gene pool is a rich reservoir of rare alleles. Therefore, efforts have been made to evaluate wild relatives to identify zero ODAP genetic resources

(Jackson and Yunus 1984). Assessment of ODAP in wild relatives indicated that none of the species is free from ODAP (Aletor et al. 1994; Hanbury et al. 1999; Siddique et al. 1996). However, on average, the ODAP concentration in *L. cicera* is lower compared to *L. sativus*. Hanbury et al. (1999) observed the lowest ODAP in *L. cicera* (0.18%) followed by *L. sativus* (0.39%) and *L. ochrus* (1.01%). Aletor et al. (1994) reported four to five times lower ODAP content in *L. cicera* (0.13%) than in *L. ochrus* (0.56%) and *L. sativus* (0.49%). Similarly, Abd-El-Moneim et al. (2000) reported ranges of 0.02–2.40% in *L. sativus*, 0.03–0.22% in *L. cicera* and 0.46–2.50% in *L. ochrus*. Eichinger et al. (2000) screened *Lathyrus* germplasm using capillary electrophoresis and found that *L. cicera* is consistently low in ODAP as compared to *L. sativus* and *L. ochrus*. Evaluation of 142 accessions of *L. cicera* at ICARDA during 2009 showed a range of 0.073–0.513% for ODAP content, which is much lower than the cultivated species. Therefore, *L. cicera* accessions hold promise as a source of low ODAP content in grass pea breeding programmes.

Protein content in the seeds of *Lathyrus* spp. ranged from 23 to 49 with a mean value of 35%. The highest content was recorded in *L. sylvestris* from the USA (49%) and the lowest in *L. cicera* from Norway (23%). Many workers reported that protein content in *Lathyrus* ranged between 23% and 31% with a mean value of 30% (Hove and King 1978; Shobhana et al. 1976; Urga et al. 2005; Granati et al. 2001; Roy and Rao 1978). Generally on average, the ODAP concentration of *L. ochrus* (6.58 mg/g) was about twice that of *L. sativus*, and *L. cicera* had the lowest ODAP concentration (1.31 mg/g) (Siddique et al. 1996).

Mondal and Puteh (2014) evaluated 30 bold and small seeded genotypes of grass pea to find out genetic variability for seed size, protein and ODAP content in seeds. The weight of 1000-seed ranged from 41 to 79 g. The protein and ODAP content ranged from 25% to 34% and 0.32% to 2.02%, respectively. A significant positive correlation between seed size and protein content ($r = 0.48$) but negative association between seed size and ODAP content ($r = -0.22$) were observed. Protein content and ODAP concentration was negatively associated ($r = -0.16$). Therefore, selection of bold seed size is the key to develop a low neurotoxin containing *Lathyrus* genotype (Mondal and Puteh 2014).

Breeding Methods for ODAP, Protein and Homoarginine Content

Significant efforts have been directed towards genetic improvement of grass pea in India, Canada, Bangladesh, Ethiopia and Nepal during the late 1970s and at ICARDA since 1989. Breeding efforts are mostly focused on three species, *L. sativus*, *L. cicera* and *L. ochrus*, and to a lesser extent *L. clymenum*, with an aim to improve grain yield, biomass and resistance to biotic and abiotic stresses and most importantly to reduce the neurotoxin from its seeds. Outcrossing percentage is very high (up to 30%) in grass pea (Ben Brahim et al. 2001; Chowdhury and Slinkard

1997; Rahman et al. 1995). Therefore, breeding methods usually employed are very similar to that of faba bean (*Vicia faba*) and pigeon pea (*Cajanus cajan*). To maintain genetic purity, different isolation techniques like covered screen houses, isolation cages and cloth begs are used. Grass pea is insect pollinated. The highest outcrossing has been found in case of coloured flowers as compared to genotypes with white flowers (9.8%) (Rahman et al. 1995). Large-sized flowers have high outcrossing rate (Kiyoshi et al. 1985). Natural outcrossing is a tremendous tool that is how the genetic variability is liberated and helps in adaptation and evolution of this species in the long run. However, there is a need to utilize this inherent system in *Lathyrus* crop improvement programmes more (Kumar et al. 2011b).

Conventional breeding approach has resulted in the development of high-yielding low ODAP varieties. In India, Pusa 24, Prateek and Mahateora, with low ODAP, were developed through intraspecific hybridization. In Bangladesh, low ODAP and high-yielding varieties BARI Khesari 1, BARI Khesari 2 and BARI Khesari 3 were developed for commercial cultivation. At ICARDA, several breeding lines with <0.1% ODAP concentration were bred, which have led to the release of BARI Khesari 3 in Bangladesh, Wasie in Ethiopia and Ali Bar in Kazakhstan. In Canada, a low ODAP (0.03%) line, LS 8246, was released for fodder and feed purposes. In Australia, two varieties, Ceora and Chalus, were released for diversification of the wheat-based system. More efforts are needed to exploit the genetic diversity existing within species of grass pea gene pools. There is a need to screen wild species for traits like protein, micronutrient, prebiotics and homoarginine contents. Due to narrow genetic variability like many other food legumes, mutation breeding has been extensively used in *Lathyrus* improvement. Both physical and chemical mutagens were used to create more genetic variability for traits including morphological traits like branching habit, leaf characters, chlorophyll mutation and ODAP content (Prasad and Das 1980; Waghmare and Mehra 2001; Waghmare et al. 2001; Talukdar 2012). Mutation breeding has also been occasionally employed to create additional genetic variability in order to develop zero/low ODAP varieties (Talukdar 2009). Two varieties, namely, Poltavskaya in the former USSR and Bina Khesari 1 in Bangladesh, were developed through mutation breeding using ethyl methanesulphonate (EMS) (0.01%) and gamma rays (250 Gy), respectively. Rybiński et al. (2006) found increased genetic variability for pod number per plant, seed number and weight per pod, reduced plant height, earliness, biomass and seed microstructure while using chemical mutagen to treat *Lathyrus sativus* varieties. The negative association between the number of seeds per pod and seed size was neutralized in 35 and 40 kR gamma-irradiated population and was validated in M3 generation also (Waghmare and Mehra 2000). A fasciated mutant commonly has broadened stem, small narrow leaves and pods reduced in size, arranged in line on the node of the upper part of the stem that was identified in M3 generation in grass pea cv. P27 following 250 Gy gamma ray treatment (Waghmare et al. 2001). This kind of physical and chemical mutagenesis should be carried out to generate the safest *Lathyrus* mutant with zero or lowest ODAP concentration.

Biotechnological approaches including plant tissue culture techniques have great potential to improve the agronomical traits through the induction of somaclonal variation or true to the type plant regeneration in *Lathyrus sativus* (Barpete et al. 2014, 2020). Valarini et al. (1997) suggested that some of the variation that arises at culture level is epigenetic and transient. But heritable changes in genome can be provoked due to in vitro stress in culture condition. However, Mohanty et al. (2008) suggested that somaclonal variation is induced due to mixoploidy following reduplication or endoduplication.

Tissue culture technique provides the great opportunity to increase the genetic variability through somaclonal variation that created new diversity, allowing the selection of desired lines with interesting agronomic traits (Roy et al. 1993; Hazrati et al. 2011). However, the development of somaclones in grass pea is limited. There are few reports of variation among in vitro regenerated grass pea plants using various culture explants, and this culture variation is used as a source of somaclonal variation for the development of low ODAP varieties (Van-Dorrestein et al. 1998; Santha and Mehta 2001). Somaclonal variation can also contribute to the development of varieties with low ODAP (Mehta et al. 1994; Mehta and Santha 1996; Santha and Mehta 2001). Ratan is released as a variety in India from selection in the somaclonal variation.

Tripathy et al. (2016) recovered a series of somaclones from four genotypes (Nirmal, P24, Nayagarh local and Dhenkanal local) and investigated their nature of genetic variation at cytological, morphological and biochemical level in grass pea. Chromosomal abnormalities and variation in the morphological traits including flower colour, leaflet length and seed colour and pod pigmentation were observed. A high-yielding low ODAP somaclone (from genotype NGOG-5) was recovered that may be promising candidate for future grass pea breeding programme (Tripathy et al. 2016).

The development of transgenic for zero ODAP *Lathyrus* is another challenging area of research. Legumes, and grass pea in particular, are very challenging to regenerate and transform, because of the problematic somatic embryogenesis or organogenesis (Iantcheva et al. 2013). Few workers have developed transformation and regeneration protocol successfully in grass pea (Zambre et al. 2002; Barik et al. 2005; Girma 2010). Girma (2010) transformed grass pea with a common bean (*Phaseolus vulgaris* L.) gene coding for additional amounts of methionine. Barik et al. (2005) developed a genetic transformation procedure expressing both a reporter gene (β-glucuronidase) and a selectable marker gene (neomycin phosphotransferase II).

A protocol has been developed of in vitro direct multiple shoot induction from mature seed embryo with two cotyledons as explants found to be a time-saving approach (Barpete et al. 2017). This protocol bypasses callus induction phase; hence, somaclonal variation is not developed (Barpete et al. 2017). The reduced level of the anti-nutritional metabolite oxalic acid (OA) in transgenic seeds of grass pea (up to 75%) was observed by the constitutive and/or seed-specific expression of an oxalate-degrading enzyme, oxalate decarboxylase (FvOXDC), of the fungus *Flammulina velutipes*. β-ODAP level of grass pea seeds had also decreased up to 73% (Kumar et al. 2016; Lambein et al. 2019). Further, in transgenic grass pea lines, seed micronutrients, such as calcium, iron and zinc, manganese and magnesium, were increased (Kumar et al. 2016; Lambein et al. 2019). Hence, there is enormous scope existing utilizing genetic transformation for nutritional improvement in grass pea.

Marker-Assisted Approaches

Genomic resources have been enriched over the years in different food legumes (chickpea, pigeon pea, lentil, field pea) (Kumar et al. 2011a; Varshney et al. 2010); however, limited genomic information was available in *Lathyrus* (Lioi et al. 2011; Shiferaw et al. 2012; Yang et al. 2014; Soren et al. 2015). This is mainly due to the large genome size and poorly characterized germplasm used for such studies. Very recently, *Lathyrus* draft genome sequence has been published (Emmrich et al. 2020); this will surely enrich marker-assisted breeding *Lathyrus*. Molecular markers have been successfully used to assess genetic diversity in *Lathyrus* (Croft et al. 1999; Chtourou-Ghorbel et al. 2001; Badr et al. 2002; Skiba et al. 2003; Belaid et al. 2006; Barik et al. 2007; Tavoletti and Iommarini 2007; Gupta et al. 2018).

Gupta et al. (2018) evaluated 118 *Lathyrus* accessions with varying β-ODAP concentrations. Genotyping data with molecular markers were analysed, and it was found that Group I consisted of 20 accessions with high β-ODAP concentration. Of these 20 accessions, 17 were wild accessions. Only one wild accession (*L. cicera*) was clustered in Group II, which was having 35 accessions in total. Interestingly, most of the Group II accessions contained low β-ODAP. Group III, which was represented by 34 accessions, and Group IV, which was comprised of 29 accessions, were mostly having very high β-ODAP concentrations.

Future Outlook

Under climate change, grass pea has emerged as climate smart crop and holds great promise of expansion in fragile agro-ecosystems, particularly in rice-based systems in South Asia, under crop-livestock systems in East Africa, North Africa, West Asia and Europe, and as a cover and forage crop in Australia and Canada. Recent studies on its nutritional value and health benefits have further helped this crop to emerge as a healthy functional food. However, observational and controlled intervention studies are required to show links between consumption of grass peas and changes in important physiological parameters, which could impact the health of the general population. Grass pea is known to reduce cholesterol, support weight management via glycaemic responses and aid digestive health. However, the evidence directly evaluating effects of controlled consumption of grass pea in a free-living environment on these markers is still very limited. Thus, there is not enough evidence to support a consumer-oriented health claim suitable for use in marketing of grass pea. The prospect of developing ODAP-free cultivars has brightened with the progress in breeding and genomics tools and technologies such as genome editing, GM technologies, genome sequencing, market assisting breeding, etc. Chances of accelerating the genetic gains in grass pea has increased manifolds with the availability of the draft genome sequence of grass pea and a large number of genome-wide markers and resequencing of genetic diversity panel.

Two research strategies could be used to obtain the scientific foundation needed to apply for health claims for grass pea. One would be to assess the health benefits of food, which is based on grass pea, to gain a unique risk reduction health claim for this specific product. The second option, which is most relevant as a follow-up to the first option, would be to aim for a health claim for grass pea more generically, as an ingredient in a variety of products, which would enable different stakeholders to use the health claim for any food that contained a sufficiently high percentage of this pulse.

References

AACC International Method 46-30.01: AACC approved methods of analysis, 11th Edition. http://methods.aaccnet.org/summaries/46-30-01.aspx

Abd El Moneim AM, van Dorrestein B, Baum M, Ryan J, Bejiga G (2001) Role of ICARDA in improving the nutritional quality and yield potential of grass pea (*Lathyrus sativus* L.) for subsistence farmers in dry areas. Lathyrus Lathyrism Newsl 2:55–58

Abd-El-Moneim AM, van Dorrestein B, Baum M, Mulugeta W (2000) Improving the nutritional quality and yield potential of grass pea (*Lathyrus sativus* L.). Food Nutr Bull 21:493–496

Adams MR, Golden DL, Register TC, Anthony MS, Hodgin JB, Maeda N, Williams JK (2002) The atheroprotective effect of dietary soy isoflavones in apolipoprotein E−/− mice requires the presence of estrogen receptor-α. Arterioscler Thromb Vasc Biol 22(11):1859–1864

Adiga PR, Padmanaban G, Rao SLN, Sarma PS (1962) The isolation of a toxic principle from *Lathyrus sativus* seeds. J Sci Ind Res (India) 21C:284–286

Adiga PR, Rao SLN, Sarma PS (1963) Some structural features and neurotoxic action of a compound from *Lathyrus sativus* seeds. Curr Sci 32:133–135

Aletor VA, Abd-El-Moneim AM, Goodchild AV (1994) Evaluation of the seeds of selected lines of three Lathyrus spp. for β-N-oxalyl amino-L-alanine (BOAA), tannins, trypsin inhibitor activity and certain in vitro characteristics. J Sci Food Agric 65:143–151

Allkin RW, Macfarlane TD, White RJ, Bisby FA, Adey ME (1983) Name and synonyms of species and subspecies in the Vicieae, Vicieae Data base project publication No. 2. Vicieae Database Project, Southampton

Allkin R, Goyder DJ, Bisby FA, White RJ (1986) Names and synonyms of species and subspecies in the Vicieae: issue 3, Vicieae Database Project, Experimental Taxonomic Information Products, Publication 7. Southampton University, Southampton

Almeida JM (1980) A cultura de leguminosas para grão, fava, tremoço, chicharos e grão de gramicha. In: Documentagao sobre Forragens e Pastagens. INIA Lisbon, Lisbon

Amara A, Coussemacq M, Geffard M (1995) Molecular detection of methionine in rat brain using specific antibodies. Neurosci Lett 185:147–150

Arslan M, Oten M, Erkaymaz T, Tongur T, Kilic M, Elmasulu S, Cinar A (2017) β-N-oxalyl-L-2,3-diaminopropionic acid, L-homoarginine, and asparagine contents in the seeds of different genotypes Lathyrussativus L. as determined by UHPLC-MS/MS. Int J Food Prop 20(Suppl):S108–S118. https://doi.org/10.1080/10942912.2017.1289961

Ayaz FA, Inceer H, Hayırlıoğlu-Ayaz S (1999) An electrophoretic analysis of the seed proteins of some Vicia L. species from northeast Anatolia (Turkey). Pak J Biol Sci 2(4):1139–1142

Badr A, Shazly HE, Rabey HE, Watson LE (2002) Systematic relationships in Lathyrus sect. Lathyrus (Fabaceae) based on amplified fragment length polymorphism (AFLP) data. Can J Bot 80:962–969

Barik DP, Mohapatra U, Chand PK (2005) Transgenic grass pea (*Lathyrus sativus* L.): factors influencing Agrobacterium-mediated transformation and regeneration. Plant Cell Rep 24:523–531. https://doi.org/10.1007/s00299-005-0957-5

Barik DP, Acharya L, Mukherjee AK, Chand PK (2007) Analysis of genetic diversity among selected grass pea (*Lathyrus sativus* L.) genotypes using RAPD markers. Z Naturforsch C (German) 62:869–874

Barpete S, Parmar D, Sharma NC, Kumar S (2012) Karyotype analysis in grass pea (*Lathyrus sativus* L.). J Food Legum 25(1):14–17

Barpete S, Aasim M, Khawar KM, Özcan SF, Özcan S (2014) Preconditioning effect of cytokinins on in vitro multiplication of embryonic node of grass pea (*Lathyrus sativus* L.). Turk J Biol 38:485–492. https://doi.org/10.3906/biy-1312-94

Barpete S, Aasim M, Ozcan SF, Khawar KM, Ozcan S (2017) High frequency axillary shoots induction in grass pea (*Lathyrus sativus* L.). Bangladesh J Bot 46:119–124

Barpete S, Gupta P, Khawar KM, Özcan S, Kumar S (2020) In vitro approaches for shortening generation cycles and faster breeding of low B-N-Oxalyl-L-A, B-diaminopropionic acid content of grass pea (*Lathyrus sativus* L.). Fresenius Environ Bull 29:2698–2706

Belaid Y, Chtourou-Ghorbel N, Marrakchi M, Trifi-Farah N (2006) Genetic diversity within and between populations of Lathyrus genus (Fabaceae) revealed by ISSR markers. Genet Resour Crop Evol 53:1413–1418

Belay A, Moges G, Solomon T, Johansson G (1997) Thermal isomerization of the neurotoxin β-N-oxalyl-L-α, β-diaminopropionic acid. Phytochemistry 45:219–223

Bell EA, O'Donovan P (1966) The isolation of a and c-oxalyl derivatives of a, c-diaminobutyric acid from seeds of *Lathyrus latifolius*, and the detection of the α-oxalyl isomer of the neurotoxin α-amino-β-oxalyl amino propionic acid which occurs together with the neurotoxin in this and other species. Phytochemistry 5:1211–1219

Ben Brahim N, Combes D, Marrakchi M (2001) Autogamy and allogamy in genus Lathyrus. Lathyrus Lathyrism Newsl 2:21–26

Berger JD, Siddique KHM, Loss SP (1999) Cool season grain legumes for Mediterranean environments: the effect of environment on non-protein amino acids in Vicia and Lathyrus species. Aust J Agric Res 50:403–412

Bisignano V, Gatta CD, Polignano GB (2002) Variation for protein content and seed weight in grass pea (Lathyrus spp.) germplasm. Plant Genet Res Newsletter 132:30–34

Bradford MM (1976) A rapid and sensitive method for the quantitation of microgram quantities of protein utilizing the principle of protein-dye binding. Anal Biochem 72:248–254

Campbell CG (1997) Grass pea (*Lathyrus sativus* L), Promoting the conservation and use of underutilized and neglected crops. Vol 18. Institute of Plant Genetics and Crop Plant Research, Gatersleben & International Plant Genetic Resources Institute, Rome, pp 1–90

Campbell CG, Mehra RB, Agrawal SK, Chen YZ, Abd-El-Moneim AM, Khawaja HIT, Yadav CR, Tay JU, Araya WA (1994) Current status and future strategy in breeding grass pea (*Lathyrus sativus*). Euphytica 73:167–175

Castiglioni S, Manzoni C, D'Uva A, Spiezie R, Monteggia E et al (2003) Soy proteins reduce progression of a focal lesion and lipoprotein oxidiability in rabbits fed a cholesterol-rich diet. Atherosclerosis 171(2):163–170

Cavada PE, Juan R, Pastor JE, Alaiz M, Vioque J (2010) Protein isolates from two Mediterranean legumes: Lathyrus clymenum and Lathyrus annuus. Chemical composition, functional properties and protein characterisation. Food Chem 122(3):533–538

Cavada EP, Juan R, Pastorb JE, Alaiza M, Vioque J (2011) Nutritional characteristics of seed proteins in 15 Lathyrus species (fabaceae) from Southern Spain. LWT Food Sci Technol 44(4):1059–1064

Chandna M, Matta NK (1994) Studies on changing protein levels in developing and germinating seeds of *Lathyrus sativus* L. J Plant Biochem Biotechnol 3:59–61. https://doi.org/10.1007/bf03321950

Chavan UD, McKenzie DB, Shahidi F (2001) Functional properties of protein isolates from beach pea (*Lathyrus maritimus* L.). Food Chem 74(2):177–187

Chowdhury MA, Slinkard AE (1997) Natural outcrossing in grasspea. J Hered 88:154–156

Chtourou-Ghorbel N, Lauga B, Combes D, Marrakchi M (2001) Comparative genetic diversity studies in the genus Lathyrus using RFLP and RAPD markers. Lathyrus Lathyrism Newsl 2:62–68

Croft AM, Pang ECK, Taylor PWJ (1999) Molecular analysis of *Lathyrus sativus* L. (grass pea) and related Lathyrus species. Euphytica 107:167–176

Datta S, Varshney RK (2009) Genomics enabled crop improvement in food legumes. In: Ali M, Kumar S (eds) Milestones in food legumes research, 1st edn. Indian Institute of Pulses Research (IIPR), Kanpur, pp 249–289

De Bruyn A, Becu C, Lambein F, Kebede N, Abegaz B, Nunn P (1994) The mechanism of the rearrangement of the neurotoxin β-ODAP to α-ODAP. Phytochemistry 36:85–89

Desroches S, Mauger JF, Ausman LM, Lichtenstein AH, Lamarche B (2004) Soy protein favorably affects LDL size independently of isoflavones in hypercholesterolemic men and women. J Nutr 134(3):574–579

Ding S, Wang M, Fang S, Xu H, Fan H, Tian Y, Zhai Y, Lu S, Qi X, Wei F, Sun G, Sun X (2018) D-dencichine regulates thrombopoiesis by promoting megakaryocyte adhesion, migration and proplatelet formation. Front Pharmacol 9:297. https://doi.org/10.3389/fphar.2018.00297

Dnyanu CU (1998) Chemical and biochemical components of Beach pea (Lathyrus maritimus L.). Doctoral (PhD) thesis, Memorial University of Newfoundland

Duke JA (1981) A handbook of legumes of world economic importance. Plenum Press, New York. https://doi.org/10.1007/978-1-4684-8151-8

Dumas JBA (1831) Procedes de L'analyse Organique. Ann Chim Phys 47:198–205

Dunlop RA, Main BJ, Rodgers KJ (2015) The deleterious effects of non-protein amino acids from desert plants on human and animal health. J Arid Environ 112:152–158

Duranti M (2006) Grain legume proteins and nutraceutical properties. Fitoterapia 77:67–82

Duranti M, Gius C (1997) Legume seeds: protein content and nutritional value. Field Crop Res 53(1–3):31–45

Duranti M, Lovati MR, Dani V, Barbiroli A, Scarafoni A et al (2004) The α′ subunit from soybean 7S globulin lowers plasma lipids and upregulates liver β-VLDL receptors in rats fed a hypercholesterolemic diet. J Nutr 134(6):1334–1339

Eichinger PCH, Rothnie NE, Delaere I, Tate ME (2000) New technologies for toxin analyses in food legumes. In: Knight R (ed) Linking research and marketing opportunities for pulses in the 21st century. Kluwer Academic Publishers, Dordrecht, pp 685–692

El Haramein FJ, Moneim AA-E, Nakkoul H (1998) Prediction of the neuro-toxin beta-Noxalyl-amino-L-alanine in Lathyrus species, using near infrared reflectance spectroscopy. J Near Infrared Spectrosc 6:A93–A96

Emmrich PMF, Rejzek M, Hill L, Brett P, Edwards A, Sarkar A, Field RA, Martin C, Wang TL (2019) Linking a rapid throughput plate-assay with high-sensitivity stable-isotope label LCMS quantification permits the identification and characterisation of low β-L-ODAP grass pea lines. BMC Plant Biol 19:489

Emmrich PMF, Sarkar A, Njaci I et al (2020) A draft genome of grass pea (*Lathyrus sativus*), a resilient diploid legume. https://doi.org/10.1101/2020.04.24.058164

Emre İ (2009) Electrophoretic analysis of some Lathyrus L. species based on seed storage proteins. Genet Resour Crop Evol 56(1):31–38

Emre I, Turgut-Balik D, Genc H, Sahin A (2006) The use of seed proteins revealed by SDS-PAGE in taxonomy of some Lathyrus L. species grown in Turkey. Pak J Biol Sci 9:2358–2361

Emre İ, Turgut-Balık D, Şahin A (2007) Electrophoretic analysis of total protein profiles of some Lathyrus L. (Sect. Cicercula) grown in Turkey. Pak J Biol Sci 10(17):2890–2894

Fikre A, Korbu L, Kuo YH, Lambein F (2008) The contents of the neuro-excitatory amino acid B-Odap (B-NOxalyl- L-A,B-Diaminopropionic acid), and other free and protein amino acids in the seeds of different genotypes of grass pea (*Lathyrus sativus* L.). Food Chem 110:422–427

Fikre A, Negwo T, Kuo YH, Lambein F, Ahmed S (2011) Climatic, edaphic and altitudinal factors affecting yield and toxicity of *Lathyrus sativus* grown at five locations in Ethiopia. Food Chem Toxicol 49(3):623–630

Fukui K, Kojima M, Tachibana N, Kohno M, Takamatsu K, Hirotsuka M, Kito M (2004) Effects of soybean β-conglycinin on hepatic lipid metabolism and fecal lipid excretion in normal adult rats. Biosci Biotechnol Biochem 68(5):1153–1155

Gatel F (1994) Protein quality of legume seeds for non-ruminant animals: a literature review. Anim Feed Sci Technol 45:317–348. https://doi.org/10.1016/0377-8401(94)90036-1

Getahun H, Lambein F, Vanhoorne M, Van der Stuyft P (2003) Food-aid cereals to reduce neurolathyrism related to grass pea preparations during famine. Lancet 362:1808–1810

Getahun H, Lambein F, Vanhoorne M, Van der Stuyft P (2005) Neurolathyrism risk depends on type of grass pea preparation and on mixing with cereals and anti-oxidants. Trop Med Int Health 10:169–178

Ghosh B, Mitra J, Chakraborty S, Bhattacharyya J, Chakraborty A, Sen SK et al (2015) Simple detection methods for antinutritive factor β-ODAP present in *Lathyrus sativus* L. by high pressure liquid chromatography and thin layer chromatography. PLoS ONE 10(11):e0140649. https://doi.org/10.1371/journal.pone.0140649

Girma D (2010) Ethiopian grass pea (*Lathyrus sativus* L.) started the genomics era: transient genetic transformation of grass pea. LAP Lambert Academic Publishing, Köln

Girma D, Korbu L (2012) Genetic improvement of grass pea (*Lathyrus sativus*) in Ethiopia: an unfulfilled promise. Plant Breed 131:231–236. https://doi.org/10.1111/j.1439-0523.2011.01935.x

Gornall AG, Bardawill CJ, David MM (1949) Determination of serum proteins by means of the biuret reaction. J Biol Chem 177:5751

Granati E, Bisignano V, Chiar D (2001) Grain quality in accessions of Lathyrus spp. Lathyrus Lathyrism Newsl 2:69–71

Grela ER, Studzinski T, Matras J (2001) Anti-nutritional factors in seeds of *Lathyrus sativus* cultivated in Poland. Lathyrus Lathyrism Newsl 2:101–104

Grela ER, Rybiński W, Klebaniuk R, Matras J (2010) Morphological characteristics of some accessions of grass pea (*Lathyrus sativus* L.) grown in Europe and nutritional traits of their seeds. Genet Resour Crop Evol 57:693–701. https://doi.org/10.1007/s10722-009-9505-4

Grela ER, Rybiński W, Matras J, Sobolewska S (2012) Variability of phenotypic and morphological characteristics of some *Lathyrus sativus* L. and Lathyrus cicera L. accessions and nutritional traits of their seeds. Genet Resour Crop Evol 59:1687. https://doi.org/10.1007/s10722-011-9791-5

Gupta P, Udupa SM, Sen Gupta DS, Kumar J, Kumar S (2018) Population structure analysis and determination of neurotoxin content in a set of grass pea (*Lathyrus sativus* L.) accessions of Bangladesh origin. Crop J 6:435–442. https://doi.org/10.1016/j.cj.2018.03.004

Gutiérrez, J, Vaquero F, Vences, F (1994) Allopolyploid vs. autopolyploid origins in the genus Lathyrus (Leguminosae). Heredity 73:29–40. https://doi.org/10.1038/hdy.1994.95

Hanbury CD, Sarker A, Siddique KHM, Perry MW (1995) Evaluation of Lathyrus germplasm in a Mediterranean type environment in south-western Australia, Occasional publication, no. 8. CLIMA, Perth

Hanbury CD, Siddique KHM, Galwey NW, Cocks PS (1999) Genotype-environment interaction for seed yield and ODAP concentration of *Lathyrus sativus* L. and L. cicera L. in Mediterranean-type environments. Euphytica 110:45–60

Hanbury CD, White CL, Mullan BP, Siddique KHM (2000) A review of the potential of *Lathyrus sativus* L. and L. cicera L. grain for use as animal feed. Anim Feed Sci Technol 87:1–27
Harrison FL, Nunn PB, Hill RR (1977) Synthesis of α- and β-ODAP and their isolation from seeds of *Lathyrus sativus*. Phytochemistry 16:1211–1215
Hazrati N, Hassanzadeh M, Shahzad J-e-s (2011) In vitro selection for salt tolerance in lentil (*Lens culinaris* Medik). Int Res J Appl Basic Sci 2:35–39
Horn MJ, Jones DB, Blum AE (1946) Colorimetric determination of methionine in proteins and foods. J Biol Chem 166:313–320
Hove EL, King S (1978) Composition, protein quality and toxins of seeds of the grain legumes, Glycine max, Lupinus spp., Phaseolus spp., Pisum sativum and Vicia faba. N Z J Agric Res 21:457–462
Iantcheva A, Mysore KS, Ratet P (2013) Transformation of leguminous plants to study symbiotic interactions. Int J Dev Biol 57:577–586. https://doi.org/10.1387/ijdb.130239pr
Jackson MT, Yunus AG (1984) Variation in the grass pea (L. sativus L.) and wild species. Euphytica 33:549–559
Jammulamadaka N, Burgula S, Medisetty R, Ilavazhagan G, Rao SLN, Singh SS (2011) β-N-oxalyl-l-α,β-diaminopropionic acid regulates mitogen-activated protein kinase signaling by down-regulation of phosphatidylethanolamine-binding protein 1. J Neurochem 118:176–186. https://doi.org/10.1111/j.1471-4159.2011.07299.x
Jiao C-J, Jiang J-L, Ke L-M, Cheng W, Li F-M, Li Z-X, Wang C-Y (2011) Factors affecting β-ODAP content in *Lathyrus sativus* and their possible physiological mechanisms. Food Chem Toxicol 49:543–549. https://doi.org/10.1016/j.fct.2010.04.050
Ke Q, Costa M (2006) Hypoxia-inducible factor-1 (HIF-1). Mol Pharmacol 70:1469–1480. https://doi.org/10.1124/mol.106.027029
Khawaja HIT (1988) A new inter-specific Lathyrus hybrid to introduce the yellow flower color into sweat pea. Euphytica 37:69–75
Khawaja HIT, Sybenga J, Ellis JR (1997) Chromosome pairing and chiasma formation in auto-polyploids of different *Lathyrus* species. Genome 40:937–944
Kiyoshi Y, Toshiyuki F, Blumenreich ID (1985) In: Kaul AK, Combes D (eds) Isozymic variation and interspecific crossability in annual species of genus Lathyrus L. Lathyrus and Lathyrism, Pau, pp 118–129
Kjeldahl J (1883) "Neue Methode zur Bestimmung des Stickstoffs in organischen Körpern" (New method for the determination of nitrogen in organic substances). Z Anal Chem 22(1):366–383
Kumar J, Pratap A, Solanki RK, Gupta DS, Goyal A, Chaturvedi SK, Nadarajan N, Kumar S (2011a) Genomic resources for improving food legume crops. J Agric Sci (Lond) 150:289–318
Kumar S, Bejiga G, Ahmed S, Nakkoul H, Sarker A (2011b) Genetic improvement of grass pea for low neurotoxin (ODAP) content. Food Chem Toxicol 49:589–600
Kumar S, Gupta P, Barpete S, Sarker A, Amri A, Mathur PN, Baum M (2013) Grass pea. In: Singh M, Upadhyaya HD, Bisht IS (eds) Genetic and genomic resources for grain legume improvement. Elsevier, Amsterdam, pp 269–293. https://doi.org/10.1016/B978-0-12-397935-3.00011-6
Kumar V, Chattopadhyay A, Ghosh S, Irfan M, Chakraborty N, Chakraborty S, Datta A (2016) Improving nutritional quality and fungal tolerance in soya bean and grass pea by expressing an oxalate decarboxylase. Plant Biotechnol J 14:1394–1405. https://doi.org/10.1111/pbi.12503
Kumari S, Vikas Kumar Jha, Diksha Kumari, Ravi Ranjan, Nimmy MS, Anand Kumar, Chandan Kishore And Vinod Kumar (2018): Protein content of *Lathyrus sativus* collected from diverse locations. J Pharmacogn Phytochem SP1: 1610-1611
Kuo YH, Khan JK, Lambein F (1993) Biosynthesis of the neurotoxin β-ODAP in developing pods of *Lathyrus sativus*. Phytochemistry 35(4):911–913
Kuo Y-H, Ikegami F, Lambein F (1998) Metabolic routes of β-(isoxazolin-5-on-2-yl)-l-alanine (BIA), the precursor of the neurotoxin ODAP (β-N-oxalyl-l-α,β-diaminopropionic acid), in different legume seedlings. Phytochemistry 49:43–48. https://doi.org/10.1016/s0031-9422(97)01001-7
Kupicha FK (1983) The infrageneric structure of Lathyrus. Notes R Bot Gard Edinb 41:209–244

Lambein F, Kuo Y-H (2004) Prevention of neurolathyrism during drought. Lancet 363:657. https://doi.org/10.1016/S0140-6736(04)15601-8

Lambein F, Khan JK, Kuo YH, Campbell CG, Briggs CJ (1993) Toxin in the seedlings of some varieties of grass pea (*Lathyrus sativus*). Nat Toxins 1:246–249

Lambein F, Haque R, Khan JK, Kebede N, Kuo YH (1994) From soil to brain: zinc deficiency increases the neurotoxicity of *Lathyrus sativus* and may affect the susceptibility for the motor-neurone disease neurolathyrism. Toxicon 32(4):461–466

Lambein F, Travella S, Kuo Y et al (2019) Grass pea (*Lathyrus sativus* L.): orphan crop, nutraceutical or just plain food? Planta 250:821–838. https://doi.org/10.1007/s00425-018-03084-0

Llorent-Martínez EJ, Zengin G, Fernández-de Córdova ML, Bender O, Atalay A, Ceylan R, Mollica A, Mocan A, Uysal S, Guler GO, Aktumsek A (2017) Traditionally used Lathyrus species: phytochemical composition, antioxidant activity, enzyme inhibitory properties, cytotoxic effects, and in silico studies of L. czeczottianus and L. nissolia. Front Pharmacol 8:83. https://doi.org/10.3389/fphar.2017.00083

Long YC, Ye YH, Xing QY (1996) Studies on the neuroexcitotoxin β-N-oxalo-L-α, β-diaminopropionic acid and its isomer α -N-oxalo-L-α, β-diaminopropionic acid from the root of Panax species. Int J Protein Pept Res 47:42–47

Lowry OH, Rosebrough NJ, Farr AL, Randall RJ (1951) Protein measurement with the Folin phenol reagent. J Biol Chem 193:265–75

Mahler-Slasky Y, Kislev ME (2010) Lathyrus consumption in late bronze and iron age sites in Israel: an Aegean affinity. J Archaeol Sci 37:2477–2485. https://doi.org/10.1016/j.jas.2010.05.008

Mehta SL, Santha IM (1996) Plant biotechnology for development of non-toxic strains of *Lathyrus sativus*. In: Arora RK, Mathur PN, Riley KW, Adham Y (eds) Lathyrus genetic resources in Asia. International Plant Genetic Resources Institute, Rome, pp 129–138

Mehta SL, Ali K, Barna KS (1994) Somaclonal variation in a food legume – *Lathyrus sativus*. Indian J Plant Biochem Biotechnol 3:73–77

Mohanty S, Panda M, Subudhi E, Nayak S (2008) Plant regeneration from callus culture of Curcuma aromatic and in vitro detection of somaclonal variation through cytophotometric analysis. Biol Plant 52:783–786

Mondal MMA, Puteh AB (2014) Spectrum of variability in seed size and content of protein and Odap in grass pea (*Lathyrus sativus* L.) germplasm. Legum Res Int J 37(5):479–482

Murti VVS, Seshadri TR, Venkatasubramanian TA (1964) Neurotoxic compound of the seeds of *Lathyrus sativus*. Phytochemistry 3:73–78

Narayan RKJ (1991) Molecular organization of the plant genome: its relation to structure, recombination and evolution of chromosomes. J Genet 70:43–61

Nunn PB, Seelig M, Zagoren JC, Spencer PS (1987) Stereospecific acute neurotoxicity of uncommon plant amino acids linked to human motor-system diseases. Brain Res 410:375–379

Ochatt SJ, Abirached-Darmency M, Marget P, Aubert G (2007) The Lathyrus paradox: 'poor men's diet' or a remarkable genetic resource for protein legume breeding? In: Ochatt S, Jain SM (eds) Breeding of neglected and under-utilised crops, spices and herbs. Science Publishers, Enfield, pp 41–60

Padmajaprasad V, Kaladhar M, Bhat RV (1997) Thermal isomerization of β-N-oxalyl-L-α, β-diaminopropionic acid, the neurotoxin in *Lathyrus sativus* during cooking. Food Chem 59:77–80

Padmanaban G (1980) Lathyrogens. In: Leiner IE (ed) Toxic constituents of plant foodstuffs. Academic Press, New York, pp 244–250

Pandey RL, Chitale MW, Sharma RN, Geda AK (1997) Evaluation and characterization of germplasm of grass pea (*Lathyrus sativus*). J Med Aromat Plant Sci 19:14–16

Pandey RL, Kashyap OP, Sharma RN, Nanda HC, Geda AK, Nair S (2008) Catalogue on grass pea (*Lathyrus sativus* L.) germplasm. Indira Gandhi Krishi Vishwavidyalaya, Raipur, p 213

Petterson DS, Sipsas S, Mackintosh JB (1997) The chemical composition and nutritive value of Australian pulses, 2nd edn. Grains Research and Development Corporation, Canberra

Piergiovanni AR, Damascelli A (2011) L-Homoarginine accumulation in grass pea (*Lathyrus sativus* L.) dry seeds. A preliminary survey. Food Nutr Sci 2:207–213. https://doi.org/10.4236/fns.2011.23028
Piergiovanni AR, Lupo F, Zaccardelli M (2011) Environmental effect on yield, composition and technological seed traits of some Italian ecotypes of grass pea (*Lathyrus sativus* L.). J Sci Food Agric 91(2011):122–129
Prasad AB, Das AK (1980) Morphological variants in Khesari. Indian J Genet Plant Breed 40:172–175
Przybylska J, Zimniak-Przybylska Z, Krajewski P (2000) Diversity of seed globulins in *Lathyrus sativus* L. and some related species. Genet Resour Crop Evol 47:239–246
Quereshi MY, Pilbean DJ, Evans CS, Bell EA (1977) The neurolathyrism of β-N-oxalyl-L-α, β-diaminopropionic acid in legume seeds. Phytochemistry 16:477–479
Rahman MM, Kumar J, Rahman MA, Afzal MA (1995) Natural outcrossing in *Lathyrus sativus* L. Indian J Genet Plant Breed 55:204–207
Rao SLN (1978) A sensitive and specific colorimetric method for the determination of a, β-diaminopropionic acid and the *Lathyrus sativus* neurotoxin. Anal Biochem 86:386–395
Rao SLN (2011) A look at the brighter facets of β-N-oxalyl-l-α,β- diaminopropionic acid, homoarginine and the grass pea. Food Chem Toxicol 49:620–622. https://doi.org/10.1016/j.fct.2010.06.054
Rao SLN, Adiga PR, Sarma PS (1964) The isolation and characterization of β-N-Oxalyl-L-α, β-Diaminopropionic acid: a neurotoxin from the seeds of *Lathyrus sativus*. Biochemistry 3:432–436
Ravindran V, Blair R (1992) Feed resources for poultry production in Asia and the Pacific. II. Plant protein sources. Worlds Poult Sci J 48:205–231. https://doi.org/10.1079/WPS19920017
Rizvi AH, Sarker A, Dogra A (2016) Enhancing grass pea (*Lathyrus sativus* L.) production in problematic soils of South Asia for nutritional security. Indian J Genet Plant Breed 76:583–592. https://doi.org/10.5958/0975-6906.2016.00074.2
Robertson LD, Abd-El-Moneim AM (1997) Status of Lathyrus germplasm held at ICARDA and its use in breeding programs. In: Mathur PN, Rao VR, Arora RK (eds) Lathyrus genetic resources network. Proceedings of a IPGRI-ICARDA-ICAR Regional Working Group Meeting, 8–10 December 1997. IPGRI, New Delhi, pp 30–41
Rosa MJS, Ferreira RB, Teixeira AR (2000) Storage proteins from *Lathyrus sativus* seeds. J Agric Food Chem 48(11):5432–5439
Ross SM, Roy DN, Spencer PS (1989) β-N-oxalyl amino-L-alanine action on glutamate receptors. J Neurochem 53:710–715
Roy DN, Rao KV (1978) Physicochemical values in different varieties of *Lathyrus sativus* and their interrelationships. J Agric Food Chem 26:687–689
Roy DN, Nagarajan V, Gopalan C (1963) Production of neurolathyrism in chicks by the injection of *Lathyrus sativus* concentrates. Curr Sci 32:116–118
Roy PK, Gupta N, Barat GK, Mehta SL (1993) 3 -N-oxalyl-L- a, 13 diaminopropionic acid in somaclones derived from internodes explants of *Lathyrus sativus*. Indian J Plant Biochem Biotechnol 2:9–13
Rybinski W, Blaszczak W, Fornal J (2006) Seed microstructure and genetic variation of characters in selected grass-pea mutants (*Lathyrus sativus* L.). Int Agrophys 20(4):317
Santha IM, Mehta SL (2001) Development of low ODAP somaclones of *Lathyrus sativus*. Lathyrus Lathyrism Newsl 2:42
Shergill-Bonner R (2013) Micronutrients. Paediatr Child Health 23(8):331–336
Shiferaw E, Porceddu E (2018) Assessment of variability in grass pea germplasm using β-ODAP content and seed protein electrophoresis. Ann Food Sci Technol 19:316–323
Shiferaw E, Pe ME, Porceddu E, Ponnaiah M (2012) Exploring the genetic diversity of Ethiopian grass pea (*Lathyrus sativus* L.) using EST-SSR markers. Mol Breed 30(2012):789–797
Shobhana S, Sangawan PS, Nainawase HS, Lal BM (1976) Chemical composition of some improved varieties of pulses. J Food Sci Technol 13:49–51

Siddique KHM, Loss SP, Herwig SP, Wilson JM (1996) Growth, yield and neurotoxin (ODAP) concentration of three Lathyrus species in Mediterranean-type environments of Western Australia. Aust J Exp Agric 36(2):209–218

Silva AB, Gonçalves L, Mecha E, Pereira F, Patto MCV, Bronze M d R (2019) An improved HILIC HPLC-MS/MS method for the determination of beta-ODAP and its alpha isomer in *Lathyrus sativus*. Molecules 24:3043

Singh SS, Rao SLN (2013) Lessons from neurolathyrism: a disease of the past & the future of *Lathyrus sativus* (Khesari dal). Indian J Med Res 138:32–37

Skiba B, Ford R, Pang EC (2003) Amplification and detection of polymorphic sequence-tagged sites in *Lathyrus sativus*. Plant Mol Biol Report 21(2003):391–404

Smartt J (1984) Evolution of grain legumes. I. Mediterranean pulses. Exp Agric 20:275–296

Soren KR, Yadav A, Pandey G, Gangwar P, Parihar AK, Bohra A, Dixit GP, Datta S, Singh NP (2015) EST-SSR analysis provides insights about genetic relatedness, population structure and gene flow in grass pea (*Lathyrus sativus*). Plant Breed 134:338–344

Spencer PS, Roy DN, Ludolph A, Hugon J, Schaumburg HH (1986) Lathyrism: evidence for role of the neuroexcitatory amino acid BOAA. Lancet 2:1066–1067

Sybenga J (1995) Meiotic pairing in autohexaploid Lathyrus: a mathematical model. Heredity 75(4):343–350

Tadesse W, Bekele E (2003) Variation and association of morphological and biochemical characters in grass pea (*Lathyrus sativus* L.). Euphytica 130:315–324

Talukdar D (2009) Recent progress on genetic analysis of novel mutants and aneuploid research in grass pea (*Lathyrus sativus* L.). Afr J Agric Res 4:1549–1559

Talukdar D (2012) Changes in neurotoxin, β-N-oxalyl-l-α,β- diaminopropionic acid (β-ODAP), level in grass pea (*Lathyrus sativus* L.) genotypes under arsenic treatments. J Appl Biosci 38:148–153

Tamburino R, Guida V, Pacifico S, Rocco M, Zarelli A, Parente A, Di Maro A (2012) Nutritional values and radical scavenging capacities of grass pea (*Lathyrus sativus* L.) seeds in Valle Agricola district, Italy. Aust J Crop Sci 6:149–156

Tavoletti S, Iommarini L (2007) Molecular marker analysis of genetic variation characterizing a grass pea (*Lathyrus sativus*) collection from central Italy. Plant Breed 126:607–611

Tavoletti S, Iommarini L, Crino P, Granati E (2005) Collection and evaluation of grass pea (*Lathyrus sativus* L.) germplasm of central Italy. Plant Breed 124:388–391

Thavarajah D, Vandenberg A, George GN, Pickering IJ (2007) Chemical form of selenium in naturally selenium-rich lentils (*Lens culinaris* L.) from Saskatchewan. J Agric Food Chem 55:7337–7341

Thavarajah D, Ruszkowski J, Vandenberg A (2008) High potential for selenium biofortification of lentils (*Lens culinaris* L.). J Agric Food Chem 56:10747–10753

Thavarajah D, Thavarajah P, Sarker A, Vandenberg A (2009) Lentils (*Lens culinaris* Medikus subspecies culinaris): a whole food for increased iron and zinc intake. J Agric Food Chem 57:5413–5419

Tripathy SK, Panda A, Nayak PK, Dash S, Lenka D, Mishra DR, Kar RK, Senapati N, Dash GB (2016) Somaclonal variation for genetic improvement in grass pea (*Lathyrus sativus* L.). Legum Res 39:329–335. https://doi.org/10.18805/1r.v0iOF.6853

Urga K, Fufa H, Biratu E, Husain A (2005) Evaluation of *Lathyrus sativus* cultivated in Ethiopia for proximate composition, minerals, β-ODAP and anti-nutritional components. Afr J Food Agric Nutr Dev 5:1–16

Valarini MJ, Otsuk IP, Carneiro Vieira ML (1997) Changes in N2 fixation in Stylosanthes scabra derived from tissue culture. Braz J Genet 20:1–6

Van Moorhem M, Decrock E, Coussee E, Faes L, De Vuyst E, Vranckx K, De Bock M, Wang N, D'Herde K, Lambein F, Callewaert G, Leybaert L (2010) l-β-ODAP alters mitochondrial Ca2+ handling as an early event in excitotoxicity. Cell Calcium 47:287–296. https://doi.org/10.1016/j.ceca.2010.01.001

van Wyk SG, Kunert KJ, Cullis CA, Pillay P, Makgopa ME, Schlüter U, Vorster BJ (2016) Review: the future of cystatin engineering. Plant Sci 246:119–127. https://doi.org/10.1016/j.plantsci.2016.02.016

Van-Dorrestein B, Baum M, Abd-El-Moneim AM (1998) Use of somaclonal variation in *Lathyrus sativus* (grass pea) to select variants with low ODAP concentration. In: Proceedings of the 3rd European Conference on grain legumes, Valladolid, Spain. AEP, Paris, p 364

Varshney RK, Glaszmann JC, Leung H, Ribaut M (2010) More genomic resources for less-studied crops. Trends Biotechnol 28:452–460. https://doi.org/10.1016/j.tibtech.2010.06.007

Waghmare VN, Mehra RB (2000) Induced genetic variability for quantitative characters in grasspea (*Lathyrus sativus* L.). Indian J Genet 60(1):81–87

Waghmare VN, Mehra RB (2001) Induced chlorophyll mutations, mutagenic effectiveness and efficiency in *Lathyrus sativus* L. Indian J Genet 61(1):53–56

Waghmare VN, Waghmare DN, Mehra RB (2001) An induced fasciated mutant in grass pea (*Lathyrus sativus* L.). Indian J Genet Plant Breed 61(2):155–157

Wang F, Chen X, Chen Q, Qin X, Li Z (2000) Determination of neurotoxin β-N-oxalyl-L-α, β-diaminopropionic acid and nonprotein amino acids in *Lathyrus sativus* by precolumn derivatization with 1-fluoro-2, 4-dinitrobenzene. J Chromatogr A883:113–118

Xiong J-L, Xiong Y-C, Bai X, Kong H-Y, Tan R-Y, Zhu H, Siddique KHM, Wang J-Y, Turner NC (2015) Genotypic variation in the concentration of β-N-Oxalyl-L-α, β-diaminopropionic acid (β-ODAP) in grass pea (*Lathyrus sativus* L.) seeds is associated with an accumulation of leaf and pod β-ODAP during vegetative and reproductive stages at three levels of water stress. J Agric Food Chem 63:6133–6141

Yan Z-Y, Spencer PS, Li Z-X, Liang Y-M, Wang Y-F, Wang C-Y, Li F-M (2005) *Lathyrus sativus* (grass pea) and its neurotoxin ODAP. Phytochemistry 67:107–121. https://doi.org/10.1016/j.phytochem.2005.10.022

Yan B, Chtourou-Ghorbel N, Marrakchi M, Trifi-Farah N (2006) Genetic diversity within and between populations of *Lathyrus* genus (Fabaceae) revealed by ISSR markers. Genet Resour Crop Evol 53:1413–1418

Yang T, Jiang J, Burlyaeva M, Hu J, Coyne CJ, Kumar S, Redden R, Sun X, Wang F, Chang J, Hao X, Guan J, Zong X (2014) Large-scale microsatellite development in grass pea (*Lathyrus sativus* L.), an orphan legume of the arid areas. BMC Plant Biol 14:65

Zambre M, Chowdhury B, Kuo Y-H, Van Montagu M, Angenon G, Lambein F (2002) Prolific regeneration of fertile plants from green nodular callus induced from meristematic tissues in *Lathyrus sativus* L. (grass pea). Plant Sci 163:1107–1112. https://doi.org/10.1016/S0168-9452(02)00319-9

Zhao L, Chen XG, Hu ZD, Li QF, Chen Q, Li ZX (1999a) Analysis of β-N-oxalyl-L-α, β-diaminopropionic acid and homoarginine in *Lathyrus sativus* by capillary zone electrophoresis. J Chromatogr 857:295–302

Zhao L, Li Z, Li G, Chen X, Hu Z (1999b) Kinetics studies on thermal isomerization of β-N-oxalyl-L-α, β-diaminopropionic acid by capillary zone electrophoresis. Phys Chem Chem Phys 1:3771–3773

Zhao R, Sun H, Mei C, Wang X, Yan L, Liu R et al (2011) The arabidopsis Ca2+-dependent protein kinase CPK12 negatively regulates abscisic acid signaling in seed germination and post-germination growth. New Phytol 192:61–73. https://doi.org/10.1111/j.1469-8137.2011.03793.x

Breeding Cowpea for Quality Traits: A Genetic Biofortification Perspective

P. Dhanasekar, J. Souframanien, and P. Suprasanna

Abstract Cowpea, *Vigna unguiculata* (L.) Walp, is an important arid legume crop cultivated widely in the arid and semiarid tropics of the world mostly by resource-poor farmers involved in subsistence farming. Also known as poor man's meat, this crop is rich in proteins and carbohydrates but does not have appreciable quantities of essential micronutrients. Micronutrient deficiency leading to malnutrition is a major concern that affects one third of the world population. Among various interventions available for alleviating malnutrition, genetic biofortification through plant breeding is considered the most viable, economical, and sustainable approach. Cowpea exhibits considerable genetic variability for important nutritional components such as protein and micronutrient levels, thus offering scope for genetic biofortification. With genetic biofortification breeding programs of primary staples attaining the intended micronutrient level targets, it is high time that similar results are replicated in secondary staples, especially pulses, and in a crop like cowpea that complement the primary staple-based diets. Breeding of cowpea quality traits from a genetic biofortification perspective is discussed with an attempt to provide a comprehensive outlook on priority biofortification traits, their genetic variability and biochemistry, and genomic and analytical tools available. The growing national and international interests of cowpea breeders for pursuing biofortification as a new, complementary intervention to address micronutrient deficiency are expected to result in the development of next-generation biofortified cowpea and ensuring a nourishing future.

Keywords Biofortification · Cowpea · Iron · Legumes · Malnutrition · Micronutrient deficiency · Zinc

P. Dhanasekar (✉) · J. Souframanien · P. Suprasanna
Nuclear Agriculture and Biotechnology Division, Bhabha Atomic Research Centre, Mumbai, India
e-mail: sekar@barc.gov.in

D. S. Gupta et al. (eds.), *Breeding for Enhanced Nutrition and Bio-Active Compounds in Food Legumes*, https://doi.org/10.1007/978-3-030-59215-8_7

Introduction

Cowpea, *Vigna unguiculata* (L.) Walp, is a proteinaceous arid grain legume crop widely cultivated in the tropical and subtropical regions of the world. They are commonly grown in the semiarid tropics between 35° N and 30° S of the equator, covering Africa, Asia, Oceania, the Middle East, Southern Europe, Central and South America, and the Southern United States (Boukar et al. 2018). The cowpea plant is a herbaceous, warm-season annual legume requiring temperatures of at least 18 °C throughout all stages of its development and having an optimal growing temperature of about 28 °C (Craufurd et al. 2010). Unlike other food legumes, this hardy crop performs well even in the drier regions. Cowpea is a dicotyledon belonging to the order Fabales, family Fabaceae, subfamily Faboideae (Syn. *Papillionoideae*), tribe Phaseoleae, subtribe Phaseolinae, genus *Vigna* Savi (Boudin and Marechal 2011). The pantropical *Vigna* is a highly variable genus encompassing 84 to 184 species (Timko et al. 2007). Cowpea belongs to the section Catiang of subgenus *Vigna* and genus *Vigna.* Section Catiang is comprised of two species, *unguiculata* and *nervosa*. The species *unguiculata* (Latin for "with a small claw," which reflects the small stalks on the flower petals) is further divided into five subspecies with all the cultivated cowpeas being found within the subspecies *unguiculata.* This subspecies is comprised of four cultivar groups: unguiculata, biflora, sesquipedalis, and textilis. All the current evidence suggests that cowpea originated in Southern Africa, although several centers of domestication such as Ethiopia, Central Africa, South Africa, and West Africa have been suggested. Presently, the wild cowpea, *Vigna unguiculata* ssp. *unguiculata* var. spontanea, is supposedly the likely progenitor of cultivated cowpea (Singh 2005). With a view to streamlining and strengthening cowpea breeding programs across the globe, the International Institute of Tropical Agriculture (IITA) was established in 1967 with a mandate to develop improved cowpea varieties for all regions. This nodal agency is maintaining more than 15,100 accessions of cultivated cowpea drawn from over 100 countries and more than 560 accessions of wild cowpeas.

The cowpea, considered to be one of the oldest domesticated crops (Chivenge et al. 2017), probably derived its name due to its use as a fodder crop for cows. It is commonly known by its indigenous or regional names such as "lobia" and "chowlee" in India; "kunde" in East Africa; "beans" and "wake" in Nigeria; "niebe" in francophone Africa; "southern pea," "crowder pea," and "black eye pea" in the United States of America; and "feijão caupe" in Brazil and also by a host of other vernacular names in different countries worldwide. Current estimates indicate that it is grown in about 14.5 million hectares with an annual production of over seven million tons on a global basis (Singh 2014). Over the last three decades, worldwide cowpea production grew at an average rate of 5%, with 3.5% annual growth in area and 1.5% growth in yield, and the area expansion accounted for 70% of the total growth during this period (Fatokun et al. 2012). India is the largest cowpea producer in Asia, and together with Bangladesh, Indonesia, Myanmar, Nepal, Sri Lanka,

Pakistan, Philippines, Thailand, and other far eastern countries, more than 1.5 million ha is under cowpea cultivation (Steele and Mehra 2009).

Cowpea is truly a multifunctional crop, providing food for man and livestock and serving as a valuable and dependable revenue-generating commodity for resource-poor farmers. It is inherently tolerant to drought and heat and has the ability to fix nitrogen (through its symbiotic relation with *Bradyrhizobium* group of nodulating bacteria, can fix 70–350 kg nitrogen per ha) even in very poor soils with a pH as low as 4–5, organic matter below 0.2%, and sand content of over 85% (Kolawale et al. 2000). Its inherent shade tolerance attribute makes it a candidate crop for intercropping with a number of cereals, root, and plantation tree crops. In addition, its quick growth and rapid ground cover has made cowpea an essential component of sustainable subsistence agriculture in marginal drier regions of the tropics where rainfall is erratic and scanty and soils are sandy with little organic matter (Carsky et al. 2001). Its plasticity toward environmental vagaries and its superior nutritional values make it a potent crop under the present context of food, nutritional security, and climate change.

Cowpea seeds provide a rich source of proteins and calories, as well as minerals and vitamins. As a legume in general, its protein content (~25%) is approximately twice that of cereals, and its amino acid (AA) profile, rich in lysine (Lys) and tryptophan (Trp), complements those of cereals, which are rich in sulfurous AAs (Nielsen et al. 1993). With very low fat content and slowly digestible starch (in comparison to cereals), cowpea is highly beneficial for human health. The grain is a rich source of an important vitamin folic acid, which helps prevent neural tube defects in unborn babies. The nutritional richness of cowpea can be comprehended in Table 1. The remnant biomass of the haulm post harvest is a source of quality fodder for ruminant livestock. Cowpea can be consumed as fresh or dry seeds, canned or frozen food, and milled flour in baked goods. In addition, cowpea has been used as an alternative to soybean for people who are allergic to soybean protein (Boukar et al. 2018). Because of its high protein content and largely being cultivated by resource-poor farmers, cowpea is aptly referred to as "poor man's meat."

Agriculture till now has been aimed at producing more calories to negate hunger, but the current scenario in most of the developing countries equally warrants the development of nutrient-rich foods to reduce hidden hunger or malnutrition. Malnutrition results from eating a diet in which one or more nutrients (calories, protein, carbohydrates, fat, vitamins, or minerals) are either not enough (undernutrition) or otherwise (overnutrition) such that the diet causes health problems. Malnutrition (often refers to undernutrition) is more predominant in developing countries with certain groups, in particular pregnant or breastfeeding women and children under 5 years of age being more susceptible. Vitamin and mineral deficiencies result in a myriad of cognitive and health impairments increasing the risk of death. In developing countries, agricultural products are the prime source of nutrients, and the nonavailability or non-affordability of nutrient-rich food grains has deprived the needy poor of these essential nutrients leading to malnutrition. Possible ways to combat those deficiencies encircle dietary diversification (healthy balanced diet), food fortification (nutrient enrichment during processing), biofortification,

Table 1 Nutritional value per 100 g of raw cowpea seeds

Nutrient	Value
Energy	336 kcal (1410 kJ)
Carbohydrates	60.03 g
Sugars	6.9 g
Dietary fiber	10.6 g
Fat	1.26 g
Protein	23.52 g
Vitamins	**Quantity**
Vitamin A equiv.	3 μg
Thiamine (B_1)	0.853 mg
Riboflavin (B_2)	0.226 mg
Niacin (B_3)	2.075 mg
Vitamin (B_6)	0.357 mg
Folate (B_9)	633 μg
Vitamin C	1.5 mg
Vitamin K	5 μg
Minerals	**Quantity**
Calcium	110 mg
Iron	8.27 mg
Magnesium	184 mg
Phosphorus	424 mg
Potassium	1112 mg
Sodium	16 mg
Zinc	3.37 mg
Other constituents	**Quantity**
Water	11.95 g

Source: USDA nutrient database

and supplementation (external nutrient-rich additives) (Ghosh et al. 2019). No single intervention can solve the problem of micronutrient malnutrition, but biofortification complements existing interventions to sustainably provide micronutrients to the most vulnerable people in a comparatively sustainable, inexpensive, and cost-effective manner (Saltzman et al. 2013). Biofortification, the process of increasing nutrient concentration in plant edible parts, can be achieved through three main approaches, namely, transgenic, conventional, and agronomic, involving the use of biotechnology, crop breeding, and fertilization strategies, respectively. Biofortification through conventional breeding is the most accepted method of biofortification. Thus, biofortification through breeding programs aims at increasing the micronutrient dietary intake without changing the diet of those targeted (Gerrano et al. 2017). A number of international initiatives have made impactful success in their sustained efforts for global redressal of malnutrition like Nutrition International, Iodine Network, iZiNCg, Iron Deficiency Project Advisory Service (IDPAS), New Partnership for Africa's Development (NEPAD), UNICEF-Micronutrients, Global

Alliance for Improved Nutrition (GAIN), Helen Keller International, CGIAR Research Program on Agriculture for Nutrition and Health (A4NH), HarvestPlus, etc. In India, various government initiatives have been launched over the years to improve the overall nutrition status in the country. These include the Integrated Child Development Services (ICDS), the National Health Mission, the Janani Suraksha Yojana, the Matritva Sahyog Yojana, the Mid-day Meal Scheme, and the National Food Security Mission, among others. However, concerns regarding malnutrition have persisted despite improvements over the years. It is in this context that the National Nutrition Strategy has been recently released (NITI Aayog, GoI 2017).

Cowpea, a crop of subsistence farming across the world, is therefore an apt crop for breeding quality traits to address the malnutrition. Quality in its broadest sense encompasses a gamut of traits that can be broadly grouped as under:

A. **Morphological and physical quality:** These traits are related to external appearance of the seed. It includes seed shape, seed size, testa color, hilum eye color, seed coat pattern, seed texture, seed weight, etc.
B. **Organoleptic quality:** These traits are related to palatability of the produce. They are easily detected and are very important in consumer preferences. It includes seed taste, aroma, flavor, softness, etc.
C. **Biological quality:** The traits included in this group define the actual usefulness of the produce, when consumed. These include protein efficiency ratio, biological value, body weight gain, bioavailability, and digestibility.
D. **Biochemical quality:** It includes protein, vitamins, minerals, carbohydrates, micronutrients, and antioxidants.
E. **Antinutritional quality:** It includes protease inhibitors, phytates, alpha-galactosides (oligosaccharides), tannins, saponins, and polyphenols.
F. **Other quality parameters**: These are important in determining the usefulness of the concerned produce. This includes cooking quality, milling quality, cooking time, and keeping quality.

However, in addressing the malnutrition through biofortification, biochemical parameters, especially micronutrient content, are of prime importance, and hence, this aspect of quality breeding will be elaborated in this chapter. Techniques to increase the total protein and mineral content of cowpea cultivars are considered as an important component of global intervention programs that are focused on alleviating human malnutrition and ensuring food security, especially in semiarid tropical areas (Santos and Boiteux 2013).

Priority Traits for Genetic Biofortification

The United Nations Food and Agriculture Organization has estimated that around 792.5 million people across the world are malnourished, out of which 780 million people live in developing countries (McGuire 2015). Apart from this, around two billion people across the world suffer from another type of hunger known as "hidden hunger," which is caused by an inadequate intake of essential micronutrients in the daily diet (Hodge 2016) despite increased food crop production (Gould 2017). With increasing incidences of protein malnutrition in developing countries and higher incidence of diabetes, heart problems, and cancer in the developed countries, the consumption of cowpea with superior nutritional quality is expected to increase. A lot of research has gone into the biofortification of primary staple crops such as rice, wheat, maize, cassava, etc., which are consumed in large quantities. Even after decades of research, the biofortified varieties in these crops are not able to meet the entire estimated average requirement (EAR) of nutrients. Hence, it is highly imperative that concerted research efforts have to be directed toward a "food basket approach," providing a range of biofortified food crop options suited to local preferences (Andersson et al. 2017). This approach allows for diversification, both on the plate and in the field. In farmers' fields, different micronutrient-dense crops can be grown in rotation to provide a steady supply of micronutrients throughout the year. The secondary staples like cowpea are usually consumed in lower quantities than primary staples. Consequently, their contribution to daily micronutrient requirements is also lower. Nevertheless, they are an important complement in daily diets and are frequently consumed together with primary staples such as rice or wheat, and any amount of biofortification levels in these crops would help realize the ultimate micronutrient target levels. Therefore, prioritizing traits for genetic biofortification in secondary staples like cowpea has to be in tandem with that of primary staples. Secondly, the target traits for genetic biofortification should be identified such that sufficient and utilizable genetic variation exists in the genetic material for the trait of interest. Thirdly, while deciding the target nutrient levels, the baseline nutrient level has to be determined, and the incremental target level has to be arrived taking into consideration the micronutrient retention after processing, the bioavailability, and the per capita consumption so that the additional percent of EAR is achieved.

The priority traits for genetic biofortification in cowpea are as follows:

(a) **Protein content and quality**: Cowpea is a rich source of proteins (23–25% in dry seeds) and carbohydrates (50–70%), which could meet the increasing consumer demand for healthier and more nutritious food. Unlike soybean, cowpea proteins do not cause allergies and are of higher quality when substituted in diets at equivalent protein contents. In recent years, there has been increasing interest in breeding cowpea cultivars with high seed protein content to improve nutritional quality. Evaluation of seed protein content in cowpea germplasm will help plant breeders select and breed high seed protein content cultivars in breeding programs. Moreover, the amino acid profile of cowpea unlike that of cereals is rich in lysine and tryptophan but lacks in sulfur-containing amino acids. Therefore, the breeding efforts need to be aimed at increasing both the protein content and the proportion of methionine and cysteine amino acids to counter protein malnutrition.

(b) **Iron content (Fe)**: Iron (Fe) is an essential micronutrient for plants and for humans, and it is a constituent of a number of important macromolecules, including those involved in respiration, photosynthesis, DNA synthesis, and metabolism (Briat 2011). Fe deficiency is ranked fifth among the top ten risk factors contributing to disease burden globally. Iron is present in all cells in the human body and has several vital functions, such as carrying oxygen to the tissues from the lungs as a key component of the hemoglobin protein, acting as a transport medium for electrons within the cells in the form of cytochromes, and facilitating oxygen enzyme reactions in various tissues. Too little iron can interfere with these vital functions and lead to morbidity and death (Centers for Disease Control and Prevention 1998). Children, premenopausal women (women of child-bearing age), and people with poor diet are most susceptible to anemia disease caused by the deficiency of Fe. The EAR of Fe in nonpregnant, nonlactating women is 1460 μg/day, while in children of 4–6 years, it is 500 μg/day. Fe retention after processing in cowpea is in the order of 90%, and bioavailability is around 2.5%.

(c) **Zinc content** (Zn): Zinc is an essential micronutrient in biological systems, which is required in small quantities. It is involved in the formation and activation of enzymes that have impact on the growth, development, and production of plants. It also affects pollen viability, flowering, and grain production. In humans, its deficiency is associated with problems of growth and learning capacity in children and increases the risk of infections, cancer, and DNA damage (Veronica et al. 2018). An estimated 17.3% of people worldwide are at risk of inadequate Zn intake (Wessells and Brown 2012), and Zn deficiency leads to estimated annual deaths of 433,000 children under the age of five (WHO 2009). It is present in around one third of the world population, which represents the sixth risk factor for diseases in developing countries (Shahzad et al. 2014). The EAR of Zn in nonpregnant, nonlactating women is 2960 μg/day, while in children of 4–6 years, it is 1390 μg/day. Zn retention after processing in cowpea is in the order of 90%, and bioavailability is around 15%.

(d) **Anti-mineral compound content:** Among the anti-minerals, antinutritional factors present in legumes like cowpea and phytic and oxalic acids are important (Liener and Kakade 1980). Phytic acid, also known as inositol hexakisphosphate (IP6), or phytate when in salt form is the principal storage form of phosphorus in many plant tissues. It is not digestible to humans or nonruminant animals, because these animals lack the digestive enzyme (phytase) required to remove phosphate from the inositol in the phytate molecule. Phytate is well documented to block absorption of not only phosphorus but also of other minerals such as calcium, magnesium, iron, and zinc (Shukkur et al. 2006). Phytic acid and oxalic acid reduce mineral bioavailability that leads to various mineral deficiency diseases, e.g., anemia, or form deleterious complexes with metal ions, e.g., calcium oxalate, which leads to renal damage. But since these antinutritional factors are mainly plant's secondary metabolites, they are involved in a variety of plant metabolic pathways. They are known to be involved in plant defense mechanism against biotic and abiotic stresses, and hence due precaution has to be taken while meddling with these compounds. The breeder has to strike a right balance as to how low these compounds could be reduced without hampering the metabolic or agronomic values of the crop. Therefore, the reduction of anti-mineral compounds leads to increased bioavailability of micronutrients or wholesomeness for consumers and could be construed as a means of biofortification.

Genetic Variability for the Target Traits

The success of genetic biofortification through recombination breeding depends on the genetic variability of the proximate contents of the various target traits. Additive genetic interaction of genes governing the target traits could lead to generation of transgressive segregants enabling the development of varieties with proximate contents greater than that of the donor parents. Therefore, it is imperative to know the extent of genetic variation for various target traits in the existing germplasm of the crop including landraces and wild species, and also the knowledge on the genetics of the trait would enable the selection of suitable breeding method to achieve the target objective. After assessing the genetic variability for the trait of interest and confirming its suitability of genetic improvement, the donor lines with these traits are identified and are used in early-stage product development and parent building. Thereafter, breeding materials with improved nutrient content and high agronomic performance as well as preferred consumer qualities are developed. If necessary, further crosses with locally adapted materials could be attempted to develop final products that meet specific traits required by local producers and consumers. When promising high-yielding, high-nutrient lines emerge, they are tested across a wide range of environments side-by-side with locally preferred varieties. If the trait is lacking in a particular crop, then genetic biofortification through biotechnological interventions like transgenics could be resorted to, provided the legal framework of the country permits.

An analysis of 1541 cowpea germplasm lines (Boukar et al. 2011) revealed that on an average cowpea has 25% protein, 38 mg Zn/kg, 53 mg Fe/kg, 1.9 g Mg/kg, 0.825 g Ca/kg, 5 g P/kg, and 15 g K/kg. The screening of 2000 lines in cowpea (Singh 2016) for studying the genetic variability for major nutritional traits showed existence of wide genetic variability for most of the traits (Table 2).

The range of protein content as reported by various researchers (Asante et al. 2006; Gupta et al. 2010; Itatat et al. 2013; Oke et al. 2015; Ravelombola et al. 2016) falls within the reported range of Singh (2016). However, Afiukwa et al. (2013), Santos and Boiteux (2013), and Dakora and Belane (2019) reported a greater variability of the total seed protein content in excess of 32% up to a maximum of 40% (South African genotype "Bengpla") in cowpea. The broad-sense heritability for seed protein was reported to range from 50.8% to 95%, in various studies (reviewed in Weng et al. 2019) indicating that seed protein content was highly heritable and selection could be rewarding for protein content. But narrow variation in amino acid

Table 2 Genetic variability for quality traits in cowpea germplasm (Singh 2016)

Parameter	Range of value	
	Min	Max
Seed size (g/100)seeds	9	27
Protein (%)	20.9	32.5
Ash (%)	2.9	3.9
Fat (%)	1.4	2.7
Carbohydrate (%)	59.7	71.6
Cooking time (m)	21.1	61.9
Iron (ppm)	51	109
Zinc (ppm)	33	51
Calcium (ppm)	581	1252
Potassium (ppm)	12,084	15,133
Magnesium (ppm)	1611	2052
Phosphorus (ppm)	3867	4922
Sulfur (ppm)	1880	2354

(AA) composition suggests a lesser possibility of improving the contents of specific AAs in cowpea (Muranaka et al. 2016). As far as Fe and Zn are concerned, the former showed more variability in comparison to the latter. Fe content in cowpea ranged from 36.5 to 150 ppm, while Zn content ranged from 33 to 61 ppm (Belane and Dakora 2011; Santos and Boiteux 2013; Singh 2016; Marappa et al. 2016). The cultivar KBC-6 from the University of Agricultural Sciences, Bangalore, was found to have the highest Fe content of 150 ppm. Incidentally, the genotypes which showed high zinc were also associated with stay green trait even after the crop maturity, thus serving as phenotypic markers (Marappa et al. 2016). The variance due to genotype was highly significant (P, 0.01) for crude protein, Fe, and Zn contents. Phytic acid contents ranged from 0.21 to 10.27 mg/g (Garinu and Ingrao 1991; Dhanasekar and Reddy 2017). However, Muranaka et al. (2016) reported phytic acid levels of up to 37 mg/g in IITA lines.

Wide genetic variation and strong correlations among crude protein, Fe, and Zn contents suggest the possibility of improving the concentrations of these nutritional factors simultaneously. There were strong positive genotypic correlations between crude protein and Fe ($r = 0.70$) and Zn ($r = 0.70$) and between Fe and Zn ($r = 0.68$) contents in cowpea (Muranaka et al. 2016). Boukar et al. (2011) also reported strong positive correlations between the contents of crude protein and Fe and of Fe and Zn in their studies with 1541 genotypes reiterating the possibility of simultaneous selection for these traits. Simple correlation coefficient values indicated that selection for high protein and mineral content does not affect grain yield and that it is feasible to obtain new biofortified cowpea cultivars by combining higher levels of protein and essential minerals (Santos and Boiteux 2013). It was also observed that the increase in levels of micronutrients in the grains also favors the agronomic performance of biofortified genotypes in soils that are naturally deficient in these minerals (Welch 2002). In addition, plants with lower concentrations of phytate improved the bioavailability of zinc and iron (Welch et al. 2000). Therefore, selection for lower levels of natural compounds that reduce the bioavailability of micronutrients in the human diet should also be a novel target for future breeding research aiming to develop biofortified cowpea cultivars.

Mutation Breeding in Genetic Biofortification

The degree of genetic variability for target traits in a crop determines the extent to which the trait of interest could be improved through combination breeding. Low genetic variability is a stumbling block in the genetic improvement, and the potential of mutation breeding in creating genetic variations during situations of low genetic variability has been demonstrated since ages. In genetic biofortification of food crops, mutations affecting various target traits have been reported. By and large, these mutants with the altered biofortification traits have been used in hybridization for transfer of these traits into elite genetic backgrounds. Maize breeders have developed quality protein maize (QPM) with high essential amino acids lysine and tryptophan by incorporating *opaque-2* (*o2*) mutant gene from naturally occurring maize into the maize cultivars (Hossain et al. 2019). Incorporation of *Or* mutant gene from orange cauliflower mutant led to increase in carotenoid level (Lopez et al. 2008). Low phytic acid accumulation is a recessive trait (Maqbool and Beshir 2019), and several losses of function mutations have been reported in various crops like rice, maize, common bean, cowpea (Neeraja et al. 2017; Cominelli et al. 2018; Dhanasekar and Reddy 2017), etc., and are being included in a range of introgressive breeding programs. Mutants could also be helpful in studying the physiological and metabolic pathways; as in maize, the mutant yellow stripe 1 (*ysl*) showed Fe deficiency due to impairment of Fe phytosiderophore uptake and that roots of iron-efficient maize mutants also absorbed more of phytosiderophore-chelated zinc (Von Wiren et al. 1996) probably owing to the involvement of nonspecific Fe transporters. Thus, mutation breeding in tandem with conventional breeding could be a great utility in realizing the biofortification goals.

Biochemistry of the Biofortification Traits

For effective genetic biofortification, knowledge on the biochemistry of the target traits is of immense importance, which would enable to maneuver the trait of interest through manipulating the genes governing the trait.

A. **Iron**: Fe is one of the most essential micronutrient that is required for the proper development of both plants and humans. Plants, as primary producers, are the gateway for iron to enter the food chain. Fe is involved in a variety of metabolic activities such as photosynthesis, mitochondrial respiration, nitrogen assimilation, hormone biosynthesis, production and scavenging of reactive oxygen species, osmoprotection, pathogen defense, and as a limiting factor for biomass production (Briat 2011). Plants obtain Fe from the soil, where Fe exists in either ferrous (Fe^{2+}) or ferric (Fe^{3+}) state. Although Fe is the fourth most abundant element in the Earth's crust, it is not readily available to plant as it binds rapidly to soil particles and forms insoluble complexes under aerobic conditions at neutral or alkaline pH (Gomez-Becerra et al. 2010). Post intake, Fe is complexed with chelators and distributed to sink tissues where it is used predominantly in the production of enzyme cofactors or components of electron transport chains. The processes of iron uptake, distribution, and metabolism are overseen by tight regulatory mechanisms, at the transcriptional and posttranscriptional level, to avoid iron concentrations building to toxic excess. Iron is also loaded into seeds, where it is stored in vacuoles or in ferritin. Iron homeostasis in plants is elaborated in detail by Connorton et al. (2017), and therefore it will be discussed only in brief in this chapter.

(a) **Fe uptake**: Plants adopt different strategies for uptake of low soluble Fe(III) oxyhydrate from the rhizosphere in higher plants: (a) Strategy I (non-Graminaceae) is reduction strategy wherein Fe^{3+} is reduced by ferric reduction oxidase 2 (FRO2) at the plasma membrane before transport across the membrane by iron-regulated transporter 1 (IRT1). In addition, plasma membrane proton pumps help acidify the rhizosphere and increase Fe^{3+} solubility. An array of metabolites including organic acids, phenolics, flavonoids, and flavins may also be exported for reduction of ferric iron. (b) Strategy II (Graminaceae) is the chelation strategy involving secretion of phytosiderophores like deoxymugineic acid (MA) which have high affinity for Fe, and the resulting chelates are imported by oligopeptide transporters like YS1. Some organisms are known to have a combination of both the strategies.

(b) **Iron distribution and storage:** Most iron enters the plant via the root and needs to be transported to the sink tissues where it is required for iron-dependent enzymes. Iron first enters the symplastic pathway through IRT1 found on the outward side of the epidermal cells of the roots. Due to its toxicity and low solubility, iron is translocated as Fe^{3+}chelated complex through a complex cascade involving xylem and phloem loading/unloading, and finally in the leaves it is reduced to Fe^{2+} mainly by FRO proteins. To facilitate this translocation, different chelators such as citrate, MAs, and nicotianamine (NA) play a crucial role. Organelle-specific iron transporters then transport a large proportion of iron into the plastids and mitochondria. Iron is then remobilized from leaf tissues with the help of oligopeptide transporter family proteins like OPT3 and reaches other sink organs through the phloem. Though present in many tissues, the terminal destination of iron is often considered to be the seed, where iron stores are important during germination before the seedling has developed a root and takes up nutrients from the soil. YSL transporters are involved in seed loading, and there is evidence that iron can be delivered to embryos as a Fe^{3+}-citrate/malate complex. Two major storage mechanisms for iron have been proposed: sequestration into vacuoles and into ferritin. The vacuolar iron transporter VIT1 was first identified in *Arabidopsis*. Genes from the *VIT* family are also known to be important for iron localization in grains. Ferritins are important iron storage proteins present across the biological kingdoms. In legume seeds, it is found that 24 subunits of ferritin form a shell capable to store up approximately 2500 Fe^{3+}ions. The proportion of total iron stored in ferritin in seeds varies among species, with approximately 60% in peas but less than 5% in *Arabidopsis* seeds. In cereal grains such as wheat and rice, most iron is present in vacuoles in the aleurone layer which is often removed during grain processing. The way in which iron is stored in seeds can affect its bioavailability when consumed, which is of great importance in biofortification studies. The iron is then used in the biosynthesis of Fe cofactors because of the toxic nature of free iron. The most common forms of iron cofactors are heme, FeS clusters, and di-iron centers.

Plants adapt their root morphology to iron-limiting conditions by increasing the density of root hairs and the number of lateral roots. The greater surface area extends contact between the epidermis and the rhizosphere, and the lateral roots help to explore fresh soil (Li et al. 2016). Great progress has been made in identifying a large number of transcriptional regulators like helix-loop-helix (bHLH) and FER-like iron deficiency-induced transcription factor (FIT) that regulate the iron deficiency response of iron homeostasis. Plants exhibit tight homeostatic control to prevent accumulation of iron where it is not needed, and this may limit iron redistribution to edible tissues such as seeds. Any successful biofortification strategy must bypass these mechanisms without causing physiological damage to the plant.

B. **Zinc:** Zn homeostasis is maintained by a tightly regulated network of low-molecular-weight ligands, membrane transport, and Zn-binding proteins, as well as regulators. Fe and Zn homeostasis interacts as a consequence of the chemical similarity between their divalent cations and the lack of specificity of the major root iron uptake transporter IRT1. A significant proportion of the Earth's arable land is considered Zn-deficient (Alloway 2009). Zn can bind tightly to soil and plant cell wall components and can form precipitates, most commonly in the form of phosphates, carbonates, or hydroxides, in the soil. Like in Fe homeostasis, Zn solubilization in the rhizosphere is thought to occur via plant-mediated acidification and secretion of low-molecular-weight organic chelators. Subsequently, Zn is taken up across the plasma membrane of root cells predominantly as a free ion in a similar fashion as that of Fe. The possible involvements of zinc-regulated transporter and iron-regulated transporter (ZRT-IRT)-like proteins (ZIPs) in cellular Zn^{2+} uptake have been established. The major root epidermal plasma membrane Fe transporter IRT1 mediates the uptake of Zn^{2+} as well as its primary substrate Fe^{2+}. In the cytoplasm of plant cells, Zn is thought to be chelated by low-molecular-weight ligands in order to prevent cytoplasmic precipitation and nonspecific binding to biomolecules. The export from cells is required for the loading of Zn into the apoplastic xylem and thus for the translocation of Zn from the root to the shoot. Zn^{2+} export from the cytoplasm and further loading into the xylem is accomplished by a subgroup of HMA proteins of the P_{1B}-type ATPase family. A subset of plant metallothioneins is likely to contribute to the buffering or storage of cytosolic Zn. Cation diffusion facilitator (CDF) family of metal cation/proton antiporters, members of which have also been named ZAT (zinc transporter of *Arabidopsis thaliana*) and MTP (metal tolerance protein or metal transport protein), acts in the removal of Zn from the cytoplasm. Inside the xylem, Zn flux into the shoot is mass flow driven. There is some evidence for the chelation of Zn by low-molecular-weight ligands inside the xylem, which could act to prevent Zn retention by metal-binding components of the surrounding cell walls or uptake into cells via Zn^{2+} transporters. Cell vacuoles are the major site for storage and detoxification of excess Zn and a source for Zn remobilization in periods of deficiency. The homeostasis of Zn has been comprehensively reviewed by Sinclair and Kramer (2012).

C. **Phytic acid (PA):** Phytic acid (*myo*-inositol-1,2,3,4,5,6-hexa*kis*phosphate, $InsP_6$) is the most abundant form of phosphorus occurring in seeds (up to 85% of total phosphorus) and other plant tissues. Due to its chemical structure (highly negatively charged at physiological pH), PA easily binds important mineral cations such as iron, zinc, potassium, calcium, and magnesium and makes them unavailable. In plants, PA biosynthesis occurs through two different routes: a "lipid-dependent" (operates in all tissues) and a "lipid-independent" pathway (predominates in seeds). PA biosynthesis begins with the production of *myo*-inositol (Ins) through a highly conserved reaction in which the enzyme d-*myo*-inositol 3-phosphate synthase (MIPS) converts d-glucose-6-phosphate to *myo*-inositol 3-phosphate ($Ins(3)P_1$). *Myo*-inositol 3-phosphate is then dephosphorylated to free Ins by inositol monophosphate phosphatase (IMP). In the "lipid-dependent" pathway, Ins is converted to phosphatidylinositol (PtdIns) by a synthase (PtdIS) and thereafter is sequentially phosphorylated by kinases. The "lipid-independent" pathway consists of sequential phosphorylation of the Ins ring to $InsP_6$, through the action of a number of specific kinases. In rice, a mutation in kinase gene (*OsPGK1*) generates an *lpa* phenotype, while overexpression increases seed $InsP_6$ content, suggesting that *OsPGK1* is a key gene for $InsP_6$ synthesis, being involved in (probably the rate-limiting) step from $InsP_1$ to $InsP_2$. Further, phosphorylation steps, required to convert $InsP_3$ into the more phosphorylated $InsP_4$, $InsP_5$, and $InsP_6$, involve at least three types of inositol kinases (for details, read Sparvoli and Cominelli 2015). Once synthesized, phytic acid is stored as globoids inside the storage vacuoles. Depending on the species, the amount and distribution of phytic acid in different parts of the seed can be quite variable. In the cereals, a large amount (80%) of phytic acid is stored in the aleurone and bran (maternal teguments), while in maize seeds 80% of phytate accumulates in the embryo and scutellum (O'Dell et al. 1972). In case of legume seeds, more than 95% is accumulated in the cotyledons (Ariza-Nieto et al. 2007), while in *Arabidopsis*, it is stored in the embryo (Otegui et al. 2002).

During germination, in order to support seedling growth, phytic acid is then degraded by phytase enzymes to remobilize the phosphorus stored as phytate salts (Raboy 2003).

Therefore, to facilitate an efficient and targeted genetic biofortification for Fe and Zn, five key steps can be addressed: (a) enhanced uptake, (b) increased translocation to seeds, (c) specialization of Fe and Zn storage toward vacuoles, (d) reduction of antinutritional compounds like phytic acid, and (e) increase of bioavailability. Either single approach or combination of multiple approaches can be applied in genetic biofortification.

Analytical Methods

The success of any biofortification program is largely dependent on robust analytical tools that can precisely and rapidly analyze the micronutrient contents for high-throughput screening of a large number of samples from segregating breeding materials in a cost-effective and efficient manner. The key to accurate measurement also depends on the chances of contamination during sample preparation and analysis. Moreover, the tools for analysis should be easily available to the breeders both cost wise and quantity wise and should be as simple as possible without the need for any special expertise. Since the micronutrient contents are very low in the order of ppm, the technology should be highly sensitive to detect accurate variations. Pfeiffer and McClafferty (2007) provide a comprehensive overview of analytical methods and diagnostic tools and also discuss other related issues, such as the varying sensitivity requirements depending on the stage of development, contamination (in the case of minerals), effects of milling/polishing, and micronutrient concentration versus content.

(a) **Protein content determination**: Protein content in cowpea has been widely determined by the age-old Kjeldahl technique or by nitrogen (N)/protein analyzer. The former involves acid digestion, distillation, and titration to determine the nitrogen content. In the latter, the nitrogen content is determined through combustion at high temperature and detection through thermal conductivity (Horneck and Miller 1998). The percent N has also been determined using mass spectrometry (Dakora and Belane 2019) or by dry oxidation (Dumas) method (Gerrano et al. 2017). In all these methods, a factor of 6.25 is typically used to calculate the crude protein content from the N content of legumes, although much lower factors, ranging from 5.32 to 6.03, have also been suggested (Sosulski and Holt 1980; Fujihara et al. 2010).

(b) **Elemental analysis of Fe and Zn**: Spectroscopic methods such as inductively coupled plasma optical emission spectroscopy (ICP-OES) and atomic absorption spectroscopy (AAS) are well established and provide accurate and sensitive results for a range of elements with analytical detection limit ranging from percent to ppm levels. The principle behind both of these methods is based on the signature spectral absorption/emission of individual elements. AAS involves volatilization of sample by passing through a flame at more than 2000 °C and studying the absorption spectra, while in ICP-OES, constituent atoms are excited at temperatures of up to 10,000 °C and by studying their emission spectra. ICP-OES has been the "gold standard" for micronutrient analysis due to the high accuracy, wide analytical detection range, capability to detect soil contamination, and expansive elemental analysis. However, it is expensive (such as equipment, high-purity reagents required, and consumables), contamination prone, and time-consuming (pre-analysis preparation). AAS is less expensive (both instrument outlay costs and analysis costs), requires greater volumes of digested plant material (compared to ICP-OES), and is generally limited to single element analysis per run (Guild et al. 2017a). ICP coupled with mass spectrometry has also been reported for elemental analysis in cowpea (Dakora and Belane 2019).

For high-throughput qualitative and quantitative elemental screening, spectrometry based on X-ray fluorescence (XRF) is also demonstrated that has proven to be cost- and time-efficient in a wide range of crops including cowpea (Guild et al. 2017b). The XRF technology is less sensitive although it is nondestructive, requires no dissolution (minimal pre-analysis preparation), and has good precision for major elements (Wobrauschek et al. 2010) making it appropriate for the analysis of large samples for multiple elements simultaneously. XRF is based on the principle of elemental excitation using X-rays and the study of secondary "fluorescent" X-ray emission during de-excitation that is characteristic and abundance of the element analyzed. Samples can be screened in either whole grain that reduces sample processing time with reduced contamination risk or flour form that improves the reproducibility and accuracy but increases likelihood of contamination and labor requirement. Therefore, it would be wise to screen a large number of samples with XRF, and later AAS or ICP-OES could be used to confirm nutrient content of the narrowed-down samples. In addition, ED (energy dispersive)-XRF analysis of cowpea indicated that when analyzing flour samples, the results were not significantly different to the reference ICP-MS analysis (average difference of ±1 mg kg^{-1} for both Fe and Zn), while whole grain analysis by XRF gave significant differences and hence is not feasible for screening grains larger than wheat (Guild et al. 2017b).

(c) **Colorimetric analysis of Fe and Zn**: An alternative to ICP and AAS analysis for elemental quantification, colorimetry is a staining technique based on color change caused by chelation of metal ion of interest with specific reagents. This technique has been shown to detect ppm levels of specific elements with added advantage of not requiring expensive equipment or pre-analysis digesting. Since the colorimetric reagent is element specific, this method is predominantly useful when screening for a particular element as in the case of biofortification trials focused on specific micronutrient (i.e., Fe or Zn). Consequently, staining techniques have been used widely to screen for genotypes with high levels of micronutrients. For Fe screening, Perls' Prussian blue (PPB) and 2,2′-dipyridyl stains have been reported, while Zn screening could be achieved by staining with dithizone (DTZ, diphenyl-thio-carbazone) and Zincon® (2-carboxy-2-hydroxy-5-sulfoformazyl benzene). The intensity of the colored chelate formed by the reaction of the stain with the metal ion determines the concentration of the metal (under optimized conditions). Consequently, it is even visually possible to differentiate nutrient-dense genotypes from those with low levels. The method has been further improved to enable semiquantitative analysis of micronutrient concentrations with the use of image analysis software such as Adobe Photoshop® and ImageJ as demonstrated by Choi et al. (2007) and Duarte et al. (2016), respectively. By using this combination of staining and image processing, it was possible to achieve results correlating color intensity with reference micronutrient analysis (ICP-OES) with $r^2 > 0.8$ for both Fe and Zn (Choi et al. 2007). This enables high-throughput screening even in basic laboratories sans costly analytical equipment.

(d) **Determination of anti-mineral compounds**: The anti-mineral compounds such as phytic acid and polyphenols have been analyzed using UV-Vis spectrophotometer through different methods. In cowpea, the polyphenols could be determined by modified Folin–Ciocalteu method (Singleton et al. 1999), while tannins could be estimated by Vanillin-HCl method as described in Dhanasekar and Reddy (2012). Phytic acid has been estimated following modified Wade's method (Dhanasekar and Reddy 2017). The phenolic compounds could also be studied both quantitatively and qualitatively using high-performance liquid chromatography (HPLC) (Moreira-Araujo et al. 2017).

Molecular Breeding Methods

Molecular markers play an important role in accelerating the pace of selection and therefore the breeding process. The utility of molecular markers depends on the availability of genomic resources in the crop or related crops. The use of molecular markers for genetic biofortification in legumes has been very limited in general and none in cowpea. Although a number of advances in cowpea genetic linkage maps and QTLs associated with some desirable traits such as resistance to *Striga*, *Macrophomina*, *Fusarium* wilt, bacterial blight, root-knot nematodes, aphids, and foliar thrips have been reported (Boukar et al. 2016). Linkage mapping provides a framework for downstream analyses including quantitative trait loci (QTL) identification, map-based cloning, diversity analysis, association mapping, and molecular breeding (Lucas et al. 2011). Now that linkage maps for cowpea with high marker density are available, there are increased opportunities for QTL resolution, map-based cloning, association mapping, and marker-assisted breeding. With the availability of improved consensus genetic linkage maps, physical maps, next-generation sequencing (NGS), and the recent publication of the whole genome sequence in cowpea (Lonardi et al. 2019), molecular markers can play a key role in the identification of QTLs and SNPs for various biofortification traits and also could be invariably put into use for marker assisted backcross selection (MABC) or marker-assisted recurrent selection (MARS). About 1100 SNPs mapped on the cowpea genome have been converted to Kompetitive allele-specific PCR (KASP) assays at IITA. Sources of cowpea genomic resources like physical maps, HarvEST:Cowpea, cowpea genomics knowledge space (CGKB), cowpea genomics initiative (CGI), microarray chip, SSR marker kit, consensus genetic linkage map, and software like "SNP selector," "KBioConverter," and "Backcross selector" have been shortlisted by Boukar et al. (2016). Several cowpea breeding programs have been exploiting these resources to implement molecular breeding, especially for MARS and MABC, to accelerate cowpea variety improvement. Molecular markers have been exploited in biofortification breeding in some of the related pulse crops. Several QTLs and/or SNP markers associated with Fe and/or Zn concentrations have been identified in peas (Ma et al. 2017; Gali et al. 2018), chickpeas (Upadhyaya et al. 2016), common beans (Blair et al. 2011), and lentils (Khazaei et al. 2017) that can be used in marker-assisted selection. The detailed discovery of a large number of QTLs for biofortification traits including Fe, Zn, selenium, carotenoids, and folates in different pulse crops is reviewed by Jha and Warkentin (2020).

In recent years, targeted gene-editing technologies using artificial nucleases, zinc finger nucleases (ZFNs), transcription activator-like effector nucleases (TALENs), and the clustered regularly interspaced short palindromic repeat (CRISPR)/CRISPR-associated protein 9 (Cas9) system (CRISPR/Cas9) have given rise to the possibility to precisely modify genes of interest and thus have potential application for crop improvement (Jaganathan et al. 2018). Recently, CRISPR/Cas9 and/or TALEN technologies have been used to generate mutant lines for genes involved in small RNA processing of *Glycine max* and *Medicago truncatula* (Curtin et al. 2018) and for disruption of symbiotic nitrogen fixation (SNF) gene activation in cowpea (Ji et al. 2019). These findings pave the way for applicability of gene-editing technologies for various traits of interest in legumes.

Future Outlook

Much progress has been made toward reaching micronutrient density targets for major primary staple food crops in Africa and Asia. Given the micronutrient malnutrition and hidden hunger among the masses in the developing countries fueled by the poor economy and low purchasing power of farmers involved in subsistence farming in these countries, it becomes highly imperative that genetic biofortification should be included as an inseparable component of national food security missions. By developing more than 150 biofortified varieties that have been released in 30 countries and being consumed by more than 20 million people in developing countries, HarvestPlus and its partners have developed strong evidence that biofortification intervention can help alleviate malnutrition.

With higher incidences of diabetes, heart problems, and cancers in the developing and developed countries, the use of cowpea with high protein content, high fiber, low glycemic index, and high levels of cancer-fighting antioxidants would become popular. Little efforts have gone into breeding for higher protein and other quality traits. However, recent screenings of cowpea germplasm have shown great variability for protein content and many health-promoting factors. Therefore, there is a need to strengthen breeding efforts to develop cowpea varieties with higher protein and minerals as well as health-promoting factors. The focus on increasing the concentration of micronutrients should go hand in hand with increasing the bioavailability of micronutrients. This can be achieved by enhancing the promoters that stimulate the absorption of minerals and by reducing the concentrations of antinutrients that interfere with absorption. A beginning has been made under the HarvestPlus Biofortification Project wherein the national partner GB Pant University of Agriculture and Technology, Uttarakhand, has successfully released four cowpea varieties with high protein (25–31%), Fe (66 to 109 ppm), and Zn contents (36 to 51 ppm) (Singh 2016), but such efforts should be further concerted and strengthened. New cowpea varieties have fairly high protein content ranging from 27% to 31%, but the cowpea protein, as in other food legumes, is deficient in sulfur-containing amino acids like methionine and cysteine. Conventional breeding for such traits having limited or no genetic variability is not tenable and can be improved through biofortification by genetic engineering. Such efforts should be diligently and unscrupulously encouraged and supported by legal federal policies to pave way for a new era of fortified and safe crop varieties. The cowpea breeders should work closely with biotechnologists to quickly transfer these traits to popularly cultivated varieties in different regions. In addition to improving cowpea varieties through genetic transformation, efforts should be made to develop markers and marker-assisted selection for accelerated genotyping. The possibility of utilizing improved genome editing tools like CRISPR for precise modification within the genome so as to target specific genes of biofortification traits should leverage rapid development of biofortified cowpea varieties. Encouraging success stories of CRISPR from other crops in tweaking the expression of genes by editing the regulatory elements of Fe homeostasis genes should help give a leeway in cowpea biofortification. In a developing country like India, where maximum people do not have sufficient access to commercially fortified foods, diversified diets, and food supplements, biofortification is an acceptable cost-effective way to eliminate malnutrition.

Looking ahead, key investments will help biofortification reach its full potential. Firstly, biofortification traits must be streamlined into conventional breeding programs of secondary staples like cowpea. High micronutrient content must be included as a core trait of breeding programs, by concatenation of micronutrient-dense parental lines. Secondly, investments in high-throughput technologies and development of molecular markers linked with biofortification traits can greatly accelerate genetic gain for these traits. Finally, more investment should be made by the government and private sectors to create awareness among the farmers and consumers to go in for the biofortified crops and products so that micronutrients get bio-concentrated in the human food chain. Given the growing national and international interest for pursuing biofortification as a new, complementary intervention to address micronutrient deficiency, it is hoped that a wider array of partners and national agricultural research systems synergize in developing the next generation of biofortified crops.

Conclusion

Micronutrients are inevitable components of both human and plant nutrition as they are essential for normal growth and development. Micronutrient deficiency leading to malnutrition is a major concern that affects one third of the world population. Among various strategies, genetic biofortification through plant breeding is considered the most viable, economical, and sustainable approach to tackle micronutrient deficiencies. This universally acclaimed potential approach can reach people living in relatively remote rural areas that have limited access to commercially marketed fortified foods and supplements. With biofortification breeding programs of primary staples attaining the intended micronutrient level targets, it is high time that similar results are replicated in secondary staples like cowpea that complement the primary staple-based diets. Moreover, nutritious crops like cowpea that are widely grown by resource-poor farmers doing sustenance farming are one of the good options for biofortification. In recent years, significant progress has been made with the release of several biofortified crop varieties that are helping to overcome micronutrient deficiencies in the target populations. Improving the nutritional profile of pulse crops like cowpeas that are an important source of protein and energy can significantly increase their consumption. Biofortification to improve the nutritional profile of pulse crops including cowpea has gained momentum in the recent past. However, there are several confrontations and challenges that require to be tackled if the consumption and cultivation of biofortified foods and crops, respectively, are to be maximized effectively.

References

Afiukwa CA, Ubi B, Kunert K et al (2013) Seed protein content variation in cowpea genotypes. World J Agr Sci 1:94–99

Alloway BJ (2009) Soil factors associated with zinc deficiency in crops and humans. Environ Geochem Health 31(5):537–548

Andersson MS, Saltzman A, Virk PS et al (2017) PROGRESS UPDATE: crop development of biofortified staple food crops under harvestplus. Afr J Food Agric Nutr Dev 17(2):11905–11935. https://doi.org/10.18697/ajfand.78.HarvestPlus05

Ariza-Nieto M, Blair M, Welch R et al (2007) Screening of iron bioavailability patterns in eight bean (*Phaseolus vulgaris* L.) genotypes using the Caco-2 cell *in vitro* model. J Agric Food Chem 55:7950–7956

Asante IK, Adu-Dapaah H, Addison P (2006) Seed weight and protein and tannin contents of 32 cowpea accessions in Ghana. Trop Sci 44:77–79

Baudoin JP, Maréchal R (2011) Genetic diversity in *Vigna*. In: Singh SR, Rachie KO (eds) Cowpea research, production and utilization. Wiley, Chichester

Belane A, Dakora F (2011) Levels of nutritionally-important trace elements and macronutrients in edible leaves and grain of 27 nodulated cowpea (*Vigna unguiculata* L. Walp.) genotypes grown in the upper west region of Ghana. Food Chem 125:99–105. https://doi.org/10.1016/j.foodchem.2010.08.044

Blair MW, Astudillo C, Rengifo J et al (2011) QTL for seed iron and zinc concentrations in a recombinant inbred line population of Andean common beans (*Phaseolus vulgaris* L.). Theor Appl Genet 122:511–521

Boukar O, Massawe F, Muranaka S et al (2011) Evaluation of cowpea germplasm lines for protein and mineral concentrations in grains. Plant Genet Resour 9:515–522. https://doi.org/10.1017/s1479262111000815

Boukar O, Fatokun CA, Huynh B-L et al (2016) Genomic tools in cowpea breeding programs: status and perspectives. Front Plant Sci 7:757. https://doi.org/10.3389/fpls.2016

Boukar O, Belko N, Chamarthi S et al (2018) Cowpea (*Vigna unguiculata*): genetics, genomics and breeding. Plant Breed 138:415–424. https://doi.org/10.1111/pbr.12589

Briat JF (2011) Iron nutrition and implications for biomass production and the nutritional quality of plant products. The molecular and physiological basis of nutrient use efficiency in crops. Wiley-Blackwell, Somerset, pp 311–334

Carsky RJ, Singh BB, Oyewole B (2001) Contribution of early season cowpea to late season maize in the savanna zone of West Africa. Biol Agric Hortic 18:303–316

Centers for Disease Control and Prevention (1998) Recommendations to prevent and control iron deficiency in the United States. Morbidity and mortality weekly report. Recommendations and reports. 47 (RR-3): 1–36

Chivenge P, Mabhaudhi T, Modi AT, Mafongoya P (2017) The potential role of neglected and underutilised crop species as future crops under water scarce conditions in sub-Saharan Africa. Int J Environ Res Public Health 12(6):5685–5711. https://doi.org/10.3390/ijerph120605685

Choi EY, Graham R, Stangoulis J (2007) Semi-quantitative analysis for selecting Fe- and Zn-dense genotypes of staple food crops. J Food Sci 20:496–505

Cominelli E, Confalonieri M, Carlessi M et al (2018) Phytic acid transport in Phaseolus vulgaris: a new low phytic acid mutant in the PvMRP1 gene and study of the PvMRPs promoters in two different plant systems. Plant Sci 270:1–12

Connorton JM, Balk J, Rodríguez-Celma J (2017) Iron homeostasis in plants – a brief overview. Metallomics Integr Biometal Sci 9(7):813–823. https://doi.org/10.1039/c7mt00136c

Craufurd PQ, Summerfield RJ, Ell RH, Roberts EH (2010) Photoperiod, temperature and the growth and development of cowpea (*Vigna unguiculata*). In: Singh BB et al (eds) Advances in Cowpea Research. Copublication Intl Inst Tropical Agric (IITA) and Japan Intl Res Center Agric Sci (JIRCAS). Sayce, Devon

Curtin SJ, Xiong Y, Michno JM et al (2018) CRISPR/Cas9 and TALENs generate heritable mutations for genes involved in small RNA processing of *Glycine max* and *Medicago truncatula*. Plant Biotechnol J 16:1125–1137

Dakora FD, Belane AK (2019) Evaluation of protein and micronutrient levels in edible cowpea (*Vigna Unguiculata* L. Walp.) leaves and seeds. Front Sustain Food Syst 3:70. https://doi.org/10.3389/fsufs.2019.00070

Dhanasekar P, Reddy KS (2012) Variation for seed protein and tannin content in elite lines of cowpea, in: abstracts of DAE-BRNS life sciences symposium on trends in plant, agriculture and food sciences, Dec 17–19, Mumbai, p 94

Dhanasekar P, Reddy KS (2017) Role of seed Raffinose family oligosaccharides and Phytic acid in better performance potential of cowpea under water stress conditions. J Bas Appl Pl Sci 1(1):106

dos Reis Moreira-Araújo RS, Sampaio GR et al (2017) Identification and quantification of antioxidant compounds in cowpea. Revista Ciência Agronômica 48(5spe):799–805. https://doi.org/10.5935/1806-6690.20170093

Duarte RF, Prom-u-thai C, Amaral DC et al (2016) Determination of zinc in rice grains using DTZ staining and image J software. J Cereal Sci 68:53–58

Fatokun CA, Boukar O, Kamara A et al (2012) Enhancing cowpea productivity and production in drought- prone areas of sub-Saharan Africa. In: Abate T (ed) Four seasons of learning and engaging small holder farmers: Progress of Phase1. International crops research Institute for the Semi-Arid Tropics, Nairobi, pp 81–112

Fujihara S, Sasaki H, Sugahara T (2010) Nitrogen-to-protein conversion factors for some pulses and soybean products. J Sci Food Agric 21:60–66

Gali KK, Liu Y, Sindhu A et al (2018) Construction of high-density linkage maps for mapping quantitative trait loci for multiple traits in field pea (*Pisum sativum* L.). BMC Plant Biol 18:172

Garinu GO, Ingrao G (1991) Gross composition, amino acid, phytic acid and trace element contents of thirteen cowpea cultivars and their nutritional significance. J Sci Food Agric 55(3):401–410

Gerrano AS, van Rensburg WSJ, Adebola PO (2017) Nutritional composition of immature pods in selected cowpea [*Vigna unguiculata* (L.) Walp.] genotypes in South Africa. Aust J Crop Sci 11(02):134–141

Ghosh A, Reja MH, Nalia A (2019) Micronutrient biofortification in pulses: an agricultural approach. Curr J Appl Sci Technol 35(4):1–12

Gomez-Becerra HF, Erdem H, Yazici A et al (2010) Grain concentrations of protein and mineral nutrients in a large collection of spelt wheat grown under different environments. J Cereal Sci 52:342–349

Gould J (2017) Nutrition: a world of insecurity. Nature 544:S7. https://doi.org/10.1038/544S6a

Guild G, Parkes E, Nutti M et al (2017a) High-throughput measurement methodologies for developing nutrient-dense crops. Afr J Food Agric Nutr Dev 17(2):11905–11935. https://doi.org/10.18697/ajfand.78.HarvestPlus05

Guild GE, Paltridge NG, Andersson MS et al (2017b) An energy-dispersive X-ray fluorescence method for analyzing Fe and Zn in common bean, maize and cowpea biofortification programs. Plant Soil 419:457–466

Gupta P, Singh R, Malhotra S et al (2010) Characterization of seed storage proteins in high protein genotypes of cowpea [*Vigna unguiculata* (L.) Walp.]. Physiol Mol Biol Plan Theory 16:53–58

Hodge J (2016) Hidden hunger: approaches to tackling micronutrient deficiencies. In: Gillespie S, Hodge J, Yosef S, Pandya-Lorch R (eds) Nourishing millions: stories of change in nutrition. International Food Policy Research Institute (IFPRI), Washington, DC, pp 35–43

Horneck D, Miller R (1998) Determination of total nitrogen in plant tissue. In: Kalra YP (ed) Handbook of reference methods for plant analysis. CRC Press, New York, pp 75–83

Hossain F, Sarika K, Muthusamy V et al (2019) Quality protein maize for nutritional security. In: Qureshi A, Dar Z, Wani S (eds) Quality breeding in field crops. Springer, Cham

Itatat SL, Nworgu EC, Ikpe EN et al (2013) Evaluation of the protein contents of selected cowpea (*Vigna unguiculata* (L.) Walp.) varieties for production in Port Harcourt. Acta Agronomica Nigeriana 13:71–75

Jaganathan D, Ramasamy K, Sellamuthu G et al (2018) CRISPR for crop improvement: an update review. Front Plant Sci 9:985

Jha AB, Warkentin TD (2020) Biofortification of pulse crops: status and future perspectives. Plan Theory 9(1):73. https://doi.org/10.3390/plants9010073

Ji J, Zhang C, Sun Z et al (2019) Genome editing in cowpea *Vigna unguiculata* using CRISPR-Cas9. Int J Mol Sci 20:2471

Khazaei H, Podder R, Caron CT et al (2017) Marker-trait association analysis of iron and zinc concentration in lentil (*Lens culinaris* Medik.) seeds. Plant Genome 10:2

Kolawale GO, Tian G, Singh BB (2000) Differential response of cowpea varieties to aluminum and phosphorus application. J Plant Nutr 23:731–740

Li G, Kronzucker HJ, Shi W (2016) The response of the root apex in plant adaptation to Iron heterogeneity in soil. Front Plant Sci 7:344. https://doi.org/10.3389/fpls.2016.00344

Liener LE, Kakade ML (1980) Protease inhibitors. In: The lectins: properties, functions, and applications in biology and medicine. Academic, New York, pp 527–552

Lonardi S, Muñoz-Amatriaín M, Liang Q et al (2019) The genome of cowpea (*Vigna unguiculata* [L.] Walp.). Plant J 98(5):767–782. https://doi.org/10.1111/tpj.14349

Lopez AB, Van Eck J, Conlin BJ et al (2008) Effect of the cauliflower or transgene on carotenoid accumulation and chromoplast formation in transgenic potato tubers. J Exp Bot 59(2):213–223. https://doi.org/10.1093/jxb/erm299

Lucas MR, Diop NN, Wanamaker S et al (2011) Cowpea-soybean synteny clarified through an improved genetic map. Plant Genome 4:218–224. https://doi.org/10.3835/plantgenome2011.06.0019

Ma Y, Coyne CJ, Grusak MA et al (2017) Genome-wide SNP identification, linkage map construction and QTL mapping for seed mineral concentrations and contents in pea (*Pisum sativum* L.). BMC Plant Biol 17:43

Maqbool MA, Beshir A (2019) Zinc biofortification of maize (*Zea mays* L.): status and challenges. Plant Breed 138:1–28. https://doi.org/10.1111/pbr.12658

Marappa N, Manoj CA, Savithramma DL et al (2016) Screening of elite cowpea genotypes for zinc, iron, protein and yield attributes. J Proteomic Bioinform 9(12 (Suppl)). https://doi.org/10.4172/0974-276X.C1.094

McGuire S. FAO, IFAD, and WFP (2015) The state of food insecurity in the world 2015: meeting the 2015 international hunger targets: taking stock of uneven progress. FAO, Rome. Adv Nutr (2015) 6(5):623–624. https://doi.org/10.3945/an.115.009936

Muranaka S, Shono M, Myoda T et al (2016) Genetic diversity of physical, nutritional and functional properties of cowpea grain and relationships among the traits. Plant Genet Res Charact Util 14(1):67–76

Neeraja CN, Ravindra BV, Ram S et al (2017) Biofortification in cereals – Progress and prospects. Curr Sci 113:1050–1057

Nielsen SS, Brandt WE, Singh BB (1993) Genetic variability for nutritional composition and cooking time of improved cowpea lines. Crop Sci 33:469–472

NITI Aayog, Government of India (2017) Nourishing India, National Nutrition Strategy. http://niti.gov.in/writereaddata/files/document_publication/Nutrition_Strategy_Booklet.pdf

O'Dell BL, de Boland AR, Koirtyohann ST (1972) Distribution of phytate and nutritionally important elements among the morphological components of cereal grains. J Agric Food Chem 20:718–721

Oke D, Tewe O, Fetuga B (2015) The nutrient composition of some cowpea varieties. Niger J Anim Prod 22:32–36

Otegui M, Capp R, Staehelin L (2002) Developing seeds of *Arabidopsis* store different minerals in two types of vacuoles and in the endoplasmic reticulum. Plant Cell 14:1311–1327

Pfeiffer WH, McClafferty B (2007) HarvestPlus: breeding crops for better nutrition. Crop Sci 47(Suppl. 3):S88–S105

Raboy V (2003) *Myo*-inositol-1,2,3,4,5,6-hexakisphosphate. Phytochemistry 64:1033–1043

Ravelombola W, Shi A, Weng Y et al (2016) Evaluation of total seed protein content in eleven Arkansas cowpea (*Vigna unguiculata* (L.) Walp.) lines. AJPS 7:2288–2296

Saltzman A, Birol E, Bouis HE et al (2013) Biofortification: Progress toward a more nourishing future. Glob Food Sec 2(1):9–17

Santos CAF, Boiteux LS (2013) Breeding biofortified cowpea lines for semi-arid tropical areas by combining higher seed protein and mineral levels. Genet Mol Res 12(4):6782–6789

Shahzad Z, Rouached H, Rakha A (2014) Combating mineral malnutrition through iron and zinc biofortification of cereals. Compr Rev Food Sci Food Saf 13(3):329–346

Shukkur MF, Abdul SE, Karthik HS et al (2006) Oxalate mediated nephronal impairment and its inhibition by c-phycocyanin: a study on urolithic rats. Mol Cell Biochem 284(1–2):95–101

Sinclair SA, Krämer U (2012) The zinc homeostasis network of land plants. Biochim Biophys Acta 1823(9):1553–1567. https://doi.org/10.1016/j.bbamcr.2012.05.016

Singh BB (2005) Cowpea (*Vigna unguiculata* (L.) Walp.). In: Singh RJ (ed) Genetic resources, chromosome engineering and crop improvement. CRC Press, Boca Raton, pp 117–162

Singh BB (2014) Cowpea: the food legume of the 21st century. Crop science Society of America. Madison, Wisconsin

Singh BB (2016) Genetic enhancement for yield and nutritional quality in cowpea [*Vigna unguiculata* (L.) Walp.]. Indian J Genet Plant Breed 76(4):568–582

Singleton VL, Orthofer R, Lamuela-Raventos RM (1999) Analysis of total phenols and other oxidation substrates and antioxidants by means of Folin-Ciocalteu reagent. In: Packer L (ed) Oxidants and antioxidants, Pt a. Academic, San Diego, pp 152–178

Sosulski FW, Holt NW (1980) Amino acid composition and nitrogen-to-protein factors for grain legumes. Can J Plant Sci 60:1327–1331

Sparvoli F, Cominelli E (2015) Seed biofortification and Phytic acid reduction: a conflict of interest for the plant? Plan Theory 4:728–755. https://doi.org/10.3390/plants4040728

Steele WM, Mehra KL (2009) Structure, evolution and adaptation to farming systems and environments in *Vigna*. In: Summerfield RJ, Bunting AH (eds) Advances in legume science. Royal Botanic Gardens, Kew, pp 393–404

Timko MP, Ehlers JD, Roberts PA (2007) Cowpea. Genome mapping and molecular breeding in plants. In: Kole C (ed) Pulses, sugar and tuber crops, vol 3. Springer, Berlin/Heidelberg, pp 49–67

Upadhyaya HD, Bajaj D, Das S et al (2016) Genetic dissection of seed-iron and zinc concentrations in chickpea. Sci Rep 6:24050

Verónica E-D, Márquez-Quiroz C, Cruz-Lázaro E et al (2018) Biofortification of cowpea beans (*Vigna unguiculata* L. Walp) with zinc: effect on yield and mineral content. Rev Mexicana cienc agric 9:4149–4160. https://doi.org/10.29312/remexca.v0i20.986

Von Wiren N, Mrschner H, Romheld V (1996) Roots of iron-efficient maize also absorb phytosiderophore-chelated zinc. Plant Physiol 111:1119–1125. https://doi.org/10.1104/p.111.4.1119

Welch RM (2002) Breeding strategies for biofortified staple plant foods to reduce micronutrient malnutrition globally. J Nutr 132:495S–499S

Welch RM, House AW, Beebe S et al (2000) Testing iron and zinc bioavailability in genetically enriched beans (*Phaseolus vulgaris* L.) and rice (*Oryza sativa* L.) in rat model. Food Nutr Bull 21:428–433

Weng Y et al (2019) Evaluation of seed protein content in USDA cowpea germplasm. HortScience 54(5):814–817. https://doi.org/10.21273/HORTSCI13929-19

Wessells KR, Brown KH (2012) Estimating the global prevalence of zinc deficiency: results based on zinc availability in national food supplies and the prevalence of stunting. PLoS One 7:e50568. https://doi.org/10.1371/journal.pone.0050568

WHO (2009) Global Health risks. Mortality and burden of disease attributable to selected major risks. Available at: http://www.who.int/healthinfo/global_burden_disease/GlobalHealthRisks_report_annex.pdf. Accessed 30 Oct 2013

Wobrauschek P, Streli C, Lindgren E (2010) Energy dispersive, X-Ray Fluorescence Analysis https://doi.org/10.1002/9780470027318.a6806.pub2

Breeding for Enhanced Nutrition in Common Bean

T. Basavaraja, Satheesh Naik S. J., Rahul Chandora, Mohar Singh, and N. P. Singh

Abstract Common bean (*Phaseolus vulgaris* L) is the most important food legume for direct human consumption, provides significant quantities of protein and energy, and is a source of vitamins and minerals including Fe and Zn. In addition to these nutritional components, common beans are rich in a variety of several phytochemicals with potential health benefits such as polyphenolic compounds, fiber, lectins, and trypsin inhibitors. Mineral deficiencies in human populations are one of the greatest health concerns given that half the current population of the world is affected by some sort of mineral deficiency. Thus, the major staples that have been targeted for mineral biofortification breeding at the international scale include mainly the seed crops of rice, wheat, maize, and common bean along with related cereals and legumes in certain more intensive national research programs that are part of the overall HarvestPlus biofortification program. Therefore, the scope of this chapter is to review the role of some bioactive compounds present in common beans, biochemistry of the biofortification traits, and their analytical methods. The main goals of mineral biofortification have been to increase the concentration of iron or zinc in certain major cereals and legumes. In humans, iron is essential for preventing anemia and for the proper functioning of many metabolic processes, whereas zinc is essential for adequate growth and for resistance to gastroenteric and respiratory infections, especially in children. This book chapter outlines the advantages and needs of mineral biofortification in common bean, starting with the steps of breeding for traits such as germplasm screening, inheritance, biochemical analytical methods, molecular approaches, and future challenges and finishing with product development in the form of new biofortified varieties.

Keywords *Phaseolus vulgaris* · Protein · Iron · Zinc · Phytic acid · Polyphenols · Genetic biofortification · Seed quality

---

T. Basavaraja (✉) · S. Naik S. J. · N. P. Singh
ICAR-Indian Institute of Pulses Research, Kanpur, India

R. Chandora · M. Singh
ICAR-NBPGR Regional Station – Shimla Phagli, Shimla, India

D. S. Gupta et al. (eds.), *Breeding for Enhanced Nutrition and Bio-Active Compounds in Food Legumes*, https://doi.org/10.1007/978-3-030-59215-8_8

Introduction

Common bean (*Phaseolus vulgaris* L.) is one of the profitable grain legume crops, third in importance after soybean and peanut but first in direct human consumption (Broughton et al. 2003). It originated in Latin America and has two primary centers of origin in the Mesoamerican and Andean regions that are easily distinguished by molecular means (Blair et al. 2006). Major bean-producing countries are Brazil and Mexico, while the United States, Canada, Argentina, and China are all exporting countries. It is vital for nutritional well-being as well as poverty alleviation among consumers and farmers with few other food or crop options. At global level, bean production is on small farms ranging from 1 to 10 ha in size. Multiple commercial seed types or horticultural classes exist based on seed color with white, yellow, cream, brown, pink, red, purple, black and mottled, pinto, or striped seed types popular in different regions of the world and with different cultures (Voysest et al. 1994; Schoonhovern and Voysest 1991). In food nutritional terms, beans are often called "poor man's meat" for their inexpensive source of protein and their rich content of minerals (especially iron and zinc) and vitamins (Beebe et al. 2000). In humans, iron is essential for preventing anemia and for the proper functioning of many metabolic processes, while zinc is essential for adequate growth and sexual maturation and for resistance to gastroenteric and respiratory infections, especially in children (Bouis 2003). Besides this, it is identified as rich in bioactive compounds such as phenolic acids, flavonoids, flavan-3-ols, condensed tannins, and anthocyanin, which are important for health. Recent scientific evidence supports the role of bioactive compounds in the prevention and treatment of major diseases such as cardiovascular diseases, stomach and prostate cancers and weight control and obesity because of their beneficial effect on health (Guzmán et al. 2002; Messina 2014; Mudryj et al. 2014).

Though historically this crop has been the main protein source in developing countries, its consumption has been declining during the last few decades as the population has adopted a western lifestyle. Its consumption has been undervalued in North America and the north of Europe (Messina 2014). The same trend is observed in the countries along the Mediterranean Sea, and this has led to an increased incidence of chronic diseases such as cancer, obesity, and cardiovascular diseases (Moreno-Franco et al. 2014). As per the global nutrition report, the problem of malnutrition remains severe: the world is not on track to achieve the targets it has set itself. Malnutrition in all its forms remains unacceptably high across all regions of the world. Despite reductions in stunting, 150.8 million children (22.2%) under five years of age were stunted while 38.3 million children under five years of age were overweight during 2017. There have been reductions in the number of children affected by stunting since 2000; overweight among children under five years of age has increased over time (Fig. 1). Moreover, the mineral concentration in major food crops has been decreased with "green revolution" varieties, where higher productivity has been achieved that has diluted some mineral constituents to a certain extent. Therefore, it is a global challenge to breed for higher mineral concentration staples combined with enhanced productivity.

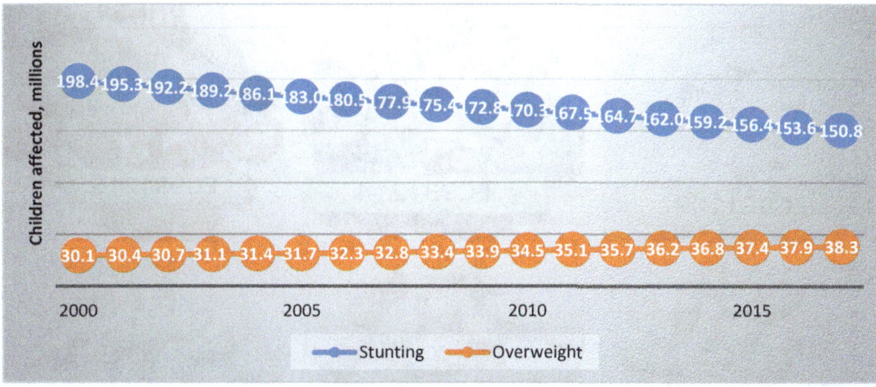

Fig. 1 Number of children affected by stunting and overweight, 2000–2017. (Source: UNICEF/World Health Organization (WHO)/World Bank Group: Joint child malnutrition estimates, https://data.unicef.org/topic/nutrition/malnutrition/)

As indicated above, common bean is a valuable source of protein, minerals, and vitamins. In terms of biofortification, enhancement of mineral content including bioactive compounds is precisely beneficial. Hence, HarvestPlus is part of the CGIAR Research Program on Agriculture for Nutrition and Health (A4NH), and its partners are developing and promoting biofortified crops (rice, wheat, common bean, pearl millet, cassava, maize, etc.) rich in vitamins and minerals (like iron and zinc) needed for good health. Accordingly, it has made the baseline for bean iron content high at 55 ppm, and it is having a challenging target on the enrichment of bean iron content, that is, an addition of approximately 44 ppm to baseline iron levels in common bean (Blair et al. 2013; Howarth 2014). However, biofortification research is instrumental in the international scientific community in developing more than 290 varieties of 12 biofortified crops, benefiting about 33 million people, in 14 countries across Africa, Asia, Latin America, and the Caribbean (Howarth 2014). Despite this progress, according to the Food and Agriculture Organization of the United Nations (FAO), the prevalence of undernourishment in the global population has decreased from 13.1% in 2007 to 10.9% in 2017, but one in nine people in the world still suffers from hunger (FAO 2018). Thus, biofortification is a global effort to develop and scale up micronutrient-rich staple food crops, including common bean. This phenomenon reduces the widespread gap between micronutrient requirements and intake by increasing the proportion of dietary vitamin A, iron, zinc, and essential bioactive compounds for public health significance globally. Therefore, efforts were underway to scale up biofortification in common bean through rigorous research, and interdisciplinary approaches could be used to counteract malnutrition by either enriching the concentration and/or bioavailability of iron in beans through agronomic, conventional plant breeding and by employing genetic engineering techniques (Fig. 2).

Fig. 2 Seed color variation in common bean genotypes

Priority Traits for Genetic Biofortification

Biofortification is considered as a cost-effective, sustainable solution that uses conventional plant breeding to increase the density of vitamins, minerals, and bioactive compounds in staple crop like common bean (Dwivedi et al. 2012). In terms of nutritional quality, common beans have large amounts of minerals (Fe, Zn) and vitamins accumulated in the seeds than in cereals (Broughton et al. 2003). It is estimated to have 4–10 times the amount of Fe and 2–3 times the amount of Zn compared to rice (Pfeiffer and McClafferty 2007). Thus, biofortification of beans is a globally accepted strategy to address micronutrient malnutrition in nutritionally insecure groups. It is an important source of minerals, especially iron, with concentrations ranging from 8.90 to 161.50 mg kg^{-1} dry matter (DM), and zinc, with concentrations ranging from 19.00 to 65.50 mg kg^{-1} DM in common bean cultivars (Talukder et al. 2010; Tryphone and Nchimbi-Msolla 2010; Silva et al. 2012). Accordingly, this crop was considered in HarvestPlus program to develop and scale up micronutrient-rich staple food crops at global level by providing micronutrients to malnourished rural populations in developing countries. In contrast to other

Table 1 Biofortified crops and their targeted micronutrients (Fe, Zn, and vitamin A) in HarvestPlus program

	Iron (parts per million, ppm)			Zinc (parts per million, ppm)			Vitamin A (parts per million, ppm)		
Biofortified crops	Base line	Target increment	Target level in crop	Base line	Target increment	Target level in crop	Base line	Target increment	Target level in crop
Rice				16	12	28			
Wheat				25	12	37			
Pearl millet	47	30	77						
Maize							0	15	15
Commnon bean	50	44	94						
Cassava							0	15	15
Lentils	40	30	70						
Sweet potato							2	30	32
Cowpea	30	33	63						
Sorghum	30	30	60	20	12	32			

Source: HarvestPlus breeding program

crops, the target Fe level of whole bean seeds is 94 ppm in HarvestPlus international research program supporting the research and development of biofortified crops (Table 1) (Blair and Izquierdo 2012; Petry et al. 2015). Like other food legumes, it contains a greater amount of essential amino acids, including lysine, which is deficient in most cereals. In addition, it is considered as highly functional food since it has great content of bioactive substances such as polyphenols, flavonoids, and anthocyanins, which contribute in a synergistic way with their therapeutic properties and may have a positive effect against some pathologies. They also serve as an excellent source of natural antioxidants for disease prevention and health promotion (Kumar and Baojun 2017; Puertas-Mejía et al. 2016).

Further, the possibility of adding nutritional value to the high agronomic performance of bean cultivars is a recent trend in breeding programs. Genetic biofortification of food crops is an important strategy to combat deficiencies of iron, zinc, and vitamins A and E in humans (Prasad 2012). Over recent decades, micronutrient deficiencies have increased due to a generalized decrease in the quality of poor people's diets in both developed and developing countries and even in areas where food is not limiting. Furthermore, micronutrient deficiencies are more widespread than deficiencies caused by inadequate consumption of bioactive substance-rich foods (Bouis 2003; Welch and Graham 1999). Therefore, reorientation of our breeding program is foremost and prioritized nutritional traits should be focused in common bean biofortification breeding program to combat micronutrient deficiencies worldwide. To develop common bean cultivars with iron and zinc concentrations in seeds that meet the specific dietary needs of bean consumers, it is necessary to study the genetic parameters of these minerals (Fig. 3).

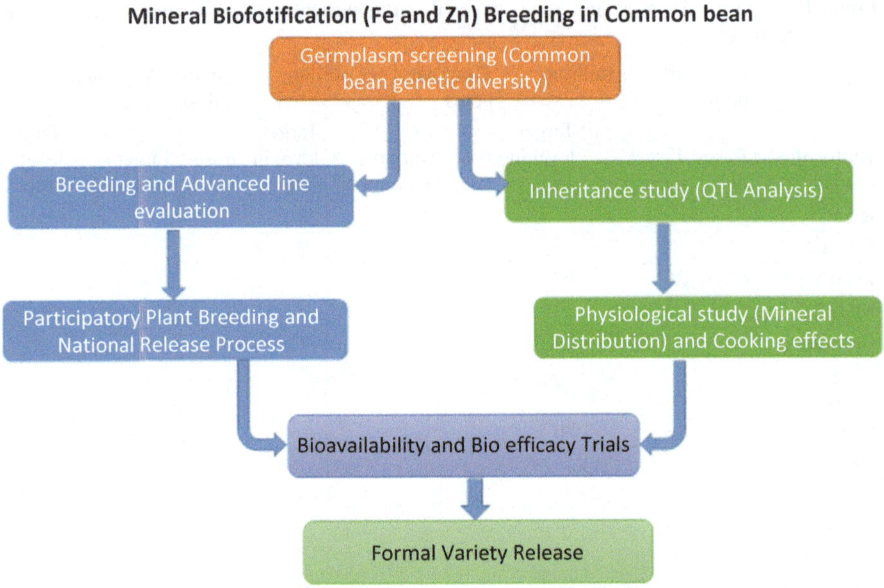

Fig. 3 Steps in the nutritional breeding pathway of common bean, including germplasm screening (common bean genetic diversity), inheritance study (quantitative trait locus [QTL] analysis for iron and zinc), physiological study (seed iron and zinc distribution) and cooking effects, breeding and advanced line evaluation, participatory plant breeding and multilocational testing, bioavailability and bioefficacy tests, and variety release (Matthew 2013)

Nutritional Value and Bioactive Compounds of Bean

Recently, common bean is gaining increasing attention as a functional or nutraceutical food, due to its high accumulation of micronutrients (Fe and Zn) in bean seed and availability of variety of phytochemicals such as fiber, polyphenolic compounds, lectins, unsaturated fatty acids, trypsin inhibitors, and phytic acid in seed. The bean cultivars may have different skin and coat colors (white, yellow, black, dark brown, red, green, and bluish gray, among others). It discerns quantitative variation in nutritional compositions of common beans such as protein, minerals, vitamins, and dietary fiber (Table 2). The presence of antioxidant compounds in foods consumed every day is extremely important for the health of human beings. Its importance consists in providing an increasing body defense mechanism against daily physical exhaustion and exhaustion originated from the resistance workout. They are also capable of reducing oxidative damage, which is believed to cause many diseases such as cancer, atherosclerosis, cardiovascular diseases, diabetes, cataracts, arthritis, and diseases related to immunodeficiency and aging (Siddhuraju 2006).

Table 2 Nutritional compositions of common beans in 100 g of edible portion (Kumar and Baojun 2017)

Nutrient	Units	Navy beans	Kidney beans	Red beans	Black beans	Pinto beans	Cranberry beans
Energy	kcal	92	92	167	464	500	257
Protein	g	6.15	5.38	22.22	14.29	10.71	22.86
Total lipid (fat)	g	0.00	0.38	0.00	21.43	21.43	0.00
Carbohydrate	g	16.15	20.77	63.89	57.14	60.71	60.00
Total dietary fiber	g	4.6	5.4	44.4	14.3	10.7	25.7
Total sugars	g	0.00	0.77	2.78	3.57	3.57	2.86
Resistant starch	g	4.2	2.0	3.8	1.7	1.9	2.6
		Minerals					
Calcium	mg	62	46	167	143	71	114
Iron	mg	1.38	1.38	7.29	3.86	2.57	5.1
Potassium	mg	300	268	222	279	96	265
Magnesium	mg	48	37	44	60	43	39
Sodium	mg	108	8	69	286	286	10
		Vitamins					
Vitamin C	mg	0.9	0.9	0.0	0.0	0.0	0.0
Folate	μg	127	115	140	128	147	124
		Lipids					
Total saturated	g	0.000	0.000	0.000	1.790	1.790	0.000
Total monounsaturated fatty acids	g	0.000	0.000	0.000	14.290	14.290	0.000
Total polyunsaturated fatty acids	g	0.000	0.000	0.000	5.360	5.360	0.000
Polyphenol	mg of gallic acid equiv./g	12.47	14.14	13.68	12.60	12.52	11.73
Flavonoids	mg of rutin equiv./g	1.78	2.59	1.55	1.28	0.98	1.65

Micronutrients (Fe and Zn)

Bean seed is a rich source of protein and minerals such as iron (Fe) and zinc (Zn) (Karen et al. 2019). Accordingly, several reports noticed that cultivars of common beans show variability for seed mineral accumulation with iron concentrations ranging from 30 to 120 ppm (Graham et al. 1999; Beebe et al. 2000; Guzman-Maldonado et al. 2003, 2004; Moraghan et al. 2002; Islam et al. 2002) and zinc concentrations ranging from 20 to 60 ppm (Moraghan and Grafton 1999; Welch et al. 2000; House et al. 2002; Hacisalihoglu et al. 2004). The range of mineral accumulation in the two gene pools of common bean (Andean and Mesoamerican) is similar, although many Andean beans or inter-gene pool hybrids have higher iron concentration than Mesoamerican beans (Islam et al. 2002). Recent studies indicated that the inheritance of iron concentration in common bean seeds meanwhile has been suggested to be quantitative in a population derived from a wild 9 cultivated cross (Guzman-Maldonado et al. 2003), while the inheritance of seed zinc concentration has been suggested to be simple although this was studied in a few specific genetic backgrounds (Forster et al. 2002; Cichy et al. 2005). Similarly, genetic studies have revealed seed Zn differences to be controlled by a single gene in two closely related navy bean genotypes, Albion and Voyager (Carolina et al. 2015). Another aspect of studies was conducted to identify potential breeding parents or analyze genetic control of mineral contents using a small number of trials. Because these traits are most likely quantitative, it is expected that the environment would have a substantial effect, and the genotype-environment interaction should be important in determining Fe and Zn content of seed. The genotype-environment interaction has been reported to have a notable effect on both Fe and Zn contents (Cichy et al. 2009; Tryphone and Msolla 2010).

Polyphenols

Studies have demonstrated that phenolic compounds are predominantly located in the seed coat of the bean than in the cotyledon and testa (López et al. 2013). The estimated content of the phenolic compound is about 145 mg/g and represents about 11% of the whole seed (Cardador-Martinez et al. 2002). The phenolic compounds in the seeds are flavones, monomers, and oligomers of flavanols, flavanones, isoflavonoids, anthocyanins, chalcones, and dihydrochalcones (Akillioglu and Karakaya 2010; López et al. 2013). However, the phenolic acids and non-flavonoid phenolic compounds (hydroxybenzoic and hydroxycinnamic acid) are mainly found in cotyledons of the bean (Ranilla et al. 2007). The color of the seed coat is based on the presence of polyphenols including anthocyanins, flavonol glycosides, and condensed tannins. Dark-colored beans normally have the highest anthocyanin content (Lin et al. 2008). In addition, red-, black-, and pink-colored varieties confer color to the bean seed coat due to their anthocyanins. The colors of light yellow or pink spot of the seed coat are generally based on the presence of condensed tannins. The seed color of beans is determined by the presence of polyphenolic compounds. The main polyphenolic compounds are flavonoids such as flavonol glycosides, anthocyanins, and condensed tannins (proanthocyanidins), and the most widely distributed group in common bean is presented in Table 3. The phenolic compounds isolation and characterization were initiated in early 1960, and four anthocyanin pigments (delphinidin 3-glucoside, petunidin 3-glucoside, malvidin 3-glucoside, and 3,5-diglucosides) were extracted from the seed coat of black violet beans (Hayat et al. 2014). Later, anthocyanins, flavonols, and tannins from the different varieties of kidney beans were isolated and characterized by many researchers. Studies have further demonstrated that wild and weedy Mexican beans are rich in anthocyanins (delphinidin, delphinidin 3-glucoside, petunidin, petunidin 3-glucoside, cyanidin,

Table 3 List of polyphenols in the common bean seeds (Kumar and Baojun 2017)

Bean name	Polyphenol class	Polyphenol subclass	Compound name
Dark bean	Flavonoids	Anthocyanins	Cyanidin 3-O-glucoside, pelargonidin 3-O-glucoside, petunidin-3-O-β-glucopyranoside, malvidin 3-O-glucoside, delphinidin acetyl-glucoside, pelargonidin acetyl glucoside, pelargonidin 3-O-malonyl glucoside, petunidin feruloyl glucose
Wild and weedy Mexican bean, pinto and black beans	Flavonoids	Anthocyanins	Peonidin, pelargonidin, cyanidin
Dark bean, wild and weedy Mexican bean	Flavonoids	Anthocyanins	Delphinidin 3-O-glucoside
Alubia, black, cranberry, dark red kidney, great northern, light red kidney, navy, pink, pinto, and small red	Flavonoids	Anthocyanins	Petunidin 3-O-(6″-acetyl-glucoside)
Dark and kidney beans, Zolfino landraces	Flavonoids	Anthocyanins	Pelargonidin 3,5-O-diglucoside
Alubia, black, cranberry, dark red kidney, great northern, light red kidney, navy, pink, pinto, and small red	Flavonoids	Anthocyanins	Delphinidin 3-O-glucosyl-glucoside
Dark bean	Flavonoids	Flavanols	(+)-Catechin, (−)-epicatechin, (+)-gallocatechin, procyanidin dimer, (−)-epigallocatechin, procyanidin dimer B2, procyanidin dimer B3, procyanidin dimer B4, procyanidin trimer, procyanidin trimer EEC, naringenin 7-glucoside
Dark bean	Flavonoids	Flavanones	Naringenin, hesperetin, naringin, naringenin 7-O-rutinoside, naringenin 7-O-glucoside, naringenin-7-methyl ether 2, hesperetin 3′-O-glucuronide, hesperetin 7-O-glucuronide, hesperetin3′,7-O-diglucuronide, hesperetin 5,7-O-diglucuronide, hesperetin7-O-rutinoside
Dark bean	Flavonoids	Flavones	Apigenin, apigenin 7-O-glucoside

(continued)

Table 3 (continued)

Bean name	Polyphenol class	Polyphenol subclass	Compound name
Brazilian bean	Flavonoids	Flavones	Chrysin
Dark bean, Brazilian bean, Mexican bean	Flavonoids	Flavonols	Kaempferol
Dark bean, Brazilian bean, Mexican bean	Flavonoids	Flavonols	Quercetin
Pinto beans, Zolfino landraces	Flavonoids	Flavonols	Kaempferol 3-O-glucosylxylose
Alubia, black, cranberry, dark red kidney, great northern, light red kidney, navy, pink, pinto, and small red	Flavonoids	Flavonols	Kaempferol 3-O-xylosyl-glucoside
Pinto beans	Flavonoids	Flavonols	Kaempferol 3-O-acetyl-glucoside
Brazilian bean	Polyphenols	Polyphenols	Coumestrol
Dark bean, pinto and black beans, Mexican bean	Phenolic acids	Hydroxybenzoic acids	Protocatechuic acid
Dark bean	Phenolic acids	Hydroxybenzoic acids	Gallic acid
Mexican bean	Phenolic acids	Hydroxybenzoic acids	Vanillic acid
Dark bean, Mexican bean	Phenolic acids	Hydroxycinnamic acids	p-Coumaric acid
Pinto and black beans	Phenolic acids	Hydroxycinnamic acids	Caffeic acid
Dark bean, Mexican bean	Phenolic acids	Hydroxycinnamic acids	Ferulic acid
Dark bean	Phenolic acids	Hydroxycinnamic acids	Sinapic acid, ferulic acid 4-glucoside
Dark bean	Stilbenes	Stilbenes	Trans-resveratrol, resveratrol 3-O-glucoside

malvidin, malvidin 3-glucoside, pelargonidin, and peonidin), present in 62 Mexican wild-type varieties (Jun et al. 2016). Thus, based on the recent prominent research efforts in common bean, it was found that common beans possess enormous quantities of polyphenols and other metabolites, have antioxidant activities, have major health-promoting effects, and protect against various diseases including diabetes, CVD, cancer, and microbial infections.

Genetic Variability for the Target Traits

Common bean is native to Latin America and is one of the five cultivated species of the *Phaseolus* genus. It has two main genetic pools: Andean (large seeds) and Mesoamerican (small seeds). Andean and inter-gene pool hybrids have higher Fe concentrations compared to Mesoamerican ones, although the range of variation is similar (Blair 2013). In any breeding program, germplasm screening for a trait of interest is an important first step to genetic improvement. In the case of biofortification, nutritional breeding also starts with assembly of parental germplasm for crosses based on the evaluation of a large amount of genetic material. Accordingly, for beans, a core collection of 1400 genotypes was the starting point for screening of mineral traits. Ranges of 30–110 ppm iron and 25–60 ppm zinc were found in the germplasm analyzed, and the high-iron genotypes, G14519 and G21242, from the FAO germplasm treaty collection held at the International Center for Tropical Agriculture were selected to initiate crosses (Blair et al. 2013). Likewise, large germplasm collection of screenings for high-Fe genotypes conducted in local and wild varieties of *P. vulgaris* has reported up to 110 ppm Fe. However, early analyses on closely related species such as *P. coccineus* and *P. dumosus* have found up to 127 ppm Fe, indicating that wild relatives might be useful (Blair et al. 2013). Though considering that high-Fe wild genetic material showed poor agronomical performance (and introgression might not be straightforward in interspecific crosses), further screening of wild genotypes is promising. Moreover, wild beans accumulate more Fe in seed coats and less in cotyledons compared to domesticated genotypes, indicating that they can contribute differently for tissue-specific biofortification (Blair et al. 2013). This variability for iron or zinc content is slightly larger than what was found in the analysis of a more limited range of genotypes. In addition, screening of advanced lines within each of the gene pools has been important for identifying potential commercial-type parents, as many of the core collection of high-iron or high-zinc lines were of noncommercial seed types. The range of mineral accumulation in the two gene pools of common bean (Andean and Mesoamerican) is similar, although many Andean beans or inter-gene pool hybrids have higher iron concentration than Mesoamerican beans. Therefore, some breeders choose a control genotype for each gene pool that is a standard variety or breeding line that can be multiplied to a large quantity, ground, and used for standardizing measurements across sites and screenings. CAL96 (Andean breeding line) has been used for this standardization in some cases. DOR390 (Mesoamerican black-seeded breeding line) is an alternative for the other gene pool as it is a variety in some parts of Central America (Blair et al. 2013).

Although the average Fe concentration in beans is high, many people still suffer from iron-deficiency anemia (IDA) due to an insufficient level of bioavailable Fe in a monotonous cereal/bean-based diet without meat (Bouis 2007). For Fe biofortification purposes, the use of common bean is advantageous because the baseline grain Fe content is high at 55 ppm and variability for the trait is great (Petry et al. 2015), ranging up to 110 ppm, allowing initial biofortification attempts to start from already high Fe levels (Blair et al. 2012; Blair 2013). Another advantage of using common beans in biofortification programs is that seeds are consumed whole after boiling. Therefore, all major components of the common bean seed could be targets of biofortification: seed coat, cotyledons, and embryo (Blair et al. 2013).

In addition, zinc deficiency (ZD) is probably as widespread globally as IDA but is not as well documented due to less testing for this nutrient (Broughton et al. 2003). ZD is possibly the leading cause of child and infant stunting; impairs immunity, vitamin A use, and vitamin D function; and leads to decreased health, higher mortality, and greater prevalence of some parasitic diseases (Singh 1999). Therefore, there is an imperative to work on iron and zinc concentrations and bioavailability in grain crops and especially in the legumes where their concentrations are higher than in the cereals (Broughton et al. 2003). As earlier mentioned, common bean has naturally 4–10 times the amount of iron as milled rice and 2–3 times the zinc (Voysest et al. 1994). A major need is also promotion of the legumes and economic policies that favor legume production. This is because legumes often cost more than cereals, and therefore, their overall consumption is more limited compared to the consumption of starchy staples. Recent reviews support that several approaches were used to breed high-iron and high-zinc bean varieties. There are three common approaches to biofortification: agronomic, conventional plant breeding, and plant breeding using genetic engineering. Agronomic biofortification provides temporary micronutrient increases through fertilizers. This approach is useful to increase micronutrients that can be directly absorbed by the plant, such as zinc, but less so for micronutrients that are synthesized in the plant and cannot be absorbed directly (Lyons and Cakmak 2012). Agronomic biofortification is particularly useful for realizing the maximum expression of biofortified traits for crops produced in micronutrient-deficient soils. Conventional plant breeding involves identifying and developing parent lines with high vitamin or mineral levels and crossing and selecting the segregants over generations to produce plants with the desired nutrient and agronomic traits (Bouis 2003).

Breeding is concentrating on both gene pools of common bean, the large-seeded Andeans and the small-seeded Mesoamericans, and bush beans as well as climbing beans. Various breeding techniques or strategies have been used for the current biofortification breeding effort, including backcrossing, recurrent selection, and various permutations of gamete and pedigree selection. Secondary characteristics of phytate and tannin content have for the most part been screened only on advanced lines due to their more expensive assays. Bioavailability tests and bioefficacy trials have been undertaken on the best bet varieties.

Biochemistry of the Biofortification Traits and Their Analytical Methods

Deficiencies of micronutrients such as iron (Fe), zinc (Zn), thiamin, and folic acid in human beings cause anemia, tiredness, weakness, alteration of immune response, impaired brain function, and even premature death. To overcome these problems, biofortification (Bouis et al. 2011) has been developed to improve the mineral content in the edible parts of plants. It can be achieved through conventional plant breeding, or modern biotechnology, agronomic practices (White and Broadley 2005). Common beans exhibit remarkable genetic variability for minerals in case of iron level, ranging from 30 to 120 ppm (Graham et al. 1999; Beebe et al. 2000; Guzman-Maldonado et al. 2003, 2004; Moraghan et al. 2002; Islam et al. 2002), Zn level, which varies from 20 to 60 ppm (Moraghan and Grafton 1999; Welch et al. 2000; House et al. 2002; Hacisalihoglu et al. 2004), and other micronutrient concentration, which is the prerequisite requirement for biofortification (White and Broadley 2005). Concentrations of iron have strongly exhibited correlation with the wild accessions of beans, the Fe concentration is higher in wild genotypes (60 μg/g) than cultivated ones (55 μg/g), and Guzmán-Maldonado et al. (2000) also reported iron concentrations much higher in wild genotypes varying from 71 to 280 μg/g. Fe content in common bean could be increased by 60–80% while Zn content around 50% through biofortification because of high heritability of Fe and Zn contents observed in common bean (Blair et al. 2009; Beebe et al. 2000). The average concentration of both minerals in bean is higher as compared to other cereals, i.e., rice, wheat, and maize (Gregorio 2002; Ortiz-Monasterio et al. 2007; Banziger and Long 2000), and is usually almost completely retained through harvest and processing (Beebe et al. 2000; Blair et al. 2010b). The inheritance of seed mineral accumulation traits is mostly to be quantitative and oligogenic, which are influenced by the environment but more specifically by genotypes (Cichy et al. 2009; Blair et al. 2009) and also associated with different candidate genes for the uptake of minerals (Vreugdenhil et al. 2004), so that breeding for high iron and zinc should be successful (Beebe et al. 2000; Blair et al. 2009). Common bean crop has been targeted through different approaches of biofortification to enhance the humane nutritional status (Table 4).

Steckling et al. (2017) reported the positive correlation between phosphorus and potassium ($r = 0.575$), copper and iron ($r = 0.729$), iron and zinc ($r = 0.641$), and copper and phosphorus ($r = 0.533$). Moderate- (Nchimbi-Msolla and Tryphone 2010; Silva et al. 2012; Pereira et al. 2014; Morais et al. 2016) and high-magnitude (Hossain et al. 2013) correlations have been reported between iron and zinc and low (Silva et al. 2012), moderate (Maziero et al. 2015), and high (Hossain et al. 2013) correlations between phosphorus and potassium in common bean grains. A positive association between minerals increases the concentration in the combined form. Correlation among the mineral concentrations suggests that selection for one mineral will in fact improve in other mineral simultaneously. Common bean iron biofortification is one of the potential strategies to combat the iron deficiency in bean-eating population, but still, many people suffer from iron deficiency due to unavailability of Fe in cereal-based diets. However, successful mineral biofortification of common bean might be limited due to the low bioavailability of particularly iron and zinc, which is associated with high concentration of anti-nutrients such as phytic acid, polyphenols, and tannins (Beninger et al. 2005; Petry et al. 2014); phytic acid (myoinositol hexakisphosphate) is the major culprit in the inhibition of Fe absorption, while polyphenols and tannins play a minor role (Petry et al. 2012, 2014). Phytic acid is positively correlated with Fe concentration in the beans, they

Table 4 Different types of common bean biofortification, status of research, variety, and concerned publications

Type of biofortification	Status	Variety/country	Papers/sources
Breeding biofortification			
High iron and zinc	Released	Rwanda: RWR 2245, RWR 2154, MAC 42, MAC 44, CAB 2, RWV1129, RWV 3006, RWV 3316, RWV 3317, RWV 2887	HarvestPlus (Rwanda)
Iron	Research		Blair et al. (2009), Gelin et al. (2006), Beebe et al. (2000)
Zinc	Research		Blair et al. (2009), Gelin et al. (2006), Beebe et al. (2000)
Agronomic biofortification			
Zinc	Research		Ibrahim and Ramadan (2015), Ram et al. (2016)
N, P, K, copper, manganese, zinc (organic + chemical fertilizers)	Research		Westermann et al. (2011)
Transgenic biofortification			
Methionine	Research		Aragao et al. (1999)

inhibit the absorption of iron in the intestinal tract, and their concentration increases with the increase of Fe concentration.

Breeders developed new bean varieties with iron concentration of about 100 ppm (Petry et al. 2015): the genotype CNFC 11948 for high iron concentration, and LEC 03–14 for high-potassium, high-phosphorus, and high-calcium concentrations in grains. Low intake of potassium can cause cardiovascular disease (He and MacGregor 2009), reduced phosphorus leads to painful bones and fatigue (Martínez-Ballesta et al. 2010), and deficiency of copper causes the problem of hypochromic anemia, neutropenia, and skeletal disturbances (Guerrero-Romero and Rodríguez-Morán 2005). These symptoms should be reduced by using biofortified common bean cultivars for increasing the concentration potassium, phosphorus, and copper in the diet.

Biofortification is of great importance in enriching seeds with micronutrient levels, and it is considered a more sustainable and cost-effective strategy than food supplementation, fortification, and diet diversification, in sorting out the problem of micronutrient malnutrition in developing countries because it targets those plant species which are daily consumed item (Dwivedi et al. 2012) and developing new micronutrient-rich genotypes. Biofortification of common bean has another advantages, that is, whole seeds are consumed after boiling and all the components of seeds (seed coat, cotyledons, and embryo) can be targeted for improvement of nutrient status (Blair et al. 2013); unlike many cereals, staple foods are polished before eating, resulting in significant loss of nutrients. HarvestPlus and their national collaborators developed biofortified bean varieties in some countries (e.g., Rwanda and Democratic Republic of the Congo). These bean varieties have shown good retention of micronutrient after processing and also good agronomic yield that farmers preferred. It could readily reach low-income households and undernourished population in remote areas which have no access to supplementation programs or commercially marketed fortified foods.

Analytical Methods

To identify elite line or varieties for Fe- and Zn-dense traits from the germplasm, we require quick and accurate analytical method to determine the level of micronutrients in the plants. The wide range of analytical methods (Table 5) are available to breeders, ranging from semiquantitative to fully quantitative and high-throughput screening techniques which have been used to quickly and accurately confirm the level of micronutrients. For precision analysis, most commonly used analytical methods such as inductively coupled plasma optical emission spectroscopy (ICP-OES), atomic absorption spectroscopy (AAS), colorimetric staining, and more recently X-ray fluorescence spectroscopy (XRF) allow the identification of wide range of micronutrients.

Traditionally, micronutrient analysis was conducted by ICP-OES and AAS techniques, inductively coupled plasma optical emission spectroscopy (ICP-OES), or atomic emission spectroscopy (ICP-AES), abbreviated as ICP techniques, and it is a powerful tool and also considered as gold standard because of its high accuracy and low detection limits to ppm for metal analysis, but atomic absorption spectroscopy is mostly preferred because of comparative ease (Table 6). The major advantages of ICP techniques are that it can be used for analysis of set of minerals all together rather than individually. This method has been used for the evaluation Fe and Zn concentration in common bean (Blair et al., 2009, 2010a, b, 2011, 2012, 2016), peanut, pea (Mojica et al. 2015), and cereals (Cakmak et al. 2010). Blair et al. (2009, 2013) reported the accumulation of Fe and Zn concentration in the RIL population of French bean from the cross DOR364 × G19833 and discover quantitative trait loci (QTLs) for 15 elements. The variability in Fe concentration (40–84.6 ppm) was higher than Zn concentration (17.7–42.4 ppm) among the RIL populations and significant correlation between two analytical methods (inductively coupled plasma spectroscopy, atomic absorption spectroscopy), between trials and between minerals (up to $r = 0.175$). Multiple heteroelements and metals such as Ca, Mg, P, and S or Cu, Cd, Co, Fe, Mn, Mo, Ni, and Zn and up to 32 elements could be analyzed in one assay (Cubadda et al. 2007). Concentration of micronutrients in common bean was calculated based on calibration curves with standard solution of Fe and Zn (AOAC International 2005):

Table 5 Analytical methods for micronutrients

	ICP-OES: inductively coupled plasma spectrometer	AAS: atomic absorption spectrometer	XRF: X-ray fluorescence spectrometer	NIRS: near-infrared reflectance spectrophotometer	Modified Perls Prussian blue	Modified 2,2'-dipyridal	Modified Zincon
Principle, Throughput and practical considerations							
Principle	Excitation and emissions at various wavelengths	Absorption	X-ray fluorescence	Absorption at wavelengths in the near infrared	Color reaction	Color reaction	Color reaction
Digestion required	Yes	Yes	No	No	No	Yes	Yes
Sample destructive	Yes		No	No	Yes	Yes	Yes
Throughput	Up to 2.5 min per sample regardless of number of elements analyzed	2.5 min per element	5–10 min per sample depending on the number of elements analyzed	~2 min per sample	~ 4 min per sample	~ 4 min per sample	~ 4 min per sample
Pros\comments	Total recovery of nutrient achieved but subject to the type of digestion procedure used; good sensitivity	Total recovery of nutrient achieved but subject to the type of digestion procedure used; good sensitivity; low cost of purchasing equipment	Total recovery of nutrient achieved; good sensitivity with more expensive models; low cost of bench-type equipment; no digestion step	Low cost	Low cost	Low cost	Low cost

Cons\comments	Gas required; high start-up cost; digestion step; destructive	Gas required; digestion step required; destructive analysis; samples require milling	Problem with Al and Ti sensitivity; sample requires milling	Calibration cost can be high; requires ongoing addition of calibration	Labor-intensive; semiquantitative	Semiquantitative	Semiquantitative
Analytical capability							
Iron	Yes	Yes	Yes	Yes	Yes, separation into high and low groups	Yes, separation into high and low groups	Yes, separation into high and low groups
Zinc	Yes	Yes	Yes	Yes	No	No	Yes, separation into high and low groups
Contamination indicators	Yes	No	Yes, but sensitivity for Al is inadequate as contaminant indicator	Not tested	No	No	No
Other elements	Yes	Yes	Yes	Yes, although data is limited in that area	No	No	No
Other relevant compounds – promoters and inhibitors	No		No	For example, protein, carotenoids, anti-nutrients	No	No	No
Precision							
Application	Plant and soil material	Plant and soil material	Plant and soil material	Plant material	Plant material	Plant material	Plant material
Accuracy of Fe	High	High	High	High	High	High	No
Accuracy of Zn	High	High	High	High	No	No	High

(continued)

Table 5 (continued)

	ICP-OES: inductively coupled plasma spectrometer	AAS: atomic absorption spectrometer	XRF: X-ray fluorescence spectrometer	NIRS: near-infrared reflectance spectrophotometer	Modified Perls Prussian blue	Modified 2,2’-dipyridal	Modified Zincon
Accuracy for carotenoids	No	No	No	High to medium, high confirmed for different crops	No	No	No
Accuracy for total carotenoids	No	No	No	High to medium, high confirmed for different crops	No	No	No
Accuracy for β-carotene	No	No	No	High to medium, high confirmed for different crops		No	No
Accuracy for minerals	Very accurate	Very accurate	Very accurate	Separates out only high and low nutrient genotypes	Separates out only high and low Fe genotypes	Separates out only high and low Fe genotypes	Separates out only high and low Zn genotypes
Economics							
Start-up costs (equipment)	\$50,000–300,000 depending on make and model	\$10,000–40,000 depending on make and model	\$50,000–350,000 depending on make and model	\$60,000–90,000		Minimal	Minimal
Running cost	Varies from lab to lab; between \$4.00 and \$7.00/sample	Varies from lab to lab; e.g., argon between \$0.22 and \$2 0.00/sample; labor the greatest cost in analysis	\$15–25 AUD/sample for XRF analysis; no cost for gas; just instrument upkeep and labor	\$0.5–2.00/samples; dramatically decreasing costs/measurement as components are measured	\$0.5–1.00/sample	\$0.5–1.00/sample	\$0.5–1.00/sample

Application in breeding							
Recommended application in breeding	Pre-screening in population development and validation of Fe- and Zn-dense genotypes identified by rapid screening techniques	Pre-screening in population development and validation of Fe- and Zn-dense genotypes identified by rapid screening techniques	Pre-screening in population development and validation of Fe- and Zn-dense genotypes identified by rapid screening techniques	Pre-screening in population development for minerals and provitamin A; precision analysis for carotenoids and β-carotene for certain crops	Pre-screening in population development	Pre-screening in population development	Pre-screening in population development

Sources: James Stangoulis, School of Agriculture and Wine, University of Adelaide, Waite Campus

Table 6 Identified genetic markers for the seed concentration, seed content, and other traits associated with the uptake of micronutrients in common bean (Hafiz et al. 2017)

Names	Lines/cultivars/ no. of accessions	Nutritional factor	Marker function(s)	Number of markers identified	References
Common bean	G2333, G19839	Phosphate-related	Root traits	19 QTLs	Ochoa et al. (2006)
Common bean	G14519, G4825	Iron	Concentration and content	8 QTLs	Blair et al. (2010)
Common bean	G14519, G4825	Zinc	Concentration and content	9 QTLs	Blair et al. (2010)
Common bean	DOR364, G19833	Phytate	Seed concentration and content	5 QTLs	Blair and Izquierdo (2012)
Common bean	AND696, G19833	Iron	Seed content	1 QTLs	Cichy et al. (2009)
Common bean	AND696, G19833	Zinc	Seed content	1 QTL	Cichy et al. (2009)
Common bean	AND696, G19833	Phosphorus	Seed content	6 linkage groups	Cichy et al. (2009)
Common bean	DOR 364, 19,833	Iron	Seed concentration	13 QTLs	Blair et al. (2009)
Common bean	DOR 364, G19833	Zinc	Seed concentration	13 QTLs	Blair et al. (2009)
Common bean	DOR364, G19833	Phosphate-related	Root hair traits	8 QTLs	
Common bean	DOR364, G19833	Phosphate-related	Root acid exudation	9 QTLs	
Common bean	G19833, DOR 364	Phosphate-	Root traits	26 QTLs	Beebe et al. (2006)
Common bean	206 accessions	Iron	Seed concentration	5 SNPs	Katuuramu et al. (2018)
Common bean	206 accessions	Zinc	Seed concentration	6 SNPs	Katuuramu et al. (2018)
Common bean	206 accessions	Iron	Seed concentration	5 SNPs	Katuuramu et al. (2018)

$$(\text{ICP mineral emission}) \times (\text{dilution factor}) / (\text{aliquot weight}) = (\text{sample mineral content} - \text{mg}/\text{kg})$$

$$(\text{Sample mineral content} - \text{mg}/\text{kg})/10 = (\text{sample mineral content} - \text{mg}/100\,\text{g})$$

$$(\text{Sample mineral content} - \text{mg}/100\,\text{g}) \times 100/(\text{sample dry weight}) = (\text{mineral dry basis content} - \text{mg}/100\,\text{g})$$

It is the more sensitive variant with mass spectrometric detection (ICP-MS). These forms of analysis require specialized equipment, high-purity reagents and highly trained analysts, and extensive sample preparation to minimize potential contamination and ensure high-quality analyses. These techniques are expensive and time-consuming both in terms of cost analysis and in time required for preparation of samples, shipment, and analysis.

Alternatively, colorimetric techniques such as Perls Prussian blue (for Fe) and dithizone (for Zn) have been developed to determine the concentration and localization of micronutrients in the seed. It is a high-throughput screening of the high level of Fe and Zn in the germplasm and reduces the cost of sample analysis (Prom-u-thai et al. 2003; Ozturk et al. 2006; Choi et al. 2007; Velu et al. 2006, 2008). It is used in selection of micronutrient-dense genotypes/RIL from germplasm/very huge population, on the basis of varying color intensity due to staining of seed. It is a more simple technique to perform than ICP-based techniques, and these methods are only semiquantitative, laborious, and not feasible when screening large numbers of samples of germplasm and are destructive process because samples must be milled. Contaminations of minerals are not determined by color-staining techniques; for this, it requires more precise analytical method.

More recently, breeders have been using X-ray fluorescence spectroscopy techniques for the analysis of micronutrients from the samples. Unlike the previously discussed methods, it is a nondestructive approach (Paltridge et al. 2012a, b), in which there is no need for digestion of sample prior to analysis, thus reducing the chances of contamination and analysis cost per sample. Additionally, it is easy to operate without the need for highly experienced analysts, specialized facilities, or additional equipment and cheaper to purchase and run than ICP-OES. While accuracy level is low in XRF, the result obtained by this technique has highly correlated ($r^2 = 0.79–0.98$) with ICP-OES techniques in small seeds, but in case of screening large grain size, the correlation between XRF intensity and the ICP reference analysis is poor (r2 = 0.33 and 0.65 for Fe and Zn, respectively) (Guild et al. 2017). It is ideal to screen large number of germplasm to identify micronutrient-dense genotypes, and if it further requires the accurate micronutrient concentration, then send it to ICP-OES.

Marker-Assisted Approaches

Understanding the nature of gene action and inheritance of seed mineral content is crucial to develop effective breeding strategies for micronutrients (Cichy et al. 2005). Very limited information has been generated on the inheritance of grain Fe and Zn content in crops. The genetic bases responsible for the uptake of some micronutrients, especially Fe uptake, in crop plants are now much better understood. Most studies have indicated multigenic inheritance of micronutrient traits (Blair and Izquierdo 2012) even while a few initial reports suggested that the inheritance of zinc concentration in common beans might be by a few genes. Germplasm evaluation shows a normal distribution for iron and zinc concentrations (Islam et al. 2002). Improving iron and zinc content is only one component of biofortification. Divalent cations like iron and zinc are conjugated by anti-nutrients like phytic acid or polyphenols, which lower bioavailability (Frossard et al. 2000). Compared to phytic acid, polyphenolic compounds have a less negative effect on bioavailability. Their effect on iron absorption ranges from moderately inhibitive to negligible (Petry et al. 2015), and their contribution to biochemical process and seed coat appearance may preclude efforts to reduce polyphenolic content for improving bioavailability. However, seeds were hard to cook and induced digestive problems in human subjects (Petry et al. 2016). Thus, further research is necessary to improve Fe bioavailability by decreasing phytate while maintaining agronomic performance and consumer preferences.

Although conventional breeding techniques have resulted in significant genetic improvement for common bean, molecular markers have recently been developed as a promising selection tool for marker-assisted selection (MAS) (Kelly and Miklas 1999; Kelly 2004). Several molecular maps have been constructed for common bean (Nodari et al. 1993; Freyre et al. 1998). Blair et al. (2003) have recently developed microsatellites that cover every chromosome in the bean genome. Quantitative trait locus (QTL) analysis provides a powerful approach to understand the genetic factors and to unravel the genes underlying the natural variation for Fe and Zn concentrations (Ghandilyan et al. 2006). The identification and tagging of major QTLs for grain micronutrients with large effects would be helpful in the selection of the QTLs in early generations with MAS technique and will greatly accelerate wheat cultivar development for improving mineral concentration in grain (Ortiz-Monasterio et al. 2007). Using various populations, many QTLs for micronutrient concentration in grain/leaf have been mapped in recent years (Table 4). The inter-gene pool study based on DOR364 × G19833 used a genetic map that covered the full common bean genome and found a large range of iron and zinc values among the recombinant inbred lines and 13 QTLs for iron content, of which five were clustered on linkage group b11 and other QTLs were identified on linkage groups b03, b06, b07, and b09 for zinc and b04, b06, b07, and b08 for iron of a total of 26 QTL detected for both minerals (Blair et al. 2009). In the Andean cross G21242 × G21078, there were nine seed mineral QTLs on five linkage groups, with the most important being new loci on b02 but with some overlapping QTL from the inter-gene pool cross on b06, b08, and b07 near *phaseolin* (Blair et al. 2011). Additional studies in the populations G21242 × G21078 derived from an Andean × Andean cross and G14519 × G4825 derived from a Mesoamerican × Mesoamerican cross have shown that some genes for zinc or iron content detected in an inter-gene pool cross (Blair et al. 2009) are also found in intra-gene pool crosses (Blair et al. 2011).

A follow-up to these studies will be the application of marker-assisted selection (MAS) for the genes and QTL that have been identified so far in an attempt to move the loci from one genetic background to another. In this regard, initial marker validation for a QTL from G14519 appears promising in a red mottled Andean grain background. In terms of further MAS research, the colocalization of QTL for seed iron and zinc would be promising for plant breeding of higher micronutrient concentration given that if the same QTL contributes jointly to both minerals, it may be easy to select for these traits simultaneously, phenotypically, and through MAS.

Future Challenges

Full proof of concept that biofortification using the conventional plant-breeding approach works will pave the way for mainstreaming and long-term sustainability. In the coming years, biofortification is expected to be increasingly integrated into international and national crop development programs, crop and food value chains, and national policies and standards. Crop development has already been initiated to develop biofortified varieties of several secondary staple crops, such as zinc and iron in common bean, sorghum, lentil, cowpea, and Irish potato and vitamin A in banana. An initial goal in the biofortification of common beans has been to produce varieties with 80% more iron content and 40% more zinc while maintaining or improving the properties that farmers and consumers require in a variety, such as adaptation to abiotic or biotic stresses and seed shape or color. Breeding is concentrating on both gene pools of common bean, the large-seeded Andeans and the small-seeded Mesoamericans, and bush beans as well as climbing beans. Various breeding techniques or strategies have been used for the current biofortification breeding effort, including backcrossing, recurrent selection, and various permutations of gamete and pedigree selection. Secondary characteristics of phytate and tannin content have for the most part been screened only on advanced lines due to their more expensive assays. Bioavailability tests and bioefficacy trials have been undertaken on the best bet varieties.

For successful and cost-effective biofortification strategy, a few facts must be satisfied:

1. Identification of micronutrient-rich germplasm/line(s) to introgress gene/QTL(s) into locally adapted varieties through breeding methods for the establishment of nutritional efficacy (biological impact under controlled conditions) and effectiveness (biological impact in real life) of a biofortified crop
2. Detection of allergenicity and toxicity
3. Withstanding of nutrient during postharvest processing
4. Acceptance of new variety by farmers and consumers for a cost-effective intervention

Simultaneously, to enhance the effectiveness of biofortification strategies, governments should recognize the benefits and consider providing structure through nutrition and agricultural policies. Furthermore, like health consequences of malnutrition, the effect of biofortified staple crops should be quick to make a difference. Even after the development of biofortified varieties, it will be essential to address various socioeconomic and sociopolitical challenges to popularize their cultivation by farmers (market price premium) and ultimately their public acceptance to combat malnutrition by biofortification. Eventually, a multitiered network and interdisciplinary research will play a pivotal role for successful breeding-based biofortification strategy to address mineral malnutrition in humans and other animals.

Concluding Remarks

Over the past 15 years, conventional breeding efforts have resulted in the development of biofortified common bean varieties with significant levels of the two micronutrients that are most limiting in diets: zinc and iron. Evidence from nutrition research has revealed that these varieties provide considerable amounts of bioavailable micronutrients, and consumption of these varieties can improve the micronutrient-deficiency status among target populations. By 2021, more than 30 countries have officially released biofortified varieties developed using the conventional plant-breeding approach, and at least an additional 20 countries have commenced the testing of these varieties. In Rwanda, for example, by 2015, about 29% of bean farmers had planted iron beans in at least one season over the past eight seasons, and about one-fifth of bean-growing farmers allocated some of their bean area to iron beans in the first season of 2015. In addition, biofortified iron beans have been demonstrated to be efficacious in Rwanda, where iron-deficient university women experienced a significant increase in hemoglobin, ferritin, and total body iron after consuming biofortified beans for 4.5 months. Despite the variety of beans grown and consumed in developed countries and the evidence about the functional potential of this legume, there is a need to develop research to further the knowledge on the nutritional value and modulation of risk factors for chronic diseases to ensure food and nutritional security.

References

Akillioglu HG, Karakaya S (2010) Changes in total phenols, total flavonoids, and antioxidant activities of common beans and pinto beans after soaking, cooking, and in vitro digestion process. Food Sci Biotechnol 19:633–639. [CrossRef]

Aragao FJL, Barros LMG, De Sousa MV, Grossi de Sa MF, Almeida ERP, Gander ES et al (1999) Expression of a methionine-rich storage albumin from the Brazil nut (Bertholletia excelsa H.B.K., Lecythidaceae) in transgenic bean plants (Phaseolus vulgaris L., Fabaceae). Genet Mol Biol 22(3):445–449. https://doi.org/10.1590/S1415-47571999000300026

Association of Official Analytical Chemists (AOAC) Method 990.08; 9.2.39. Minerals quantification (ICP) – Official Methods of Analysis of AOAC International. 18th edn. AOAC International; Gaithersburg, 2005. p 46

Banziger M, Long J (2000) The potential of increasing the iron and zinc density of maize through plant breeding. Food Nutr Bull 21:397–400

Beebe S, Gonzalez AV, Rengifo J (2000) Research on trace minerals in the common bean. Food Nutr Bull 21:387–391. https://doi.org/10.1177/156482650002100408

Beebe S, Rojas-Pierce M, Yan X, Blair MW, Pedraza F, Muñoz F, Tohme J, Lynch JP (2006) Quantitative trait loci for root architecture traits correlated with phosphorus acquisition in common bean. Crop Science 46:413–423

Beninger M, Setimela PS, Hodson H, Vivek H (2005) Breeding for improved abiotic stress tolerance in maize adapted to southern africa. Agric Water Manag 80:212–224

Blair MW (2013) Mineral biofortification strategies for food Staples: the example of common bean. J Agric Food Chem 61:8287–8294. https://doi.org/10.1021/jf400774y

Blair MW, Izquierdo P (2012) Use of the advanced backcross-QTL method to transfer seed mineral accumulation nutrition traits from wild to Andean cultivated common beans. Theor Appl Genet 125:1015–1031. https://doi.org/10.1007/s00122-012-1891-x

Blair MW, Pedraza F, Buendia HF, Gaitán-Solís E, Beebe SE, Gepts P, Tohme J (2003) Development of a genome-wide anchored mi-crosatellite map for common bean (*Phaseolus vulgaris* L.). Theor Appl Genet 107:1362–1374

Blair MW, Giraldo MC, Buendia HF, Tovar E, Duque MC, Beebe SE (2006) Microsatellite marker diversity in common bean (*Phaseolus vulgaris* L.). Theor Appl Genet 113(1):100–109

Blair MW, Astudillo C, Grusak MA, Graham R, Beebe SE (2009) Inheritance of seed iron and zinc concentrations in common bean (Phaseolus vulgaris L.). Mol Breed 23(2):197–207. https://doi.org/10.1007/s11032-008-9225-z

Blair MW, Knewtson SJB, Astudillo C, Li C-M, Fernandez AC, Grusak MA (2010a) Variation and inheritance of iron reductase activity in the roots of common bean (*Phaseolus vulgaris* L.) and association with seed iron accumulation QTL. BMC Plant Biol 10:215. https://doi.org/10.1186/1471-2229-10-215

Blair MW, Medina JI, Astudillo C, Rengifo J, Beebe SE, Machado G et al (2010b) QTL for seed iron and zinc concentration and content in a Mesoamerican common bean (*Phaseolus vulgaris* L.) population. Theor Appl Genet 121:1059–1070. https://doi.org/10.1007/s00122-010-1371-0

Blair MW, Astudillo C, Rengifo J, Beebe SE, Graham R (2011) QTL analyses for seed iron and zinc concentrations in an intra-genepool population of Andean common beans (*Phaseolus vulgaris* L.). Theor Appl Genet 122:511–521. https://doi.org/10.1007/s00122-010-1465-8

Blair MW, Herrera AL, Sandoval TA, Caldas GV, Filleppi M, Sparvoli F (2012) Inheritance of seed phytate and phosphorus levels in common bean (*Phaseolus vulgaris* L.) and association with newly-mapped candidate genes. Mol Breed 30:1265–1277. https://doi.org/10.1007/s11032-012-9713-z

Blair MW, Izquierdo P, Astudillo C, Grusak MA (2013) A legume biofortification quandary: variability and genetic control of seed coat micronutrient accumulation in common beans. Front Plant Sci 4:275. https://doi.org/10.3389/fpls.2013.00275

Bouis HE (2003) Micronutrient fortification of plants through plant breeding: can it improve nutrition in man at low cost? Proc Nutr Soc 62:403–411. https://doi.org/10.1079/PNS2003262

Bouis HE (2007) The potential of genetically modified food crops to improve human nutrition in developing countries. J Dev Stud 43:79–96. https://doi.org/10.1080/00220380601055585

Bouis HE, Hotz C, McClafferty B, Meenakshi JV, Pfeiffer WH (2011) Biofortification: a new tool to reduce micronutrient malnutrition. Food Nutr Bull 32(Suppl. 1):S31–S40

Broughton WJ, Hernandez G, Blair MW, Beebe SE, Gepts P, Vanderleyden J (2003) Beans (*Phaseolus* spp.) – model food legumes. Plant Soil 252:55–128

Cakmak I, Kalayci M, Kaya Y, Torun AA, Aydin N, Wang Y, Arisoy Z, Erdem H, Gokmen O, Ozturk L, Horst WJ (2010) Bio fortification and localization of zinc in wheat grain. J Agric Food Chem 58:9092–9102

Cardador-Martinez A, Castano-Tostado E, Loarca-Pina G (2002) Antimutagenic activity of natural phenolic compounds present in the common bean (*Phaseolus vulgaris*) against aflatoxin B1. Food Addit Contam 19:62–69. [CrossRef] [PubMed]

Carolina AR, Andrea C. Fernandez, Karen A. Cichy (2015) Transcriptome Characterization of Developing Bean (*Phaseolus vulgaris* L.) Pods from Two Genotypes with Contrasting Seed Zinc Concentrations. PLOS ONE. https://doi.org/10.1371/journal.pone.0137157

Choi EY, Graham RD, Stangoulis JCR (2007) Rapid semi-quantitative screening methods for determination of iron and zinc in grains and staple foods. J Food Compos Anal 20:496–505

Cichy KA, Forster S, Grafton KF, Hosfield GL (2005) Inheritance of seed zinc accumulation in navy bean. Crop Sci 45(3):864–870. https://doi.org/10.2135/cropsci2004.0104

Cichy KA, Caldas GV, Snapp SS, Blair MW (2009) QTL analysis of seed iron, zinc, and phosphorus levels in an Andean bean population. Crop Sci 49:1742–1750

Cubadda RE, Carcea M, Marconi E, Trivisonno MC (2007) Influence of protein content on durum wheat gluten strength determined by the SDS sedimentation test and by other methods. Cereal Foods World 52(1):273–277

Dwivedi SL, Sahrawat KL, Rai KN, Blair MW, Andersson M, Pfieffer W (2012a) Nutritionally enhanced staple food crops. Plant Breed Rev 34:169–262. https://doi.org/10.1002/9781118358566.ch

Dwivedi SL, Sahrawat KL, Rai KN, Blair MW, Andersson M, Pfieffer W (2012b) Nutritionally enhanced staple food crops. Plant Breed Rev 34:169–262. https://doi.org/10.1002/9781118358566.ch3

FAO (2018) World food and agriculture – statistical pocketbook 2018. Rome, p 254. Licence: CC BY-NC-SA 3.0 IGO

Forster SM, Moraghan JT, Grafton KF (2002) Inheritance of seed Zn accumulation in navy bean. Annu Rep Bean Improv Coop 45:30–31

Freyre R, Skroch PW, Geffroy V, Adam-Blondon A-F, Shirmohamadali A, Johnson WC, Llaca V, Nodari RO, Pereira PA, Tsai SM, Tohme J, Dron M, Nienhuis J, Vallejos CE, Gepts P (1998) Towards an integrated linkage map of common bean: 4. Dev Core Linkage Map Alignment RFLP Maps Theor Appl Genet 97:847–856

Frossard E, Bucher M, Mächler F, Mozafar A, Hurrell R (2000) Potential for increasing the content and bioavailability of Fe, Zn and ca in plants for human nutrition. J Sci Food Agric 80(7):861–879

Gelin JR, Forster S, Grafton KF, McClean P, Rojas-Cifuentes GA (2006) Analysis of seed-zinc and other nutrients in a recombinant inbred population of navy bean (Phaseolus vulgaris L.). Crop Sci 47:1361–1366. https://doi.org/10.2135/cropsci2006.08.0510

Ghandilyan A, Vreugdenhil D, Aats MGM (2006) Progress in the genetic understanding of plant iron and zinc nutrition. Physiol Plant 126:407–417

Graham RD, Senadhira D, Beebe SE, Iglesias C, Ortiz-Monasterio I (1999) Breeding for micronutrient density inedible portions of staple food crops: conventional approaches. Field Crops Res 60:57–80

Gregorio GB (2002) Progress in breeding for trace minerals in staple crops. J Nutr 132:500S–502S

Guerrero-Romero F, Rodríguez-Morán M (2005) Complementary therapies for diabetes: the case for chromium, magnesium, and antioxidants. Arch Med Res 36:250–257

Guild GE, Paltridge NG, Andersson MS et al (2017) An energy-dispersive X-ray fluorescence method for analysing Fe and Zn in common bean, maize and cowpea biofortification programs. Plant Soil 419:457

Guzmán-Maldonado SH, Acosta-Gallegos J, Paredes-López O (2000) Protein and mineral content of a novel collection of wild and weedy common bean (*Phaseolus vulgaris* L.). J Sci Food Agric 80:1874–1881

Guzmán MSH, Gallegos JAA, Muñoz MDLÁÁ, Delgado SG, Piña GL (2002) Calidad alimentariay potencial nutracéutico del frijol (*Phaseolus vulgaris* L.) [food quality and nutraceutical potential of common bean (*Phaseolus vulgaris* L.)]. Agric Téc Méx 28:159–173

Guzman-Maldonado SH, Martinez O, AcostaGallegos JA, Guevara-Lara F, Paredes-Lopez O (2003) Putative quantitative trait loci for physical and chemical components of common bean. Crop Sci 43(3):1029–1035

Guzman-Maldonado SH, Acosta-Gallegos J, Paredes-Lopez O (2004) Protein and mineral content of a novel collection of wild and weedy common bean (Phaseolus vulgaris L.). J Sci Food Agric 80:1874–1881

Hacisalihoglu G, Osturk L, Cakmak I, Welch RM, Kochian L (2004) Genotypic variation in common bean in response to zinc deficiency in calcareous soil. Plant Soil 259:71–83. https://doi.org/10.1023/B:PLSO.0000020941.90028.2c

Hafiz A, Riaz T, Shakoori FR (2017) Metabolic profile of a stored grain pest Trogoderma granarium exposed to deltamethrin Pakistan: J Zool 49(1):183–188

Hayat I, Ahmad A, Masud T, Ahmed A, Bashir S (2014) Nutritional and health perspectives of beans (Phaseolus vulgaris L.): an overview. Crit Rev Food Sci Nutr 54:580–592. [CrossRef] [PubMed]

He FJ, MacGregor GA (2009) A comprehensive review on salt and health and current experience of worldwide salt reduction programmes. J Hum Hypertens 23:363–384. https://doi.org/10.1111/j.1399-3054.2006.00646.x

Hossain KG, Islam N, Jacob D, Ghavami F et al (2013) Interdependence of genotype and growing site on seed mineral compositions in common bean. Asian J Plant Sci 12:11–20

Howarth B (2014) Biofortification progress briefs, Aug 2014

Ibrahim EA, Ramadan WA (2015) Effect of zinc foliar spray alone and combined with humic acid or/and chitosan on growth, nutrient elements content and yield of dry bean (Phaseolus vulgaris L.) plants sown at different dates. Sci Hortic 184:101–115. https://doi.org/10.1016/j.scienta.2014.11.010

Islam FMA, Basford KE, Jara C, Redden RJ, Beebe SE (2002) Seed compositional and disease resistance differences among gene pools in cultivated common bean. Genet Resour Crop Evol 49:285–293. https://doi.org/10.1023/A:1015510428026

Jun S, Shin S, Joung H (2016) Estimation of dietary flavonoid intake and major food sources of Korean adults. Br J Nutr 115:480–489. [CrossRef] [PubMed]

Karen V, Armendáriz F, Ibeth MHH, Muñoz-Márquez E, Esteban S (2019) Characterization of Bioactive Compounds, Mineral Content, and Antioxidant Activity in Bean Varieties Grown with Traditional Methods in Oaxaca, Mexico. Antioxidants, 8:26. https://doi.org/10.3390/antiox8010026

Katuuramu D, Hart J, Porch T, Grusak M, Glahn R, Cichy K (2018) Genome-wide association analysis of nutritional composition-related traits and iron bioavailability in cooked dry beans (Phaseolus vulgaris L.). Mol Breed 38:44. https://doi.org/10.1007/s11032-018-0798-x

Kelly JD (2004) Advances in common bean improvement: some case histories with broader applications. Available at www. actahort.org/books/637/637_11.htm (verified 12 Apr 2007). Acta Hortic (ISHS) 637:99–122

Kelly JD, Miklas PN (1999) Marker-assisted selection. In: Singh SP (ed) Developments in plant breeding, vol 7. Common bean improvement in the twenty-first century. Kluwer Academic, Boston

Kumar G, Baojun X (2017) Polyphenol-rich dry common beans (Phaseolus vulgaris L.) and their health benefits, review. Int J Mol Sci 18:2331. https://doi.org/10.3390/ijms18112331

Lin LZ, Harnly JM, Pastor-Corrales MS, Luthria DL (2008) The polyphenolic profiles of common beans (Phaseolus vulgaris L.). Food Chem 107:399–410. [CrossRef] [PubMed]

López A, El-Naggar T, Dueñas M, Ortega T, Estrella I, Hernández T, Gómez-Serranillos MP, Palomino OM, Carretero ME (2013) Effect of cooking and germination on phenolic composition and biological properties of dark beans (*Phaseolus vulgaris* L.). Food Chem 138:547–555. [CrossRef] [PubMed]

Lyons G, Cakmak I (2012) In: Bruulsema TW, Heffer P, Welch RM et al (eds) Agronomic biofortification of food crops with micronutrients. In fertilizing crops to improve human health: a scientific review. International Plant Nutrition Institute, Paris, pp 97–122

Martínez-Ballesta MC, Dominguez-Perles R, Moreno DA, Muries B et al (2010) Minerals in plant food: effect of agricultural practices and role in human health. A Rev Agron Sustain Dev 30:295–309

Matthew WB (2013) Mineral biofortification strategies for food Staples: the example of common bean. J Agric Food Chem 61:8287–8294. https://doi.org/10.1021/jf400774y

Maziero SM, Ribeiro ND, Storck L (2015) Simultaneous selection in beans for architecture, grain yield and minerals concentration. Euphytica 205:369–380. https://doi.org/10.1007/s10681-015-1392-5

Messina V (2014) Nutritional and health benefits of dried beans. Am J Clin Nutr 100(Suppl 1):437S–442S. [CrossRef] [PubMed]

Mojica M, Chen K, González de Mejía E (2015) Impact of commercial precooking of common bean (Phaseolus vulgaris) on the generation of peptides, after pepsin-pancreatin hydrolysis, capable to inhibit dipeptidyl peptidase-IV. J Food Sci 80(1):H188–H198. https://doi.org/10.1111/1750-3841.12726
Moraghan JT, Grafton K (1999) Seed zinc concentration and the zinc-efficiency trait in navy bean. Soil Sci Soc Am J 63:918–922
Moraghan JT, Padilla J, Etchevers JD, Grafton K, Acosta-Gallegos JA (2002) Iron accumulation in seed of common bean. Plant Soil 246:175–183. https://doi.org/10.1023/A:1020616026728
Morais NM, Ribeiro ND, Storck L, dos Santos PRF et al (2016) Selection of common bean land cultivars based on agronomic performance, cooking time, and mineral concentration. Semina Cienc Agrar 37:1255–1266
Moreno-Franco B, León-Latre M, Andrés-Esteban EM, Ordovás JM, Casasnovas JA, Peñalvo JL (2014) Soluble and insoluble dietary fibre intake and risk factors for cvd and metabolic syndrome in middle-aged adults: the AWHS cohort. Atherosclerosis 235:e279–e280. [CrossRef]
Mudryj AN, Yu N, Aukema HM (2014) Nutritional and health benefits of pulses. Appl Physiol Nutr Metab 39:1197–1204
Nchimbi-Msolla S, Tryphone GM (2010) The effects of the environment on iron and zinc concentrations and performance of common bean (Phaseolus vulgaris L.) genotypes. Asian J Plant Sci 9:455–462
Nodari RO, Tsai SM, Gilbertson RL, Gepts P (1993) Towards an integrated linkage map of common bean: 2. Development of an RFLP-based linkage map. Theor Appl Genet 85:513–520
Ortiz-Monasterio JI, Palacios-Rojas N, Meng E, Pixley K, Trethowan R, Pena RJ (2007) Enhancing the mineral and vitamin content of wheat and maize through plant breeding. J Cereal Sci 46:293–307
Ozturk L, Yazici MA, Yucel C, Tornn A, Cekic C, Bagci A, Ozkan H, Braun HJ, Sayers Z, Cakmak I (2006) Concentration and localization of zinc during seed development and germination in wheat. Physiol Plant 128:144–152
Paltridge NG, Milham PJ, Ortiz-Monasterio JI, Velu G (2012a) Energy-dispersive X-ray fluorescence spectrometry as a tool for zinc, iron and selenium analysis in whole grain wheat. Plant Soil 361:261–269
Paltridge NG, Palmer LJ, Milham PJ, Guild GE (2012b) Energy-dispersive X-ray fluorescence analysis of zinc and iron concentration in rice and pearl millet grain. Plant Soil 361:251–260
Pereira HS, Del Peloso MJ, Bassinello PZ, Guimarães CM et al (2014) Genetic variability for iron and zinc content in common bean lines and interaction with water availability. Genet Mol Res 13:6773–6785
Petry N, Egli I, Gahutu JB, Tugirimana PL, Boy E, Hurrell R (2012) Stable iron isotope studies in Rwandese women indicate that the common bean has limited potential as a vehicle for iron biofortification. J Nutr 142:492–497. https://doi.org/10.3945/jn.111.149286
Petry N, Egli I, Gahutu JB, Tugirimana PL, Boy E, Hurrell R (2014) Phytic acid concentration influences iron bioavailability from biofortified beans in Rwandese women with low iron status. J Nutr 144:1681–1687. https://doi.org/10.3945/jn.114.192989. [PubMed] [CrossRef]
Petry N, Boy E, Wirth JP, Hurrell RF (2015) Review: the potential of the common bean (Phaseolus vulgaris) as a vehicle for iron biofortification. Nutrients 7(2):1144–1173. https://doi.org/10.3390/nu7021144
Petry N, Rohner F, Gahutu JB, Campion B, Boy E, Tugirimana PL et al (2016) In Rwandese women with low iron status, iron absorption from low-phytic acid beans and biofortified beans is comparable, but low-phytic acid beans cause adverse gastrointestinal symptoms. J Nutr 146:970–975. https://doi.org/10.3945/jn.115.2
Pfeiffer WH, McClafferty B (2007) HarvestPlus: breeding crops for better nutrition. Crop Sci 47:S88–S105. https://doi.org/10.2135/cropsci2007.09.0020IPBS
Prasad R (2012) Micro mineral nutrient deficiencies in humans, animals and plants and their amelioration. P Natl Acad Sci USA 82:225–233
Prom-u-thai C, Dell B, Thomson G, Rerkssem B (2003) Easy and rapid detection of iron in rice seed. Sci Asia 29:314–317

Puertas-Mejía MA, Mosquera-Mosquera N, Rojano B (2016) Study of in vitro antioxidant capacity of *Phaseolus vulgaris* L. (frijol) by microwave assisted extraction. Rev Cubana Plant Med 21:42–50

Ram H, Rashid A, Zhang W, Duarte AP, Phattarakul N, Simunji S et al (2016) Biofortification of wheat, rice and common bean by applying foliar zinc fertilizer along with pesticides in seven countries. Plant Soil 1(403):389–401. https://doi.org/10.1007/s11104-016-2815-3

Ranilla LG, Genovese MI, Lajolo FM (2007) Polyphenols and antioxidant capacity of seed coat and cotyledon from Brazilian and Peruvian bean cultivars (Phaseolus vulgaris L.). J Agric Food Chem 55:90–98. [CrossRef] [PubMed]

Schoonhovern A, Voysest O (eds) (1991) Common beans: research for crop improvement. C.A.B. Int, Wallingford

Siddhuraju P (2006) The antioxidant activity and free radical-scavenging capacity of phenolics of raw and dry heated moth bean (*Vigna aconitifolia*) (Jacq.) Marechal seed extracts. Food Chem 99:149–157. [CrossRef]

Silva CA, Abreu AFB, Ramalho MAP, Maia LGS (2012) Chemical composition as related to seed color of common bean. Crop Breed Appl Biotechnol 12(2):132–137. https://doi.org/10.1590/S1984-70332012000200006

Singh SP (ed) (1999) Common bean improvement for the twenty-first century. Kluwer Academic Publishers, Dordrecht

Steckling S Dem, Ribeiro ND, Arns FD, Mezzomo HC, Possobom MTDF (2017) Genetic diversity and selection of common bean lines based on technological quality and biofortification. Genet Mol Res 16(1):gmr16019527. http://dx.doi.org/10.4238/gmr16019527

Talukder ZI, Anderson E, Miklas PN, Blair MW, Osorno J, Dilawari M, Hossain KG (2010) Genetic diversity and selection of genotypes to enhance Zn and Fe content in common bean. Can J Plant Sci 90(1):49–60. https://doi.org/10.4141/CJPS09096

Tryphone GM, Msolla N (2010) Diversity of common bean (Phaseolus vulgaris L.) genotypes in iron and zinc contents under greenhouse conditions. Afr J Agric Res 5:738–747

Tryphone GM, Nchimbi-Msolla S (2010) Diversity of common bean (Phaseolus vulgaris L.) genotypes in iron and zinc contents under screenhouse conditions. J Agric Res 5(8):738–747. https://doi.org/10.5897/AJAR10.304

Velu G, Kulkarni VN, Rai KN, Muralidharan V, Longvah T, Sahrawat KL, Raveendran TS (2006) A rapid method for screening grain iron content in pearl millet. Int Sorghum Millets News 47:158e161

Velu G, Bhattacharjee R, Rai KN, Sahrawat KL, Longvah T (2008) A simple and rapid screening protocol for grain Zn content in pearl millet. J SAT Agric Res 48:5e8

Voysest O, Valencia M, Amezquita M (1994) Genetic diversity among Latin American Andean and Mesoamerican common bean cultivars. Crop Sci 34:1100–1110

Vreugdenhil D, Aarts MGM, Koornneef M, Nelissen H, Ernst WHO (2004) Natural variation and QTL analysis for cationic mineral content in seeds of Arabidopsis thaliana. Plant Cell Environ 27:828–839

Welch RM, Graham RD (1999) A new paradigm for world agriculture: productive, sustainable and nutritious food systems to meet human needs. Field Crops Res 60:1–10. https://doi.org/10.1016/S0378-4290(98)00129-4

Welch RM, House WA, Beebe S, Cheng Z (2000) Genetic selection for enhanced bioavailable levels of iron in bean (Phaseolus vulgaris L.). J Agric Food Chem 48:3576–3580. https://doi.org/10.1021/jf0000981

Westermann DT, Teran H, Munoz-Perea CG, Singh SP (2011) Plant and seed nutrient uptake in common bean in seven organic and conventional production systems. Can J Plant Sci 91:1089–1099. https://doi.org/10.4141/cjps10114

White PJ, Broadley MR (2005) Biofortifying crops with essential mineral elements. Trends Plant Sci 10:586–593

Pulses: Milling and Baking Applications

Clifford Hall

Abstract The application of pulses in bakery items is an ideal complement to cereal and starch ingredient use in gluten-free and non-gluten-free products. The addition of pulses to cereal-based products improves nutrition by providing complimentary amino acid to cereal grains and increasing fiber and protein of gluten-free products. This review highlights milling and bakery applications of pulses. In some cases, the information available was reported three decades ago. However, many of these published documents are still relevant today and will serve as a starting point for those interested in milling and incorporation of pulses into bread. The application data for cakes and cookies is relatively new compared to milling information. In general, pulses can be milled effectively using pin, hammer, and roller mills. The resulting flours can be incorporated into bakery products as a whole flour or protein, starch, or fiber fraction. This review highlights some applications. Information regarding particle size effects of pulses in cakes and cookies has been provided. There is no general trend about the impact of particle size on bakery products given that baking systems evaluated impact how particle size influences product quality. The level of pulse fortification also impacts quality, and thus no general recommendation can be made with regard to the usage level for all bakery products. However, pulse fortification of 10% appears to produce acceptable pan breads, while 100% pulse flour can be used in cookies. Therefore, the usage level will be system dependent and research to identify optimal percentages may be needed. The applications presented in this review focus on pea, chickpea, lentil, and beans such as navy, pinto, and black. However, the use of other pulses may be suitable for bakery applications.

Keywords Food legumes · Gluten-free · Flour quality · Pulses protein · Bread · Dough · Cookies · Healthy food

---

C. Hall (✉)
Department of Dairy and Food Science, South Dakota state University, Brookings, SD, USA
e-mail: clifford.hall@sdstate.edu

D. S. Gupta et al. (eds.), *Breeding for Enhanced Nutrition and Bio-Active Compounds in Food Legumes*, https://doi.org/10.1007/978-3-030-59215-8_9

Introduction

Pulses as ingredients in baking systems have not been extensively evaluated. The primary reason for this relates to the availability of pulse ingredients. Recent trends in the food industry have resulted in increased interest in pulse ingredients; however, much of the information on ingredient production is held confidential by the ingredient supplier. Pulse ingredients include whole or split flours, starch and protein flours, and fiber. Furthermore, precooked and extruded flours are also available in limited supplies.

Growing interest in pulses is due to their high nutrient density and health-promoting compounds. Pulses have approximately double the protein content of cereal grains and are high in the amino acid, lysine (Udahogora 2012). Lysine is the limiting amino acid in cereals, and thus, combinations of pulses with cereals can create a complete protein. Unlike soy, pulses are not a primary food allergen and do not need to be listed on food packages. Furthermore, food manufacturers are looking to replace soy in their formulations (Eigenmann et al. 2008). Pulses, although not a one to one replacement, may be applicable in bakery items as a complete or partial replacement for soy or wheat.

The level of replacement in baking applications will depend on the desired functionality. Milling and fractionation can be used as a means of providing flours with varying functionality. This chapter highlights milling and fractionation approaches in the context of baking applications. Where applicable, information on other modifications such as extrusion will be provided.

Milling

Milling of pulses has been sporadically evaluated over the last three to four decades. Much of the early research on pulse processing focused on fractionation of pulse flours. For example, Vose et al. (1976) used pin milling to reduce the particle size of dry pea to less than 44 μm followed by air classification. The resulting fractions consisted mainly of protein and starch. The nature of the pin milling and degree of air classification dictate the composition of the fractions. Sosulski and Youngs (1979) reported that air fractionation of great northern bean flour was effective in producing protein and starch fractions. Protein increased from 24 to 53.5% in high protein fractions, whereas the starch increase was smaller, from 40 to 51.5% in the high starch fractions. Patel et al. (1980) were able to enhance the protein from 30.4% to 60–63% by air fractionation of navy beans. Tyler et al. (1981) reported similar increases in protein content of great northern and navy bean flours subjected to air classification. These authors did report that further fractionation of the primary fractions resulted in a reduction in protein content from 61% to 49%. This suggests that simple fractionation protocols could be useful in developing food ingredients; however, extensive fractionation may not be warranted.

Tyler et al. (1981) reported the adaptability of pin milling and air classification using eight different pulses. They noted that only cowpea and lima beans were not well suited for fractionation. Greater classifier speeds resulted in smaller particles and less protein yields (Pelgrom et al. 2013). This observation supports those of prior researchers (Tyler et al. 1981; Bergthaller et al. 2001). Kon et al. (1977) reported that air classification enhanced the lipid content in higher protein fractions. Furthermore, ash and crude fiber tended to concentrate in the protein fractions. Pelgrom et al. (2015a) also found high fiber content in the protein-rich fraction of yellow peas. Electrostatic separation of yellow pea flour allowed for greater separation of protein and fiber components (Pelgrom et al. 2015b). In contrast, Vose et al. (1976) observed higher fiber content in the starch fractions. Thus, treatment of the pulses prior to milling and air classification can impact composition and may account for differences observed by researchers.

Roasting of navy and pinto beans and chickpeas resulted in about 2.5 times more protein in the protein fraction of these pulses compared to the non-roasted versions (Han and Khan 1990a). Of the three pulses, only the chickpea fractionation process was considered inefficient. Both roasting and fractionation impacted functional characteristics of navy and pinto beans and chickpeas. Han and Khan (1990b) reported decreased nitrogen solubility and foaming capacity in roasted pulses; however, water-holding capacities and cold paste viscosity were higher in roasted samples compared to non-roasted pulse flours. The fractionation of pulses leads to the mixed functional results. Some fractions had significantly higher functional properties (e.g., water-holding capacity), while other fractions had functional properties similar to the control (Han and Khan 1990b). The starch fraction had higher water-holding capacities and cold paste viscosity compared to the control flour, while emulsification and foamability were greater in the protein fraction compared to the control.

The extrusion expansion index of the high starch and protein fractions of navy and pinto beans varied with fractions. The high starch fractions of both navy and pinto beans produced expansion indexes that were similar to blends of corn flour and bean flour fractions (Gujska and Khan 1991). However, blending of high protein fractions into the high starch fractions resulted in significant reductions in expansion index values. The extrusion of a high-starch pinto bean flour resulted in significantly lower hot and cold paste viscosities compared to a whole pinto flour or a non-extruded high-starch pinto bean flour (Simons et al. 2017). The abovementioned reports are just a few of the fractionation studies that support the combination of milling and fractionation as a means to obtain ingredients with desired functional characteristics.

In contrast to fractionation, milling to produce flour with multiple nutrient compositions and functionalities is desired. However, limited studies have been reported on milling of whole pulses without the intention of fractionation. Maskus et al.

(2016) reported the impacts of milling methods to compositional and functional properties of yellow pea flour. They found that stone milling produced the largest mean particle followed by the hammer mill, roller mill, and fine pin milling. These authors also found minor but significant differences in protein, starch, and fiber contents among flours from the different mills. The roller mill tended to produce flours with higher peak and final viscosities compared to other milling methods (Maskus et al. 2016). Research in the corresponding author's laboratory supported these findings. In this research, roller-milled pea flour had greater peak, hot paste, and final viscosities than flour obtained by a hammer mill (Hall 2017). The higher viscosities observed in roller-milled flour were attributed to higher starch damage.

The level of starch damage within hammer-milled flours increased with decreasing mill screen aperture size (Hall 2017). The flour with the greatest starch damage and highest viscosities was obtained from a hammer mill using a 0.84 mm screen, while the opposite was found in flours milled through a 9.53 mm screen. Another factor affecting flour properties during hammer milling is the rotor speed. Regardless of screen size, pasting viscosities were always higher in flours obtained at a rotor speed of 102 m/s (7200 rpm) compared to 34 m/s (2400 rpm). Furthermore, the smaller particles obtained from milling at the higher rotor speeds likely contributed to the observed starch damage and pasting properties (Hall 2017). In addition to functional properties, flow characteristics of a flour are impacted by milling conditions.

Understanding flow properties of flours will allow food manufactures to determine the impact pulse flours might have on the flow of wheat flour in conveying systems. The angle of repose and angle of slide for flours are two measures used to define flow properties. A low angle of repose indicates that a flour flows well, while a high angle suggests that it easily clumps. Angle of slide indicates an interaction between the flour and the surface. A flour with a low angle indicates that the flour flows smoothly over the surface, while a flour with a high angle of slide indicates possible flow issues through a hopper or tubing. Kallenbach et al. (2017) reported that flow properties were significantly affected by hammer speed, screen size, and flour moisture.

The moisture of milled peas was a significant factor for only the angle of slide on a stainless steel surface. The flour from the peas with an 11% moisture had a significantly greater angle of slide compared to the flour from a pea with 9% moisture. Regardless of moisture, flours from the hammer mill obtained at a rotor speed of 102 m/s had angles of slide and angles of repose that were greater than flours obtained from 34 m/s rotor speeds. This suggests that increasing rotor speeds produced flours with significantly worse flow properties. Smaller screens produced flours with significantly larger angles of repose. Screen sizes smaller than 2.0 μm had significantly larger angles of repose compared to screens larger than 2.0 μm (Kallenbach et al. 2017). In general, smaller particles had the worse flow properties. The difference in flow properties might be due to greater surface adhesion of the flour with higher moisture and smaller particles. Understanding particle flow and flour properties will allow millers to produce flours suitable for desired applications.

Dough and Bread

The relative recent interest in using pulses in bread stems from the desire to create a complete protein. However, utilization of the appropriate pulse may require extensive reformulation in products compared to wheat alone. Furthermore, the quality and functionality of the pulse flour and protein can be affected by a pretreatment, such as extrusion or cooking, of the pulse or pulse cultivar (Kaur and Sandhu 2010; Nosworthy et al. 2017a,b). The particle size of the flour can also impact functionality of the pulse flour (Borsuk et al. 2012; Luhovyy et al. 2017).

The addition of pulse protein or their respective protein fractions is widely accepted as the approach to enhance protein content in wheat flour blends. Fenn et al. (2010) found that protein content increased by 8–32% in yellow pea-fortified wheat blends, while an 8–29% increase in protein was found in chickpea-fortified wheat blends. As expected, protein content increased from 13.5% in bread without texturized pinto bean protein to 17.4% with the addition of 15% texturized pinto bean protein (Simons et al. 2014). Furthermore, a 139% increase in lysine was reported for the bread containing 15% texturized pinto bean protein compared to the control.

The addition of pulse flours or ingredients in baking systems generally impacts dough rheology and bread quality. Pulse flour addition to wheat flour tends to increase peak height of flours. D'Appolonia (1977) observed that 5% addition of pulse (i.e., lentil, faba, mung, navy, or pinto bean) flour to hard red spring wheat (HRSW) flour increased peak height. The HRSW flour had a peak height of 470 B.U., while the pulse-fortified flours had values between 500 and 590 B.U., with lentil having the most significant impact of peak height. The addition of pulse flours beyond 5% (i.e., 10 and 20%) did not further increase the peak heights. However, the addition of 10 and 20% pinto flours caused reduced peak heights and setback values. Setback for the other pulse-flour-fortified blends was greater than HRSW control flour. Mohammed et al. (2014) observed lower peak and final viscosities with increasing chickpea flours compared to the control wheat flour.

Farinograph properties are commonly measured to characterize wheat flour. Water absorption, dough development time, and stability are measures used to compare flours. Navy and pinto bean flours tended to increase the water absorption of HRSW, while faba beans and lentils had no effect on water absorption (D'Appolonia 1977). Mungbean generally caused a reduction in water absorptions of HRSW flour. Mohammed et al. (2012) observed increasing water absorption values with increasing chickpea flour supplementation. Fenn et al. (2010) observed increased water absorptions of several different wheat cultivars fortified with yellow pea or chickpea protein concentrates (60–72% protein). They did observe that flour blend with 2% of the pulse protein had water absorption values comparable to the control flours. However, the 5 and 8% addition of pulse protein caused significant increases in water absorptions of the control wheat flour. This trend was observed in all four of the cultivars tested. Water absorption properties of defatted and glycated cowpea flour were higher than raw cowpea (Campbell et al. 2016). The addition of pea fiber also impacts water absorption. Blends of wheat with insoluble cotyledon fiber of pea, lentil, and chickpea had water absorptions that were higher compared to wheat blends with hull and soluble cotyledon fiber (Dalgetty and Baik 2006).

Dough development time and stability were significantly reduced by the addition of pulse flour compared to the HRSW control (D'Appolonia 1977). Mohammed et al. (2012) observed increased dough development times and decreased dough stability with increasing levels of chickpea up to 30%. Stability of the flour blends also decreases with increasing pulse protein (Fenn et al. 2010). However, the degree of stability reduction was wheat cultivar dependent. The pretreatment of chickpea, pea, and lentil impacted mixograph mixing times differently (Baik and Han 2012). However, with few exception, the mixograph mixing time was lower than those for the HRSW flour. In contrast, longer mixing times were observed in flour containing texturized pinto bean protein (Simons et al. 2014), but stability was less than the dough without pinto protein. Dalgetty and Baik (2006) observed increasing mixing times of wheat dough with increasing levels of pea and chickpea fiber additions, while this relationship was not observed for lentil fiber. However, regardless of the pulse, wheat flour with soluble cotyledon fiber had the shortest mixing time.

As expected, pan bread quality was impacted by pulse addition. The general trend was that the level of pulse flour or fraction impacted the specific volume of bread more so than the pulse itself (D'Appolonia 1977; Fenn et al. 2010; Kasprzak and Rzedzicki 2010; Mohammed et al. 2012). The addition of 2% yellow pea or chickpea protein produced breads with comparable specific volumes to control breads, but the addition of 5 and 8% caused noticeable differences in the bread volume (Fenn et al. 2010). Loaf volumes were approximately half for 70:30 HRSW-pulse blends compared to the HRSW control (Baik and Han 2012). These authors also observed that precooked pulse flour caused the greatest reduction in loaf volume when compared to the control. Bread fortified with 10% extruded black bean flour and all wheat control bread had similar specific volumes, while bread with 15% extruded black bean flour had significantly lower specific volume compared to the control (Batista et al. 2011). Specific volume of bread fortified with 5% texturized bean proteins was not significantly different from the control bread, while breads with 10 and 15% texturized pinto bean protein had specific volumes that were significantly lower from the control bread (Simons et al. 2014). The lower starch content of the texturized pinto bean protein may have limited the usage level to 5% given that 10% usage level of black bean flour resulted in specific volumes comparable to the control. Loaf volume of bread made with defatted and glycated cowpea flour was similar to bread made with raw cowpea (Campbell et al. 2016). Dalgetty and Baik (2006) reported that increasing (from 3 to 7%) addition of pea hull fiber and cotyledon fiber significantly affected loaf volume. The soluble cotyledon had the least negative impact on loaf volume, although lower levels were used in the bread formula compared to the insoluble fibers. Kasprzak and Rzedzicki (2010) observed a reduction in bread volume with the addition of pea hull. The reduction observed was influenced by the particle size (0.28 and 0.83 mm) of the milled hull; the larger the particle size, the greater was the loaf volume reduction. Increasing percentages of fiber from lentils and chickpea also caused loaf volume reduction (Dalgetty and Baik 2006). In contrast, 0.75 and 1% pea pod fiber and broad bean pod fibers increased loaf volume compared to the control (Fendri et al. 2016).

The crumb characteristics were also affected by pulse flour and protein addition. D'Appolonia (1977) observed that the 5% addition of navy or pinto bean flour to HRSW produced comparable crumb grain; however, additions beyond 5% resulted in poor crumb scores for all pulses tested. A significant reduction in crumb texture and grain scores was observed beyond a 10% chickpea fortification (Mohammed et al. 2012). Fenn et al. (2010) also reported lower bread crumb scores with increasing amounts of pea or chickpea protein addition to wheat flour. However, only yellow pea caused a significant increase in crumb firmness. The firmness of bread with pea fiber was lower than the control bread as measured during a 72-hour storage study (Gomez et al. 2003). Furthermore, the overall acceptability of bread with pea fiber was rated comparable to the control bread by a sensory panel.

Processing of pulses prior to use in bread applications has focused on reducing the antinutrients prior to use of pulse flours in bread applications. D'Appolonia (1978) observed that roasting of navy beans prior to milling into flour tends to alter the functionality compared to raw flour. When blended with HRSW, roasted navy bean flour had higher water absorption and dough stability values compared to raw navy bean flour blends. The loaf volume and crumb grain scores were greater for breads made with roasted navy bean flour compared to the breads made with raw flour. The pretreatment of chickpea, pea, and lentil impacted dough and bread properties differently (Baik and Han 2012). Cooking of the chickpea, pea, and lentil tended to have the greatest impact on dough and bread qualities compared to roasting and fermentation. Flour from germinated pulses has been evaluated in baking applications. Dough with germinated pulse flours tends to have longer mixing times and similar or slightly lower water absorptions compared to wheat flour (Morad et al. 1980; Hsu et al. 1982). In contrast, Sadowska et al. (2003) observed slightly higher water absorption and stability values in dough containing germinated pea flour compared to dough with raw pea flour. However, quality parameters were slightly better for bread made with raw pea flour compared to bread with germinated pea flour. Slight to no reductions in bread loaf volumes were observed in bread with 3–5% germinated pulse flour, but as levels increased to 20%, a significant reduction in bread volume was observed (Morad et al. 1980; Hsu et al. 1982). The starch structure and composition can be affected by germination (Morad et al. 1980), which could be responsible for some of the observed affect in bread systems that include germinated pulse flour.

The variability of results published on pulse utilization in baking relates to a number of factors. The usage level of the pulse and modification to the pulse prior to use in bread systems appear to be the most significant contributors to their impact on dough and bread quality. To a lesser extent, the pulse type can influence bread baking performance.

Gluten Free

Gluten-free (GF) bread baking suffers from the lack of gluten functionality. Many GF applications rely on gums to provide gluten-like functionality. However, recent interest in label-friendly ingredients has resulted in a shift to flours and fractions from commodities such as pulses.

Mariotti et al. (2009) reported that the highest water absorption values in a GF bread flour occurred when a combination of pea isolate and psyllium fiber was used at the 6 and 2% in the mix, respectively. The combination of pea isolate and psyllium in the flour also produced the best dough handling properties, which took on a wheat dough-like consistency compared to a liquid batter that is commonly found in GF batters. Marco and Rosell (2008a) reported that water absorption was dependent on level of protein in a rice flour model. Combination of soy and pea isolates, respectively, at concentrations of 25 and 1%, 1 and 25%, and 25 and 25% produced water absorptions of approximately 85, 39, and 150%, respectively. The synergistic benefit of soy and pea protein in contrast was not observed in mechanical properties of the dough. Separately, dough with increasing levels of soy or pea protein had increasing springiness values, while the combination of soy and pea resulted in springiness values that were lower than the dough containing the individual proteins (Marco and Rosell 2008a).

The specific volume of GF bread made with chickpea flour was significantly higher than specific volumes of bread made with pea isolate, soy flour, or carob germ flour (Minarro et al. 2012). Specific volumes of GF bread made with chickpea flour also were higher than bread made with tiger nut flour (Aguilar et al. 2015). Bread volume of a lupine-protein-fortified bread was higher than the volume for the pulse-free control, while pea-protein-fortified bread had a lower volume than the control (Ziobro et al. 2016). Muffins made with pea protein isolate had similar volumes to muffins made with soy protein isolate (Matos et al. 2014). However, muffins with pea protein isolate were softer and more elastic than muffins with soy protein isolate. Aprodu and Banu (2015) observed increased specific volumes in GF bread formulated with pea fiber and 0.1% glucose oxidase compared to the pea fiber alone. They also observed slightly lower crumb firmness values in the GF bread containing pea fiber and 0.1% glucose oxidase.

The rate of staling was slower in GF bread made with chickpea flour (Minarro et al. 2012) and combinations of chickpea flour with tiger nut flour (Aguilar et al. 2015). The complexing of starch, lipid, and protein through the emulsification activity of the chickpea flour was proposed as the reason for the better bread quality compared to other flours. However, Marco and Rosell (2008b) did not observe an increase in emulsification activity of rice flour when pea protein was added at 5%. Furthermore, 10% pea protein did not inhibit staling in a corn and potato starch-based bread formulation (Ziobro et al. 2016). The type and level of pulse or pulse ingredient appear to affect product quality. Thus, adjustments must be made for the desired application. Jeradechachai (2012) used response surface methodology to optimize yellow pea flour utilization in GF bread. The optimal conditions resulted in a high loaf specific volume (2.6 ml/g), soft crumb (174.2 g_f), bright crumb color (L* value =68.2), and small cell diameter (3.81 mm). The precooking of pea flour at 156.9 °C, water addition of 523.8 g, and proof time (18.0 min) were deemed the optimal conditions. Water addition had the greatest effects on the quality of pea flour GF breads. Loaf volume and cell diameter increased with increasing water additions.

Tortillas, Pita, and Flat Breads

Tortillas are flat bread used as a carrier for other ingredients such as beans and meat. Unlike pan breads, tortillas do not have as extensive a structure. Thus, non-wheat flours can be used at higher rates without significantly affecting the structure of the baked tortilla.

Anton et al. (2008) compared the flours of four different bean market classes as ingredients in tortillas. Regardless of market class, bean flour blended at rates of 15, 25, and 35% with wheat flour had higher water absorptions than the wheat flour alone. However, dough stability decreased significantly from the control with increasing levels of bean flour. The same results were observed for tortillas with pinto bean flour (Anton et al. 2009). However, Sharma et al. (1995) found lower water absorption and stability increased with the addition of chickpea and pigeon pea flours in a flat bread dough.

Increasing levels of bean flour reduced the firmness and cohesiveness of the tortillas compared to the all wheat control (Anton et al. 2008). However, only the tortillas with 25 and 35% bean flours had diameters that were greater than the control. Flat breads supplemented with chickpea and pigeon pea were firmer than control (Sharma et al. 1995). However, chickpea-fortified flat bread at 20% and pigeon pea addition at 10% produce better sensory quality than the unsupplemented flat bread. Firmness of pinto bean-fortified tortillas maintained lower firmness values than the control tortillas over a 7-day storage at both 4 and 25 °C conditions (Anton et al. 2009). Pita bread made with pinto bean flour had the best texture among pita breads made with flours of navy bean and lentil (Borsuk et al. 2012). These authors also reported coarse particles produced pita bread with greater water absorptions.

Cakes

Unlike bread, cake structure is driven by starch in the formulation. As a result, greater flexibility in the formulations can be used provided that there is sufficient structure-forming components available. Pea flour, pea protein, and pea hull have been used in cake formulations (Hillen 2016; Hoang 2012; Kaack and Pedersen 2005) and represent the diverse application of pulses in cake production.

Gómez et al. (2008) evaluated chickpea-wheat flour blends in cake formulations. Cakes with chickpea flour had pasting viscosities that were significantly lower than the wheat alone. They attributed the observation to the reduction in starch in the chickpea-wheat blends. The type of chickpea flour and cultivar of chickpea used to make the flour for blending did not impact viscosity parameters. Singh et al. (2015) observed significantly higher cake batter viscosity of navy bean flours with high protein compared to the control and lower protein navy bean flours. Gómez et al. (2012) reported similar sponge cake batter specific volumes between batters made with a 50:50 (wt %) blend of wheat and pin-milled pea flour and 100% wheat flour. However, 100% pin-milled pea flour had significantly lower batter specific volumes compared to the control. In contrast, all pea flour containing layer cake formulas had batter specific volumes that were greater than the 100% wheat flour control. Likewise, replacement of wheat flour with pea starch and protein fractions produced batter specific volumes that were greater than the wheat flour-only batter. However, the protein fraction did cause a reduction in batter specific volume compared to the control (i.e., wheat only) batter, while the starch fraction produced batters that had specific volumes similar to the control (Gómez et al. 2012). The densities of layer cake and sponge cake batters formulated with lentil or lentil-wheat (1:1) were lower and higher than the density of the control cake (100% wheat flour), respectively (de la Hera et al. 2012). In the 100% lentil batters, the batter made with the coarse (>140 um particles) flour had lower batter density than batters made with fine (<140 um particles) flour. No trend in cake batter density was observed for navy bean flours with different particles; however, flours with medium protein levels tended to produce batters with the lowest densities (Singh et al. 2015).

In both layer and sponge cakes, the batter viscosity of the pin-milled pea flour and starch fraction blends was not significantly different from the control batter viscosity. However, the pea protein fraction produced batters that were more viscous than the control, regardless of the level of protein fraction used in the batter formulation (Gómez et al. 2012). The addition of lentil flour significantly increased batter viscosity in both layer and sponge cake formulas (de la Hera et al. 2012). The particle size of the flours affected the batter viscosity differently. In layer cakes, the fine lentil flours had batter viscosities that were lower than batters made with coarse lentil flours. However, in sponge cakes, no relationship between particle size and batter viscosity was observed (de la Hera et al. 2012).

Cake volume was affected by the cultivar of chickpea used in the flour blends (Gómez et al. 2008). A 50% replacement of wheat flour with pea protein resulted in a significant reduction in both sponge and layer cake volumes compared to the cake made with 100% wheat flour (Gómez et al. 2012). Hoang (2012) also observed pea protein isolate caused a reduction in cake volume. Modification of cowpea proteins by defatting or glycation had little impact on cake volume but did reduce firmness of cakes (Campbell et al. 2016). Varying results were reported for replacement of wheat with pin-milled pea flour or a pea starch fraction (Gómez et al. 2012). In layer cakes, the starch fraction caused a reduction in cake volume while in sponge cakes only the cake with 50% starch fraction had lower cake volumes than the control. Cakes with pin-milled pea flour had cake volumes similar to the control. Layer cakes containing either 50 or 100% lentil flour had volumes lower than the control (i.e., 100% wheat flour) cakes (de la Hera et al. 2012). The opposite outcome was observed in sponge cakes. Furthermore, cakes made with fine lentil flours had volumes that were greater than cakes made with coarse lentil flours (de la Hera et al. 2012). In contrast, cakes made with whole navy bean flour tended to have greater cake volumes than cakes made with coarse or fine navy bean flours (Singh et al. 2015). The volume of muffins made with 30 and 60% pea fiber replacement for sugar was significantly lower than the control muffins. The reduction in volume was most pronounced in the muffin containing pea fiber at the 60% sugar replacement level (Struck et al. 2016).

Sponge cake characteristics tended to be affected to a greater degree than in layer cakes made with chickpea four. For example, sponge cake firmness increased significantly in cakes containing 50 and 100% chickpea flours compared to the cake with 100% wheat flour (Gómez et al. 2008). de la Hera et al. (2012) generally observed increased firmness in layer cakes made with lentil flours compared to the control cakes. The effect was pronounced in the cakes made with 100% lentil flour and with coarse lentil flours. In sponge cake, cakes made with lentil and wheat blends (1:1) had volumes similar to the control, while cakes made with 100% of the coarse lentil flours had significantly higher firmness values compared to the control (de la Hera et al. 2012). Whole navy bean flours tended to produce less firm cakes compared to cake flour, while coarse navy bean flours generally produced cakes with slightly greater firmness (Singh et al. 2015). All cakes made with navy bean flour had less springiness compared to the control cake. Gómez et al. (2012) also observed higher firmness values in both sponge and layer cakes made with a pea protein fraction compared to the control. In contrast, a pea starch fraction did not significantly impact firmness of the cakes, while the whole pin-milled pea flour at the 100% addition level produced cakes with higher firmness values than the control cakes. Unlike starch fractions, the firmness of muffins made with pea fiber as a replacement for sugar was significantly higher than the control muffins (Struck et al. 2016). The potential for proteins to denature during batter preparation generally is beneficial, but early (i.e., at lower temperatures) denaturation during baking can reduce the starch gelatinization and prevent expansion during baking, resulting in stiffening of the cake (Lee and Boonsupthip 2014). Ozkahraman et al. (2016) observed that the gelatinization degree (%) decreased from approximately 95% in the cake flour control to 79, 64, and 60% in cakes made with blends of 90% cake flour and 10% lentil, chickpea, and pea, respectively. Therefore, much of the observed effects on cakes likely is related to the effects of components on starch, which serves as a structure-forming component in cake systems.

Chickpea, pea, lentil, and bean flours were evaluated as a rice flour replacement (i.e., 50%) in GF cake formulas. Bean flours had the most significant effect on batter viscosity followed by chickpea, lentil and, pea flours (Gularte et al. 2012). Cakes made with lentil flour had the greatest specific volume followed by bean and pea and finally chickpea, which had the same specific volume as the control. Softer more elastic cakes were made with lentil flour compared to the all-rice control cake (Gularte et al. 2012). Cakes with other pulses tended to have greater hardness values compared to the control cake. Hillen (2016) pretreated pea flour with ethanol as a means to reduce flavor compounds prior to application in GF cakes. The cakes made with pea flour treated with ethanol had higher cake height values and lower firmness values than the cake made with raw pea. The sensory panel also rated the flavor, texture, and overall acceptability of cakes made with the ethanol-treated pea flour better than the cake made with raw pea flour.

Cookies and Crackers

Cookies and crackers fall under the sweet and savory snack food categories. Most of these products tend to be high in calories, fats, and easily digestible carbohydrates. However, recent trends indicate an interest in making these products healthy. One approach has been to incorporate more whole grains and to a lesser extent pulses.

Cookies are considered a shortened bakery product because a fat source (e.g., shortening) is a major part of the formula. The added shortening interferes with gluten formation and results in a product with a soft texture. However, overcoming starch retrogradation in cookie products is necessary to prevent staling or unwanted firmness. The incorporation of pulse flours into cookies and other food items has been recognized as an approach to improve the nutrient composition (Anton et al. 2008).

Cookies have served as a model system for pulse flour addition for over three decades. In the context of this review, the study by Patel and Rao (1995) represents studies predating the year 2000, and thus, the chapter author encourages readers to review the highlighted publication for additional information prior to 2000. Patel and Rao (1995) reported significant reductions in cookie diameters and diameter-to-spread ratio with increasing incorporation of blackgram bean flour. Of the black bean flours tested, flour made from germinated black beans had the smallest cookie diameters. At comparable black bean percentages in the blends, cookies made with untreated black bean had the highest diameter-to-spread ratio and germinated the lowest (Patel and Rao 1995). Cookie weights were highest for cookies made with germinated black bean flour. Cookies with germinated black bean flours also had weights greater than all-wheat control cookie. The hardness of cookies increased with increasing black bean incorporation into cookies. Cookies made with roasted black bean flour tended to have the lowest hardness values at comparable black bean percentages in the blends. Cookies made with germinated black bean flour had the greatest hardness value. The lowest hardness value likely was attributed to the higher water absorption values observed in the roasted black bean flour (Patel and Rao 1995).

Green lentil, navy bean, and pinto bean flours milled into a coarse and fine particles were evaluated in a cookie. The interaction between pulse (i.e., flour), substitution level, and particle size (i.e., degree of milling) for thickness indicated significant differences among the control cookies and cookies made from pulses

(Zucco et al. 2011). However, a difference in cookie thickness was observed between cookies made with differing pulses at each level of substitution. In general, cookies made with fine green lentil flours had the greatest thickness. The spread of cookies made with navy and pinto bean flours tended to be less than for cookies made with green lentils. Cookies made with fine-particle flours generally had less spread than cookies made from coarse particles, regardless of pulse flour used (Zucco et al. 2011). The higher damaged starch levels were believed to be the reason for lower cookie spread in cookies made from navy and pinto bean flours. Cookie hardness was affected by flour type. Cookies made with lentil flours tended to have higher hardness values compared to cookies made with navy and pinto bean flours. Furthermore, beyond 25% substitution, fine pulse flour produced harder cookies than coarse flours.

Siddiq et al. (2013) observed greater cookie diameter and height for cookies made with extruded navy bean flour compared to steam-treated navy bean flour. In contrast, no difference in cookie quality was observed in thermally processed pinto bean flour. However, sensory evaluation supported extrusion over steam cooking in terms of flavor, texture, and overall acceptability. Extrusion tends to remove some of the off-flavors associated with beans. Simons and Hall (2017) reported no significant differences in cookie spread or hardness for cookies made with 40% raw, cooked, or germinated pinto bean flours, which is in agreement with results reported by Siddiq et al. (2013). Although not significant, cookies made with cooked pinto flour had flavor and acceptability scores greater than other treatments. The studies by Siddiq et al. (2013) and Simons and Hall (2017) support the pretreatment of bean flours prior to incorporation into products such as cookies due to reduction in off-flavors.

Hillen (2016) reported improved sensory flavor, texture, and overall acceptance scores in cookies made from ethanol-extracted pea flour compared to cookies containing raw pea flour. Cookies made with ethanol-extracted pea flours had greater cookie heights but lower cookie diameters than the cookies with raw pea flour. Combinations of rice flour, corn starch, and pea protein resulted in cookies with varying cookie thicknesses and widths, but not spread. The only treatment that had significantly different spread was the treatment with 60% corn starch and 40% rice flour (Mancebo et al. 2016). Pea protein negatively impacted both thickness and width as observed by decreasing values for parameters with increasing pea protein concentrations. The authors suggested that the glass transition temperature of the protein or the higher dough viscosities resulting from protein addition might have been responsible for the observed effects (Mancebo et al. 2016). In addition, the ability of proteins to bind water was the likely reason for the lower hardness values of protein-fortified cookies. The cookies containing pea protein did have the highest overall acceptability of the cookies evaluated by a sensory panel. Hoang (2012) reported no significant differences in overall acceptability of the control cookie and cookie containing pea protein isolate. However, cookies made with pea protein treated with transglutaminase had sensory scores significantly lower than the control.

Unlike cookies, crackers follow a manufacturing process similar to flatbreads and breads. However, sheeting of the cracker dough is important for texture development and is the most common manufacturing method. The dough is produced through a chemical leavening step or sponge and dough process followed by sheeting (Han et al. 2010). Flavor develops during the dough development stage and baking step of the process. Therefore, the addition of ingredients that interfere with sheeting and the dough development will affect the cracker quality. Limited data is available regarding pulse usage in crackers.

Han et al. (2010) systematically evaluated several pulses and fractions as ingredients in gluten-free crackers. Chickpea flour produced a cracker with the greatest peak force followed by lentils and crackers made with pea fiber. The least firm cracker was made with navy bean flour. However, sensory panel rated crackers with pea protein or pea fiber crispier than with the pulse flours, although values were not significant. Crackers made with chickpea flour had the highest crispiness rating among pulse flours. The addition (i.e., 40%) of broad bean and green and yellow pea flour to a wheat formulation resulted in crackers with significantly more protein than the wheat control cracker (Millar et al. 2017). The dietary fiber content of the pulse-containing crackers was also higher than control. The weights of pulse-containing crackers were less than those of the control. However, the difference likely is due to the higher moisture content of the control cracker. Both cracker volume and hardness were lower in crackers with pulses compared to the control (Millar et al. 2017).

Conclusion

The interest in using pulses for nutritional enhancement fits well in bakery products due to the complementary nature of the protein to wheat and it being a source of folate, which is typically added to cereal products. The recent interest in optimizing milling methods stems from the general lack of pulse milling knowledge. Several researchers have documented that particle size can influence product quality. For example, fine particles allowed cakes to expand more, resulting in greater cake volume. In contrast, fine particles were found to inhibit cookie spread. Additional research is needed to further define the role of pulse particle size in product quality. Pulses appear to be less impactful on cookie and cake quality in comparison to bread. In bread, substitution levels of 10% or less generally can be used without significantly impacting bread quality. However, levels beyond 10% clearly impact bread quality. Limited studies have been completed using combinations of commercially available dough conditioners or vital wheat gluten and pulses. Alasino et al. (2011) recommended the use of sodium stearoyl lactylate in formulas containing 10% pea flour. However, this could be an area of research to explore. Furthermore, few researches have investigated the functionality of pulses in crackers. The protein and fiber of pulses make pulses an ideal ingredient in crackers, which typically lack high protein and fiber contents unless specifically targeted during manufacturing.

References

Aguilar N, Albanell E, Minarro B, Capellas M (2015) Chickpea and tiger nut flours as alternatives to emulsifier and shortening in gluten-free bread. LWT-Food Sci Tech 62(1):225–232

Alasino M, Osella C, De LTM, Sanchez H (2011) Use of sodium stearoyl lactylate and azodicarbonamide in wheat flour breads with added pea flour. Int J Food Sci Nut 62:385–391

Anton AA, Ross KA, Lukow OM, Fulcher G, Arntfield SD (2008) Influence of added bean flour (Phaseolus vulgaris L.) on some physical and nutritional properties of wheat flour tortillas. Food Chem 109(1):33–41

Anton AA, Lukow OM, Fulcher RG, Arntfield SD (2009) Shelf stability and sensory properties of flour tortillas fortified with pinto bean (Phaseolus vulgaris L.) flour: effects of hydrocolloid addition. LWT-Food Sci Tech 42(1):23–29

Aprodu I, Banu I (2015) Influence of dietary fiber, water, and glucose oxidase on rheological and baking properties of maize based gluten-free bread. Food Sci Biotech 24(4):1301–1307

Baik BK, Han IH (2012) Cooking, roasting, and fermentation of chickpeas, lentils, peas, and soybeans for fortification of leavened bread. Cereal Chem 89(6):269–275

Batista KA, Prudencio SH, Fernandes KF (2011) Wheat bread enrichment with hard-to-cook bean extruded flours: nutritional and acceptance evaluation. J Food Sci 76(1):S108–S113

Bergthaller W, Dijkink BH, Langelaan HC, Vereijken JM (2001) Protein from pea mutants as a co-product in starch separation – isolates from wet and dry separation: yield, composition and solubility. Nahrung/Food 45:390–392

Borsuk Y, Arntfield S, Lukow OM, Swallow K, Malcolmson L (2012) Incorporation of pulse flours of different particle size in relation to pita bread quality. J Sci Food Agric 92(10):2055–2061

Campbell L, Euston SR, Ahmed MA (2016) Effect of addition of thermally modified cowpea protein on sensory acceptability and textural properties of wheat bread and sponge cake. Food Chem 194:1230–1237

D'Appolonia BL (1977) Rheological and baking studies of legume-wheat flour blends. Cereal Chem 54(1):53–63

D'Appolonia BL (1978) Use of untreated and roasted navy beans in bread making. Cereal Chem 55(6):898–907

Dalgetty DD, Baik BK (2006) Fortification of bread with hulls and cotyledon fibers isolated from peas, lentils, and chickpeas. Cereal Chem 83(3):269–274

De la Hera E, Ruiz-Paris E, Oliete B, Gomez M (2012) Studies of the quality of cakes made with wheat-lentil composite flours. LWT-Food Sci Tech 49(1):48–54

Eigenmann PA, Beyer K, Burks AW, Lack G, Liacouras CA, Hourihane JO, Sampson HA, Sodergren E (2008) New visions for food allergy: an iPAC summary and future trends. Pediatr Allergy Immunol 19:26–39

Fendri LB, Chaari F, Maaloul M, Kallel F, Abdelkafi L, Chaabouni SE, Ghribi-Aydi D (2016) Wheat bread enrichment by pea and broad bean pods fibers: effect on dough rheology and bread quality. LWT-Food Sci Tech 73:584–591

Fenn D, Lukow OM, Humphreys G, Fields PG, Boye JI (2010) Wheat-legume composite flour quality. Int J Food Prop 13(2):381–393

Gomez M, Ronda F, Blanco CA, Caballero PA, Apesteguia A (2003) Effect of dietary fibre on dough rheology and bread quality. Eur Food Res Tech 216(1):51–56

Gomez M, Oliete B, Rosell CM, Pando V, Fernandez E (2008) Studies on cake quality made of wheat-chickpea flour blends. LWT-Food Sci Tech 41(9):1701–1709

Gomez M, Doyague MJ, de la Hera E (2012) Addition of pin-milled pea flour and air-classified fractions in layer and sponge cakes. LWT-Food Sci Tech 46(1):142–147

Gujska E, Khan K (1991) Functional-properties of extrudates from high starch fractions of navy and pinto beans and corn meal blended with legume high protein-fractions. J Food Sci 56(2):431–435

Gularte M, Gomez M, Rosell CM (2012) Impact of legume flours on quality and in vitro digestibility of starch and protein from gluten-free cakes. Food Bioprocess Technol 5:3142–3150

Hall C (2017) Pulse flour milling and pulse protein isolation. Pulse product innovation – pathway to future foods session. American Association of Cereal Chemist's International annual meeting, San Diego
Han JY, Khan K (1990a) Physicochemical studies of pin-milled and air-classified dry edible bean fractions. Cereal Chem 67(4):384–390
Han JY, Khan K (1990b) Functional-properties of pin-milled and air-classified dry edible bean fractions. Cereal Chem 67(4):390–394
Han J, Janz JAM, Gerlat M (2010) Development of gluten-free cracker snacks using pulse flours and fractions. Food Res Int 43(2):627–633
Hillen C (2016) Sensory and quality attributes of deodorized pea flour used in gluten-free food products. A Thesis Submitted to the Graduate Faculty of the North Dakota State University of Agriculture and Applied Science. ProQuest Dissertations Publishing. https://search-proquest-com.ezproxy.lib.ndsu.nodak.edu/docview/1853477020?pq-origsite=primo
Hoang H (2012) Evaluation of Pea protein and modified Pea protein as egg replacers A Dissertation Submitted to the Graduate Faculty of the North Dakota State University of Agriculture and Applied Science. ProQuest Dissertations Publishing. https://library.ndsu.edu/ir/handle/10365/26825
Hsu DL, Leung HK, Morad MM, Finney PL, Leung CT (1982) Effect of germination on electrophoretic, functional, and bread-baking properties of yellow pea, lentil, and faba bean protein isolates. Cereal Chem 59(5):344–350
Jeradechachai T (2012) Application of response surface methodology in the development of gluten-free bread with yellow pea flour addition. A Thesis Submitted to the Graduate Faculty of the North Dakota State University of Agriculture and Applied Science. ProQuest Dissertations Publishing. https://library.ndsu.edu/ir/handle/10365/26829
Kaack K, Pedersen L (2005) Low energy chocolate cake with potato pulp and yellow pea hulls. Eur Food Res Tech 221(3–4):367–375
Kallenbach J, Kaiser A, Fujiwara N, Hall C, Manthey F (2017) Flow properties of flour obtained from hammer milled peas. Pulses for sustainable agriculture session. American Association of Cereal Chemist's International annual meeting. San Diego, CA, Oct. 8–11, 2017. Poster Abstract 1316
Kasprzak M, Rzedzicki Z (2010) Effect of pea seed coat admixture on physical properties and chemical composition of bread. Int Agrophys 24(2):149–156
Kaur M, Sandhu KS (2010) Functional, thermal and pasting characteristics of flours from different lentil (*Lens culinaris*) cultivars. J Food Sci Tech- Mysore 47(3):273–278
Kon S, Sanshuck DW, Jackson R, Huxsoll CC (1977) Air-classification of bean flour. J Food Proc Pres 1:69–77
Lee SJ, Boonsupthip W (2014) Physical, chemical, and sensory properties of antioxidant-enriched raw and cooked rice by vacuum drying impregnation in a semidry state. Cer Chem 91(5):445–452
Luhovyy BL, Hamilton A, Kathirvel P, Mustafaalsaafin H (2017) The effect of navy bean flour particle size on carbohydrate digestion rate measured in vitro. Cer Foods World 62(5):208–213
Mancebo CM, Rodriguez P, Gomez M (2016) Assessing rice flour-starch-protein mixtures to produce gluten free sugar-snap cookies. LWT-Food Sci Tech 67:127–132
Marco C, Rosell CM (2008a) Functional and rheological properties of protein enriched gluten free composite flours. J Food Eng 88(1):94–103
Marco C, Rosell CM (2008b) Effect of different protein isolates and transglutaminase on rice flour properties. J Food Eng 84(1):132–139
Mariotti M, Lucisano M, Pagani MA, Ng PKW (2009) The role of corn starch, amaranth flour, pea isolate, and *Psyllium* flour on the rheological properties and the ultrastructure of gluten-free doughs. Food Res Int 42(8):963–975
Maskus H, Bourre L, Fraser S, Sarkar A, Malcolmson L (2016) Effects of grinding method on the compositional, physical, and functional properties of whole and split yellow pea flours. Cer Foods World 61(2):59–64

Matos ME, Sanz T, Rosell CM (2014) Establishing the function of proteins on the rheological and quality properties of rice based gluten free muffins. Food Hydrocoll 35:150–158
Millar KA, Barry-Ryan C, Burke R, Hussey K, McCarthy S, Gallagher E (2017) Effect of pulse flours on the physiochemical characteristics and sensory acceptance of baked crackers. Int J Food Sci Tech 52(5):1155–1163
Minarro B, Albanell E, Aguilar N, Guamis B, Capellas M (2012) Effect of legume flours on baking characteristics of gluten-free bread. J Cer Sci 56(2):476–481
Mohammed I, Ahmed AR, Senge B (2012) Dough rheology and bread quality of wheat-chickpea flour blends. Ind Crops Prod 36(1):196–202
Mohammed I, Ahmed AR, Senge B (2014) Effects of chickpea flour on wheat pasting properties and bread making quality. J Food Sci Tech- Mysore 51(9):1902–1910
Morad MM, Leung HK, Hsu DL, Finney PL (1980) Effect of germination on physicochemical and bread-baking properties of yellow pea, lentil, and faba bean flours and starches. Cereal Chem 57(6):390–396
Nosworthy MG, Franczyk A, Zimoch-Korzycka A, Appah P, Utioh A, Neufeld J, House JD (2017a) Impact of processing on the protein quality of pinto bean (*Phaseolus vulgaris*) and buckwheat (*Fagopyrum esculentum* Moench) flours and blends, as determined by in vitro and in vivo methodologies. J Agric Food Chem 65(19):3919–3925
Nosworthy MG, Franczyk AJ, Medina G, Neufeld J, Appah P, Utioh A, Frohlich P, House JD (2017b) Effect of processing on the in vitro and in vivo protein quality of yellow and green split peas (*Pisum sativum*). J Agric Food Chem 65(35):7790–7796
Ozkahraman BC, Sumnu G, Sahin S (2016) Effect of different flours on quality of legume cakes to be baked in microwave-infrared combination oven and conventional oven. J Food Sci Tech-Mysore 53(3):1567–1575
Patel MM, Rao GV (1995) Effect of untreated, roasted and germinated black gram (*Phaseolus mungo*) flours on the physicochemical and biscuit (cookie) making characteristics of soft wheat-flour. J Cer Sci 22(3):285–291
Patel KM, Bedfor CL, Younds CW (1980) Amino-acid and mineral profile of air-classified navy bean flour fractions. Cereal Chem 57:123–125
Pelgrom PJM, Vissers AM, Boom RM, Schutyser MAI (2013) Dry fractionation for production of functional pea protein concentrates. Food Res Int 53(1):232–239
Pelgrom PJM, Boom RM, Schutyser MAI (2015a) Functional analysis of mildly refined fractions from yellow pea. Food Hydrocoll 44:12–22
Pelgrom PJM, Wang J, Boom RM, Schutyser MAI (2015b) Pre- and post-treatment enhance the protein enrichment from milling and air classification of legumes. J Food Eng 155:53–61
Sadowska J, Blaszczak W, Fornal J, Vidal-Valverde C, Frias J (2003) Changes of wheat dough and bread quality and structure as a result of germinated pea flour addition. Eur Food Res Tech 216(1):46–50
Sharma S, Sekhon KS, Nagi HPS (1995) Legume supplemented flat bread: nutritive value, textural and organoleptic changes during storage. J Food Proc Pres 19:207–222
Siddiq M, Kelkar S, Harte JB, Dolan KD, Nyombaire G (2013) Functional properties of flour from low-temperature extruded navy and pinto beans (*Phaseolus vulgaris* L.). LWT-Food Sci Tech 50(1):215–219
Simons CW, Hall C (2017) Consumer acceptability of gluten-free cookies containing raw cooked and germinated pinto bean flours. Food Sci Nut 6(1):77–84
Simons CW, Hunt-Schmidt E, Simsek S, Hall C, Biswas A (2014) Texturized pinto bean protein fortification in straight dough bread formulation. Plant Foods Human Nut 69(3):235–240
Simons C, Hall C, Biswas A (2017) Characterization of pinto bean high-starch fraction after air classification and extrusion. J Food Proc Pres 41(6):e13254
Singh M, Byars JA, Liu SX (2015) Navy bean flour particle size and protein content affect cake baking and batter quality. J Food Sci 80(6):E1229–E1234
Sosulski F, Youngs CG (1979) Yield and functional properties of air-classified protein and starch fractions from eight legume flours. J Amer Oil Chem Soc 56:292–295

Struck S, Gundel L, Zahn S, Rohm H (2016) Fiber enriched reduced sugar muffins made from iso-viscous batters. LWT-Food Sci Tech 65:32–38

Tyler RT, Youngs CG, Sosulski FW (1981) Air classification of legumes. 1. Separation efficiency, yield, and composition of the starch and protein-fractions. Cer. Chem 58(2):144–148

Udahogora M (2012) Health benefits and bioactive compounds in field peas, faba beans, and chickpeas. In: Shahidi F, Yu L, Tsao R (eds) Cereals and pulses nutraceutical properties and health benefits. Wiley, Ames, pp 199–216

Vose JR, Basterrechea MJ, Gorin PAJ, Finlayson AJ, Youngs CG (1976) Air classification of field peas and horsebean flours - chemical studies of starch and protein-fractions. Cer Chem 53(6):928–936

Ziobro R, Juszczak L, Witczak M, Korus J (2016) Non-gluten proteins as structure forming agents in gluten free bread. J Food Sci Tech – Mysore 53(1):571–580

Zucco F, Borsuk Y, Arntfield SD (2011) Physical and nutritional evaluation of wheat cookies supplemented with pulse flours of different particle sizes. LWT-Food Sci Tech 44(10):2070–2076

Bioactives and Nutraceuticals in Food Legumes: Nutritional Perspective

Faruk Toklu, Debjyoti Sen Gupta, Tolga Karaköy, and Hakan Özkan

Abstract Peas, common bean, cowpea, chickpeas, and lentils are the major protein sources in the grain legume group and have an important place in terms of human and animal feeding. High protein content and other nutrient compositions like micronutrients, dietary fiber, prebiotics, and folates make leguminous seeds an important food source. Low glycemic index (GI) content of food legumes is important for long-term benefit of health. Isoflavones within legumes (isoflavones = an active substance that stabilizes the estrogen hormone) play an important role in the resistance mechanism of the plant, in the formation of nodules in the root, and have human health benefits. Proteins, glycosides, tannins, saponins, alkaloids, and many other bioactive substances in grain legume seeds increased the importance of legumes for human consumption. In developing and underdeveloped countries, because of the high cost of animal proteins, people meet their everyday protein requirements from cheaper legumes. In addition, they get additional supply of the abovementioned bioactives for a better health and nutrition.

Keywords Bioactives · Nutrients · Anti-nutrients · Protein · Glycemic index · Dietary fiber

F. Toklu (✉) · H. Özkan
Faculty of Agriculture, Field Crops Department, Çukurova University, Adana, Turkey
e-mail: fapet@cu.edu.tr

D. Sen Gupta
ICAR-Indian Institute of Pulses Research, Kanpur, UP, India

T. Karaköy
Faculty of Agricultural Sciences and Technology, Sivas University of Science and Technology, Sivas, Turkey

D. S. Gupta et al. (eds.), *Breeding for Enhanced Nutrition and Bio-Active Compounds in Food Legumes*, https://doi.org/10.1007/978-3-030-59215-8_10

Introduction

Today, the world population is around 7.2 billion, and it is expected that this number will reach 9.7 billion in 2050 and 11.2 billion in the year 2100. Today, the current population in developing and underdeveloped countries could not get enough and healthy food. For this reason, to boost agricultural production in general, promoting sustainable agricultural practices and ensuring high-quality and adequate food production are among the main topics that engage researchers. Grain legumes such as dry beans, chickpeas, lentils, and peas are important sources of human nutrition, starch, and mineral substance (Iqbal et al. 2006; Karaköy et al. 2012). On the other hand, legume biomass and straw are rich in mineral matter and carbohydrates, making them important animal feed (Muehlbauer et al. 2006). Especially in underdeveloped and developing countries, people could not get enough animal protein or usually animal protein is not part of the regular diet. People who live here receive a significant portion of the protein requirement from the edible grain legumes (Upadhyaya et al. 2011; Blair 2013; Toklu et al. 2015). The other issue that makes grain legumes important is the rhizobium-type bacteria that live as symbionts in their roots; because of their ability to fix the free nitrogen in the soil, it has an important place in terms of food safety and sustainable agriculture. At the same time, it is one of the products that must be included in the production systems as an integral part of organic agricultural production. Due to their unique amino acid composition, they have an important place in diets of humans and feeds of animals. They are especially rich in terms of mineral substances that are important for the development of children, such as iron (Fe) and zinc (Zn). In recent years, some of the substances found in the seeds are also very important for human health in medical terms which has been revealed by various researchers (WHO 1999; Mayer et al. 2008; White and Broadley 2009).

In this chapter, the role of grain legumes will be evaluated with its nutritional value, bioactives, and nutraceutical properties.

Nutrient Contents of Edible Legumes

Mineral elements play important physiological roles in plants and in the human body. The human body requires more than 22 minerals that can be supplied by an appropriate diet (White and Broadley 2005), and the most important minerals are phosphorus (P), potassium (K), calcium (Ca), magnesium (Mg), iron (Fe), zinc (Zn), copper (Cu), and manganese (Mn). Dietary deficiencies in mineral elements can have significant negative impacts, such as learning disabilities in children, increased morbidity and mortality, low worker productivity, and high healthcare costs. The most common micronutrient deficiencies are Fe, Zn, and I, but certain populations may also suffer from deficiencies in Mg, Ca, and Se (White and Broadley 2005). It has been estimated that nearly 3.7 billion people worldwide are

Fe-deficient (60%) and that 54% of these 3.7 billion people are severely deficient (Yang et al. 2007). Zn deficiency ranks the 11th among the 20 most important nutritional deficiencies worldwide and the fifth among the 10 most important deficiencies in developing countries (Cakmak 2008). Hotz et al. (2004) reported that Zn deficiency affects about one-third of the world population and that its incidence ranges from 4% to 73% depending on the country. Micronutrient deficiencies mainly result from their low concentrations in the daily diet. The concentrations of most minerals in most plant foods are not enough to meet daily dietary requirements when these foods are consumed in typical amounts. Hence, there has been an interest in increasing the mineral concentrations of various seed crops. Although food supplements were traditionally used to treat mineral deficiencies, agricultural strategies for increasing micronutrient density in foods are now being assessed as sustainable and long-term solutions.

Legumes (*Leguminosae* or *Fabaceae*) are among the greatest three plant families; the family includes about 700 taxon and 18,000–20,000 species. The family generally includes annual or perennial plants in the forms of herb, shrub, and tree. The family is quite widespread throughout the world, except for the poles. Besides being a significant family to be used in human food and animal feed, the family also includes several ornamental species. There are also some kind of plants in this family used in lumber, flooring, and dye industries.

Edible legumes include beans, peas, lentils, cowpeas, chickpeas, and broad beans, and they all have been used in human diets for thousands of years. Evidences indicated that the history of edible legumes went back to 3000 BC and that they were used by the Mediterranean, Mesopotamians, Egyptians, Hungarians, Trojans, and ancient Britons. Legumes were used to be known by farmers as "antique" nutrients or "old-fashion nutritional habit" until quite recently. However, such a perception has changed with the introduction of rice, bread, and meat-like "modern" basic nutrients. While local consumptions are stable in developing countries, the demand for legumes is increasing in the USA and developed Western European countries. The reason for such increasing demands was expressed as quite distinctive nutritional attributes of the legumes (Devos 1988).

With regard to nutritional properties, the general characteristics of edible legumes can be specified with its high protein contents, lysine-rich essential amino acids, poor in methionine and cysteine essential amino acids, perfect complementary protein sources for cereal grains, zero cholesterol levels, and hard-to-digest because of anti-nutritional substances (Kostyra 1996; Duranti 2006).

Proteins

In general, grain legumes are moderate to good sources of protein, containing 150 to 400 g/kg crude protein (Hedley 2001). The predominant protein fraction in legume seeds is made of globulins (60%–90%), which are storage proteins rich in arginine, glutamic acid, aspartic acid, and their amides. However, legume seeds are deficient in sulfur-containing amino acids (Wang et al. 2003). The deficiency of these amino acids, however, does not pose a problem in commercial feed manufacturing because of the availability and low cost of crystalline methionine. The deficiency of methionine and cystine could also be overcome, in part, by mixing the legume seeds with cereal proteins (Shewry and Tatham 1999). This is the reason why rice or wheat-based foods with cooked pulses are part of the diet in many countries in Asia.

Edible grain legumes are cheap and high-quality protein sources. Legume grains contain twice as much protein than cereal grains, and protein contents usually vary between 20% and 25%. When used to compensate animal-originated protein in diets, high plant-originated protein contents of edible legumes generally reduce blood cholesterol levels (Anderson et al. 1999). Edible legume protein is quite rich in amino acid lysine, which exists at quite low levels in cereals, and regarding this essential amino acid, they are almost equal to red meat. Protein contents of bean (*Phaseolus vulgaris* L.) grains at dry-ripe conditions vary between 14.6% and 35.1%. Bean grain proteins are composed of mixture of several proteins. Of these proteins, 62%–95% are dissolved in water, about 13% are dissolved in soda, and about 2%–25% are dissolved in salts. Protein contents and amino acid composition generally vary based on growing conditions and sowing times. Bean protein contains lysine (74.4%), threonine (27.9%), leucine (43.7%), phenylalanine (71.2%), tryptophan (0.6%), and methionine and cysteine (71.4%) amino acids. Bean grains contain about 45.0%–58.0% nitrogen-free substances. The high starch content of the grains increases nutritional value of the beans (Şehirali 1988). Peas (*Pisum sativum* L.) are used in human nutrition either as green or dry grains. Green grains are not used as a significant source of protein, but protein contents of dry grains vary between 18.3% and 28.4%. Protein contents of broad bean (faba bean) (*Vicia faba* L.) grains in dry-matter basis vary between 25.5% and 36.05%. Protein concentrations of summer broad bean species are genetically greater than the protein concentrations of winter species. Small-grained ones have greater protein contents than large-grained ones, and late-ripening ones have greater protein contents than early-ripening ones. Protein quality of broad beans is largely related to lysine contents. Previous studies revealed that broad beans had the second place among edible legumes after soybean regarding lysine contents. However, there is a negative correlation between protein content and lysine content of broad beans. Therefore, selections in breeding lines in terms of high protein contents generally reduce protein quality. Cowpea (*Vigna sinensis* L.) is another edible legume species. Fresh pods have protein contents of between 2.0% and 4.3%, and fresh grains have protein contents of between 4.5% and 5.0%. However, protein concentrations of dry-ripened cowpea grains vary between 20.42% and 34.60% based on cultivars and environmental conditions. Chickpeas (*Cicer arietinum* L.) are among the mostly cultivated edible legumes worldwide, and chickpea grains have protein content between 16.4% and 31.2% (Erdemci et al. 2017). Chickpea protein contains isoleucine (1.95 mg), leucine (1.54 mg), lysine (1.44 mg), methionine (276 mg), phenylalanine (1.012 mg), threonine (739 mg), tryptophan (710 mg), and valine (1.025 mg) amino acids. Even at quite low levels, tryptophan reduces the protein quality of chickpea grains. Lysine concentrations decrease, but histidine and cysteine concentrations increase with the progress of ripening. Tryptophan contents initially increase but then gradually decrease with the progress of ripening. The other amino acid quantities generally increase toward the ripening period. Lentils (*Lens culinaris* Medik.) are among the most significant edible legumes worldwide, and protein contents of lentil grains generally vary between 20.40% and 30.90%. Lentil protein contains tryptophan (0.22 mg), isoleucine (1.32 mg), lysine (1.53 mg), phenylalanine (1.41 mg), methionine (0.18 mg), leucine (1.76 mg), threonine (0.89 mg), and valine (1.36 mg) amino acids. The straw and pod residues left after harvest and threshing of lentils contain about 10.2% moisture, 1.8% fat, 4.4% protein, 50% carbohydrate, 21.4% cellulose, and 12.2% ash; thus, they have significant place in animal feeding.

On the other hand, except for cowpeas, edible legume protein is quite poor in methionine amino acid as compared to cereal protein (Tiwari and Singh 2012). Except for broad beans, the digestibility of edible legume proteins usually varies between 71% and 94% based on the cultivars (Grabner and Hofer 1985). Trypsin inhibitor is the primary reason for the low digestibility of broad bean protein. As compared to cereals, edible legumes are quite rich in tryptophan, lysine, and aspartic acid like amino acids, but they contain less methionine, cysteine, and glutamic acid. Therefore, mixtures of lentil and chickpea with wheat and rice almost compensate such a deficiency of legumes and provide a balanced diet (Sharma 1988). To compensate deficiencies of legumes in basic amino acids, they should be combined with the other food stuff rich in those amino acids. For instance, the combination of broad beans with cereals improves protein quality of both species.

Lipids

Edible legumes usually have low fat ratios (about 0.8–1.55), and they do not contain cholesterol. Such an attribute makes them a perfect heart health-friendly source of fat, and thus, they are largely used as an alternative option in the prevention of cardiovascular diseases. Except for soybean and peanut (not counted as pulses), most of the legumes are quite poor in fats. Fat contents of peas, lentils, broad beans, and beans vary between 1% and 2%. The fat content of chickpea varies between 4% and 5%. Legume fats are usually polyunsaturated and have high linoleic acid levels. Such a case indicates quite high nutritional value. Fats present are quite less influenced by processing practices (Devos 1988; Pekşen and Artık 2005).

Vitamins

Raw (uncooked) legumes are quite rich in B-group vitamins and generally deficient in A-, C-, and E-group vitamins. Peeling usually increases vitamin content of legumes. Cooking, on the other hand, reduces vitamins, especially vitamin B1, B2, and C contents. Excessive cooking negatively influences vitamin B contents. Since B-group vitamins dissolve in water, they are largely lost through cooking water. Pressure-cooking or autoclave-type cooking are the best way of cooking without losing the vitamins (Devos 1988; Pekşen and Artık 2005).

Bean vitamins vary based on growing conditions and sowing periods. Bean grains are quite rich in provitamin A (carotene) and ascorbic acid (vitamin C), but carotene contents decrease, and vitamin C contents increase with the progress of ripening (Pekşen and Artık 2005). Pea grains are also rich in vitamins B and C. However, green-ripened grains are richer in these vitamins than dry-ripened grains.

Folates (Vitamin B9 and Salts)

Folic acid plays a significant role in healthy cell formation, and it is a water-dissolved vitamin B. Water-dissolved does not necessarily mean that the vitamin stayed in the body for longer durations. Therefore, the vitamin should be taken daily to prevent damages on nerve cells. Vitamin B9 needs of the body increases during pregnancy and fetus development periods (Margaret et al. 2001). Edible legumes, especially beans, are a significant source of vitamin B and folate and meet almost half of daily needs of humans at every meal. Growing environment greatly affects concentration of folates in seeds. It was found that folate concentration in lentil cultivars ranged from 216 to 290 μg/100 g, in chickpea from 42 to 125 μg/100 g, in yellow field pea from 41 to 55 μg/100 g, and in green field pea from 50 to 202 μg/100 g (Sen Gupta et al. 2013). Jha et al. (2015) studied another set of different pulses and found that the total folate concentration ranged from 351 to 589 μg/100 g in chickpea, 165 to 232 μg/100 g in common bean, 136 to 182 μg/100 g in lentil, and 23 to 30 μg/100 g in pea.

Minerals

Same as the other unrefined nutrients, legumes are also quite rich in minerals, especially in potassium (K), phosphorus (P), calcium (Ca), and iron (Fe). Peeling reduces mineral contents, and cooking mixes minerals into cooking water. Mineral contents of 100 g bean grain are specified as follows: K (1.2–1.9 g), P (0.49–0.58 g), Ca (0.1–0.2 g), Mg (0.15–0.20 g), S (0.05–0.23 g), Fe (0.012–0.008 g), and Mn (0.002 g) (Gebhardt and Thomas 2002). These values largely vary with the growing conditions and sowing times of beans. Similarly, in lentils, concentrations of Fe, Zn, Ca, Cu, and Mg in raw seeds varied from 26 to 92, 17 to 51, 97 to 536, 3 to 12, and 272 to 892 mg/kg, respectively (Sen Gupta et al. 2016). Analytical methods used also affect estimations.

Carbohydrates and Dietary Fibers

Carbohydrates are composed of sugar, starch and other polysaccharides. Starch is the major component of legumes, and starch contents vary between 35% and 53% in lentils and between 37% and 50% in chickpeas (Devos 1988). Carbohydrates of legumes play a significant role in processing practices. The basic functional characteristics of carbohydrate can be specified as water absorption, swelling, solubility, gelatinization and stickiness, fat absorption, and structural characteristics (Bressani and Elias 1988). Cooking and pressure-cooking facilitate carbohydrate digestion. Dietary fibers, the undigested organic portions of foods, constitute the most significant portion of carbohydrates. Dietary fibers are composed of cellulose, hemicellulose, pectin, and lignin. The first three of them are carbohydrates but lignin is not. Legumes are quite rich in dietary fibers. Such a rate is around 18% in peas, lentils, and chickpeas and 28% in beans. The majority of the dietary fibers are concentrated within the testa. Therefore, peeling significantly reduces fiber contents (Devos 1988). In 1970s, the rise of several diseases such as constipation, diverticulosis, hemorrhoid, diabetes, obesity, intestinal cancer, and cardiovascular diseases, also called as "civilization diseases," was due to greater consumption of refined food stuff and less fiber intake (Trowell et al. 1985; Pekşen and Artık 2005).

Soluble Dietary Fibers (Pectin, Gums, and Some Hemicelluloses)

Edible legumes are great sources of soluble dietary fibers. They usually contain about 3%–7% fiber. Previous studies indicated that soluble dietary fibers had significant positive impacts on cardiovascular diseases through reducing total serum and LDL (low density lipoprotein = bad cholesterol) cholesterol levels (Glore et al. 1994). Clinical studies also revealed that soluble dietary fibers were quite effective in reducing postprandial blood sugar, insulin, and blood serum lipid levels; thus, it is quite effective in type II diabetes (Tabatabai and Li 2000; Pekşen and Artık 2005).

Insoluble Dietary Fibers (Lignin, Cellulose, and Some Hemicelluloses)

Edible legumes also contain about 11% insoluble dietary fibers, and these fibers are quite good for intestinal health because of laxative properties. Soluble and insoluble dietary fibers have positive impacts on nutrition and weight loss (Anderson and Bryant 1986; Marlett et al. 2002). Consumption of insoluble dietary fibers may reduce the risks of intestinal cancer and cardiovascular diseases (Hughes 1991; Marlett et al. 2002; Pekşen and Artık 2005).

Bioactive Molecules of Grain Legumes

Legume seeds contain many chemicals that are important for human health and are generally referred to as bioactive substances. Bioactive substances in legume seeds are very different from each other in terms of biochemical structures and properties. Many molecules like proteins, glycosides, tannins, saponins, and alkaloids belong to this group (Singh et al. 2017). The extraction and detection methods of these bioactive molecules are different from each other, and the amounts and structures vary among different legume seeds. The physiological effect of each bioactive molecule is also different (Carbonaro et al. 2012; Pedrosa et al. 2012). Some of the bioactive substances found in leguminous seeds are used as a source of energy in the defense mechanism of plants against biotic and abiotic factors and some as accumulation in seed as reserve substances (Roberts and Wink 1998). The processing of legume seeds changes the nutrient profile and increases the digestibility of protein and starch molecules in the structure of seeds. Molecules such as protease inhibitors and lectins, which are found in leguminous seeds, are heat-labile, and heat treatment in the consumption phase eliminates the negative effects of these molecules (Chau and Cheung 1997). On the other hand, tannins, saponins, and phytic acid molecules show a stable response to temperature, and their effects can be reduced by peeling, wetting, germination, or fermentation applications (Knorr 1999). In recent years, research on bioactive molecules has led to a better understanding of their importance in human nutrition. Scientific research on these substances emphasizes the possibilities of their use as probiotic intestinal, metabolic, and hormonal regulators (Muzquiz et al. 2012). Nowadays, in human nutrition, the concept of eating has left its place to consume balanced and beneficial foods and to maintain a healthy life (Dillard and German 2000). In the researches, it is reported that the calories required by people in the daily nutrition program are 15% protein, 60% carbohydrate, and 25% fats (Nishida et al. 2004). Food grains have become a focus of attention as a food source with their balanced protein, starch, fiber, vitamin, and mineral contents (Leterme 2002). Legume seeds have many bioactive compounds with different biochemical structures. These include proteins (protease inhibitors, α-amylases, lectins), glycosides (α-galactosides, vicine and convicine), tannins, saponins, or alkaloids (Muzquiz et al. 2000). The acquisition, identification, and quantification of bioactive molecules are quite different. Not all bioactive molecules are detected in all grain legumes, and their physiological effects are different. For example, pyrimidine, glycosides, vicine, and convicine are seen in *Vicia faba* and cause favism. Nonprotein amino acid β-N-oxalyl-L-α and β-diaminopropionic acid (β-ODAP) are present in *Lathyrus* spp. It is reported to cause lathyrism, also known as neurolathyrism. α-Galactosides found in many other legumes cause various health problems (Muzquiz et al. 2012).

Oligosaccharides

The most common oligosaccharides in plants are α-galactosides, of which raffinose group oligosaccharides are prevalent (Kadlec et al. 2000). Raffinose group oligosaccharides are raffinose (a trisaccharide), stachyose (a tetrasaccharide), verbascose (a pentasaccharide), and ajugose (a hexasaccharide). The oligosaccharides of the raffinose group are present in each section of the legumes, but they accumulate in seeds and roots during the development of plants. These oligosaccharides significantly vary in amounts (0.4%–16.1% dry matter) in lentils, beans, and peas (Rao and Belavady 1978; Martinez-Villaluenga et al. 2008). It has been reported that raffinose oligosaccharides are effective in drought resistance (Arora 1983). Raffinose and stachyose play a role in frost resistance (Castonguay et al. 1995). Stachyose content ranged from 0.45 (*Vicia ervilia*) to 59.08 (*Lupinus albus* var. Multolupa) mgg^{-1} and verbascose are mostly found in the *Pisum* and *Vicia* species. The bean seeds contain high levels of verbascose, while the ajugose is found only in *V. faba* seeds. Ciceritol is also an α-galactoside oligosaccharide but not from the raffinose family. Ciceritol is an α-D-galactoside (Quemener and Brillouet 1983) which was first detected in chickpea and was found in lentil, pea, and some *Vicia* species. The rate of ciceritol in desi-type chickpea seeds is reported to vary between 21.1 and 38.3 mg/gm (Pedrosa et al. 2012).

Protein Anti-Nutrients

Protein anti-nutrients, which are the most important ones, are enzyme inhibitors (pancreatic proteases and α-amylases) and lectins.

Protease Inhibitors

Seed protease inhibitors in leguminous seeds show a great effect on seed nutritional value as they inhibit the function of digestive enzymes such as trypsin and chymotrypsin by recombinant binding. These protease inhibitors do not contain carbohydrates and show two different constructs from the Kunitz and Bowman-Birk families. Legume seeds have protease inhibitors of both families (Lajolo et al. 2004; Srinivasan et al. 2005). The protease inhibitors of both families can inhibit trypsin and chymotrypsin. Since protease inhibitors are heat-labile, legumes are generally not harmful to humans when they are consumed as cooked food. Genetic engineering is now being used to reduce protease inhibitors.

In recent years, the effects of the protease inhibitors have been focused in relation to human health. It has adverse effects on the development of animals not only by preventing protein digestion in the intestine but also by leading to the development of free amino acids in food (Lajolo et al. 2004). Kunitz and Bowman-Birk inhibitors cause pancreatic growth in rodents and poultry. This is due to the reduction of endogenous hormones rich in sulfur content and consequently develop depression. Proteins found in leguminous seeds are generally poor in terms of sulfur-containing amino acids (Lajolo and Genovese 2002).

In recent years, protease inhibitors have been associated with health and are called natural bioactive. It has been reported that protease inhibitors have an anticarcinogenic effect (Clemente et al. 2004). Therefore, BBI has been included in the drug group to be investigated by the Food and Drug Administration (FDA) in the USA, and studies on humans have not demonstrated the toxic effect of BBI (Kennedy and Wan 2002). In contrast, this effect was not observed in autoclaved BBI, which indicates that the activity of protease inhibitors is very important for anticarcinogenic effects (Kennedy et al. 2002).

α-Amylase Inhibitors

α-Amylases (α-1,4-glucanohydrolases) are endo-amylases that catalyze α-D-(1–4) glycosidic bonds in starch and related molecules. α-Amylases are produced by the synthesis of glucose and other sugars, which are energy sources in humans and animals, and play an important role in carbohydrate metabolism. Among the α-amylase inhibitors (αAI) found in plants, attention is paid to the α-amylase inhibitors found in legumes, especially in beans (Lajole and Genovese 2002; Muzquiz et al. 2012).

According to few studies conducted, α-amylase inhibitors found in legumes were 11.6 inhibitory units g^{-1} with 51.4 inhibitory units g^{-1} in chickpea (Mulimani et al. 1994), in Solara pea cultivar were 16.8 inhibitory units g^{-1} (Alonso et al. 1998), in Pinto bean cultivar were 2.26 inhibitory units g^{-1} (Marzo et al. 2002). It has been reported that the main function of α-amylase inhibitors contained in plants is to protect it from the digestive enzymes of insects; these enzymes also prevent pancreatic amylases in humans and animals (Le Berre-Anton et al. 1997). α-Amylase inhibitors show a harmful or anti-metabolic effect to many harmful pests in plants. Transgenic plants obtained by transferring these genes to some plants have been reported to be more resistant to insect damage (Gatehouse et al. 2011). As a result, α-amylase inhibitors reduce the amylase activity and starch digestion in the intestine (Singh et al. 1982) when administered to humans by mouth, and it has been found to be beneficial in the fight against obesity or diabetes as it prevents the increase in blood glucose after meals.

Lectins

Lectins (hemagglutinins) are specific sugars on the surface of cells in the intestinal wall and glycoproteins which are able to reverse the glycoproteins, thereby reducing the breakdown and absorption of nutrients. These results were obtained in the form of agglutination of red blood cells of some organisms in vitro. Lectins are measured and expressed by the activity of hemaggluten in their contents (Grant 1991). It has been reported that high levels of legume lectins and weakness in skeletal muscles and lipid and glycogen content decrease the stomac acid (Bardocz et al. 1996). In addition to bubble, lectins are used to treat obesity, which partially inhibits tumor formation by affecting bowel function (Pusztai et al. 2008).

Saponins

Saponins are named after the stable soap-water structure that bubbles in the liquid. They have complex and different chemical structures. Chemically, saponins consist of a steroidal or triterpene aglycone bound to one, two, or three saccharide chains of varying size and complexity through ester and ether linkages (Fenwick et al. 1991; Muzquiz et al. 2012). Saponins are found in many legumes, such as, lentils, chickpeas, peas and soybeans (Shi et al. 2004). In saponins, soy sapogenol B shows water solubility (Price et al. 1988; Tava et al. 1993). It has been reported that sapogenol B is present in varying proportions,.and the highest sapogenol B content is in beans (Burbano et al. 1999).

Nutraceutical Properties of Edible Grain Legumes

The term nutraceutical was first derived from the "Nutrition" and "Pharmaceutical" is a term coined by Dr. Stephen De Felice in 1989, founder and chairman of the Foundation for Innovation in Medicine. In his catalog, nutraceutical was defined in 1994 as a nutritional supplement with proven healing or protective effect against nontoxic hazards (Scarafoni et al. 2005). Due to their nutritious properties, the grained grain legumes have been among the most important foods of humanity for many years. In addition to the nutrients of food stuffs, legumes have been shown to be of some benefit in recent years. Thus, regular and scheduled consumption of edible legumes in the daily nutrition program was found to be very important in terms of leading a healthy life (Krauss et al. 2001). Experimental and clinical studies have shown that some biological molecules in the legume seeds have a positive effect on the function and health of the human body. Alkaloids, oligosaccharides, phenols, etc. are the main components of legumes, with the high storage protein in the leguminous seeds.

Some studies have shown that some bioactive substances of grain legumes have balanced blood sugar and thus reduce the risk of cardiovascular diseases. This effect is associated with high fiber content of edible legumes, low glycemic index, and smaller contents of phytosterols, saponins, and oligosaccharides. The US Food and Drug Administration reported in 1999 that the consumption of 6.25 g of soy protein in a portion (about 25 g per day) reduced the risk of heart disease in one portion (PDA 1999/3).

7S globulin in soy is reported to reduce blood cholesterol in rats by 35% in experimental studies, which means that the rate of cholesterol in blood is significantly decreasing. The 7S globulin subunit alpha molecule was found to have a direct enhancing-stabilizing effect on LDL receptors (Manzoni et al. 1998; Duranti et al. 2004; Scarafoni et al. 2005). It has been reported that the regular consumption of perennial fertilizers decreases coronary heart disease and cardiovascular disorder, coronary heart disease risk is 22% and cardiovascular disease risk is 11% lower than those who consume legumes once a week for those who consume 4 or more beans a week (Bazzano et al. 2001; Flight and Clifton 2006). Anderson and Major (2002) reported that reducing intake of saturated fat and cholesterol-enhancing nutrients and increasing the consumption of nutrients such as legumes that have high fiber content and cholesterol-lowering effect are important in preventing heart diseases. In experimental observations, they reported that the effect of legumes other than soy on cholesterol ratio in serum is important, and this effect is due to the fiber content of other legumes, plant protein, oligosaccharides, isoflavones, phospholipids, fatty acids, saponins, and other factors. Therefore, regular consumption of pulses reduces blood pressure, the risk of cardiovascular diseases, diabetes and obesity. Some carbohydrate foods cause a lower increase in blood sugar compared to others. The glycemic index (GI) of a nutrient is defined as an increase in the rate of sugar caused by a reference food. It is reported that the special starch composition of legumes causes this situation. Because of these properties, grain legumes are included in the food group with low glycemic index, and it has been reported that it plays an important role in the regulation of blood sugar in diabetic individuals (Lang et al. 1999; Scarafoni et al. 2005). In addition, edible grain legumes have been reported to be effective in the prevention of insulin resistance, which is a preliminary symptom of type II diabetes (Duranti 2006; Venn and Mann 2004). Research on the effects of foodstuffs on cancer has been conducted for a long time. According to the results of this research, protease inhibitors, saponins, phytosterols, isoflavones, and pythates are shown in the seeds of legumes (Mathers 2002). Some studies have reported inverse relationship between mortality rate of colon, stomach, pancreatic, and prostate cancer and the rate of legume consumption; in other words, high level of legume consumption has been reported to lead to a decrease in the number of deaths from cancer (Jain et al. 1999). It has been reported that legumes are an important source of vitamin B, which plays an important role in the prevention of some types of cancer (Kim 2003).

Conclusion

Legumes are an important nutrient for all people and animals due to their high protein content, folic acids, vitamins, minerals, and fiber contents. Besides the nutritional properties of legumes, some bioactive compounds have made legumes more important for human nutrition. Enzyme inhibitors, lectins, pythates, oligosaccharides, and phenolic compounds are different bioactive compounds present in many legumes. In addition, some compounds that are found in legumes may be undesirable or neutral in terms of nutrition but may have clinical values. It has been reported by many investigators that the consumption of legumes can be effective in reducing some chronic diseases (lifestyle diseases).

References

Alonso R, Orue E, Marzo F (1998) Effects of extrusion and conventional processing methods on protein and antinutritional factor contents in pea seeds. Food Chem 63:505–512

Anderson JW, Bryant CA (1986) Dietary fiber: diabetes and obesity. Am J Gastroenterol 81(10):898–906

Anderson JW, Major AW (2002) Pulses and lipaemia, short-and long-term effect: potential in the prevention of cardiovascular disease. Br J Nutr 88(S3):263–271

Anderson JW, Smith BM, Washnock CS (1999) Cardiovascular and renal benefits of dry bean and soybean intake. Am J Clin Nutr 70(3):464s–474s

Arora SK (1983) Legume carbohydrates [Includes cyclitols, galactosides of pinitol]. In: Arora SK (ed)Chemistry and biochemistry of legumes

Bardocz S, Grant G, Pusztai A, Franklin MF, Carvalho ADF (1996) The effect of phytohaemagglutinin at different dietary concentrations on the growth, body composition and plasma insulin of the rat. Br J Nutr 76(4):613–626

Bazzano LA, He J, Ogden LG, Loria C, Vupputuri S, Myers L, Whelton PK (2001) Legume consumption and risk of coronary heart disease in US men and women: NHANES I epidemiologic follow-up study. Arch Intern Med 161(21):2573–2578

Blair MW (2013) Mineral biofortification strategies for food staples: the example of common bean. J Agric Food Chem 61(35):8287–8294

Bressani R, Elias LG (1988) Seed quality and nutritional goals in pea, lentil, faba bean and chickpea breeding. In: World crops: cool season food legumes. Springer, Dordrecht, pp 381–404

Burbano C, Muzquiz M, Ayet G, Cuadrado C, Pedrosa MM (1999) Evaluation of antinutritional factors of selected varieties of Phaseolus vulgaris. J Sci Food Agric 79(11):1468–1472

Cakmak I (2008) Enrichment of cereal grains with zinc: agronomic or genetic biofortification? Plant and Soil 302(1–2):1–17

Carbonaro M, Maselli P, Nucara A (2012) Relationship between digestibility and secondary structure of raw and thermally treated legume proteins: a Fourier transform infrared (FT-IR) spectroscopic study. Amino Acids 43(2):911–921

Castonguay Y, Nadeau P, Lechasseur P, Chouinard L (1995) Differential accumulation of carbohydrates in alfalfa cultivars of contrasting winter hardiness. Crop Sci 35(2):509–516

Chau CF, Cheung PK (1997) Effect of various processing methods on antinutrients and in vitro digestibility of protein and starch of two Chinese indigenous legume seeds. J Agric Food Chem 45(12):4773–4776

Clemente A, MacKenzie DA, Johnson IT, Domoney C (2004) Investigation of legume seed protease inhibitors as potential anti-carcinogenic proteins. Pub-Eur Assoc Animal Prod 110:137–142
Devos P (1988) Mercimek ve nohutun besin değeri ve proses sırasındaki değişiklikler. Herkes İçin Mercimek Sempozyumu, pp.29–30
Dillard CJ, German JB (2000) Phytochemicals: nutraceuticals and human health. J Sci Food Agric 80(12):1744–1756
Duranti M (2006) Grain legume proteins and nutraceutical properties. Fitoterapia 77(2):67–82
Duranti M, Lovati MR, Dani V, Barbiroli A, Scarafoni A, Castiglioni S, Ponzone C, Morazzoni P (2004) The α′ subunit from soybean 7S globulin lowers plasma lipids and upregulates liver β-VLDL receptors in rats fed a hypercholesterolemic diet. J Nutr 134(6):1334–1339
Erdemci I, Aktas H, Nadeem MA (2017) Effect of fertilization and seed size on nodulation, yield and yield components of chickpea (Cicer arietinum L.). Appl Ecol Environ Res 15(1):563–571
Fenwick GR, Price KR, Tsukamoto C, Okubo K (1991) Saponins. In: D'Mello JPF, Duffus CM, Duffus JH (eds) Toxic substances in crop plants. The Royal Society of Chemistry, Cambridge, pp 285–327
Flight I, Clifton P (2006) Cereal grains and legumes in the prevention of coronary heart disease and stroke: a review of the literature. Eur J Clin Nutr 60(10):1145–1159
Gatehouse AMR, Ferry N, Edwards MG, Bell HA (2011) Insect-resistant biotech crops and their impacts on beneficial arthropods. Philos Trans R Soc B Biol Sci 366(1569):1438–1452
Gebhardt SE, Thomas RG (2002) Nutritive value of foods. U.S. Department of Agriculture, Agricultural Research Service, Home and Garden Bulletin 72
Glore SR, Van Treeck D, Knehans AW, Guild M (1994) Soluble fiber and serum lipids: a literature review. J Am Diet Assoc 94(4):425–436
Grabner M, Hofer R (1985) The digestibility of the proteins of broad bean (*Vicia faba*) and soya bean (*Glycine max*) under in vitro conditions simulating the alimentary tracts of rainbow trout (*Salmo gairdneri*) and carp (*Cyprinus carpio*). Aquaculture 48(2):111–122
Grant G (1991) Lectins. In: D'Mello JPF, Duffus CM (eds) Toxic substances in crop plants. The Royal Society of Chemistry, pp 49–67
Hedley CL (2001) Introduction. In: Hedley CL (Ed) Carbohydrate in grain legume seeds, pp. 1–11
Hotz C, Brown KH, Rivera JA (2004) Assessment of the risk of zinc deficiency in populations and options for its control. Food and Nutrition Bulletin 25(1, supplement 2):S94–S204
Hughes JS (1991) Potential contribution of dry bean dietary fiber to health. Food Technol 45:122–126
Iqbal A, Khalil IA, Ateeq N, Khan MS (2006) Nutritional quality of important food legumes. Food Chem 97(2):331–335
Jain PC, Kushwaha PS, Dhaked US, Khan H, Trivedi SK (1999) Response of chickpea (*Cicer arietinum* L.) to phosphorus and biofertilizer. Indian J Agron 22:241–244
Jha AB, Ashokkumar K, Diapari M, Ambrose SJ, Zhang H, Tar'an B et al (2015) Genetic diversity of folate profiles in seeds of common bean, lentil, chickpea and pea. J Food Compost Anal 42:134–140
Kadlec P, Kaasova J, Bubnik Z, Pour V (2000) Effect of germination and microwave treatment on chemical composition of pea. In: Proceedings of International Conference on Microwave Chemistry. Francie, Antibes, pp 241–243
Karaköy T, Erdem H, Baloch FS, Toklu F, Eker S, Kilian B, Özkan H (2012) Diversity of macro- and micronutrients in the seeds of lentil landraces. Sci World J 2012
Kennedy AR, Wan XS (2002) Effects of the Bowman-Birk inhibitor on growth, invasion, and clonogenic survival of human prostate epithelial cells and prostate cancer cells. Prostate 50(2):125–133
Kennedy AR, Billings PC, Wan XS, Newberne PM (2002) Effects of Bowman-Birk inhibitor on rat colon carcinogenesis. Nutr Cancer 43(2):174–186
Kim I (2003) Role of folate in colon cancer development and progression. J Nutr 133:3731–3739
Knorr D (1999) Novel approaches in food-processing technology: new technologies for preserving foods and modifying function. Curr Opin Biotechnol 10(5):485–491

Kostyra H (1996) Food proteins—evolution and nutritional aspects. In: Bardocz S, Gelencsér E, Pusztai A, (ed) Effects of antinutrients on the nutritional value of legume diets. COST98, vol. 1. European Commission Directorate-general XII; 1996. p. 86

Krauss RM, Eckel RH, Howard B, Appel LJ, Daniels SR, Deckelbaum RJ, Erdman JW, Kris-Etherton P, Goldberg IJ, Kotchen TA, Lichtenstein AH, Mitch WE, Mullis R, Robinson K, Wylie-Rosett J, St. Jeor S, Suttie J, Tribble DL, Bazzarre TL (2001) Revision 2000: a statement for healthcare professionals from the nutrition Committee of the American Heart Association. J Nutr 131:132–146

Lajolo FM, Genovese MI (2002) Nutritional significance oflectins and enzyme inhibitors from legumes. J Agric Food Chem 50(22):6592–6598

Lajolo FM, Genovese MI, Pryme IF, Dale TM (2004) Beneficial (antiproliferative) effects of different substances. Pub Eur Assoc Animal Prod 110:123–136

Lang V, Bornet FR, Vaugelade P, van Ypersele de Strihou M, Luo J, Pacher N, Rossi F, La Droitte P, Duée PH, Slama G (1999) Euglycemic hyperinsulinemic clamp to assess posthepatic glucose appearance after carbohydrate loading. 2. Evaluation of corn and mung bean starches in healthy men. Am J clin Nutr 69(6):1183–1188

Le Berre-Anton V, Bompard-Gilles C, Payan F, Rougé P (1997) Characterization and functional properties of the α-amylase inhibitor (α-AI) from kidney bean (Phaseolus vulgaris) seeds. Biochimica et Biophysica Acta (BBA)-Protein Struct Mol Enzymol 1343(1):31–40

Leterme P (2002) Recommendations by health organizations for pulse consumption. Br J Nutr 88(S3):239–242

Manzoni C, Lovati MR, Gianazza E, Morita Y, Sirtori CR (1998) Soybean protein products as regulators of liver low-density lipoprotein receptors. II. α– α 'rich commercial soy concentrate and α 'deficient mutant differently affect low-density lipoprotein receptor activation. J Agric Food Chem 46(7):2481–2484

Margaret AH, Leonard JP, Mathews TJ, Erickson JD, Wong LC (2001) Impact of folic acid fortification of the US food supply on the occurrence of neural tube defects. JAMA 285:2981–2986

Marlett JA, McBurney MI, Slavin JL (2002) Position of the American Dietetic Association: health implications of dietary fiber. J Am Diet Assoc 102(7):993–1000

Martinez-Villaluenga C, Frias J, Vidal-Valverde C (2008) Alpha-galactosides: antinutritional factors or functional ingredients? Crit Rev Food Sci Nutr 48(4):301–316

Marzo F, Alonso R, Urdaneta E, Arricibita FJ, Ibanez F (2002) Nutritional quality of extruded kidney bean (Phaseolus vulgaris L. var. pinto) and its effects on growth and skeletal muscle nitrogen fractions in rats. J Anim Sci 80:875–879

Mathers JC (2002) Pulses and carcinogenesis: potential for the prevention of colon, breast and other cancers. Br J Nutr 88(S3):273–279

Mayer JE, Pfeiffer WH, Beyer P (2008) Biofortified crops to alleviate micronutrient malnutrition. Curr Opin Plant Biol 11(2):166–170

Muehlbauer FJ, Cho S, Sarker A, McPhee KE, Coyne CJ, Rajesh PN, Ford R (2006) Application of biotechnology in breeding lentil for resistance to biotic and abiotic stress. Euphytica 147(1–2):149–165

Mulimani VH, Rudrappa G, Supriya D (1994) Alpha-amylase inhibitors in chick pea (Cicer arietinum L.). J Sci Food Agric 64:413–415

Muzquiz M, Burbano C, Cuadrado C, Martin M (2000) Analytical methods for determination of compounds with no nutritive value. In: Handbook on common bean related laboratory methods. Galicia, Spain, pp 11–26

Muzquiz M, Varela A, Burbano C, Cuadrado C, Guillamón E, Pedrosa MM (2012) Bioactive compounds in legumes: pronutritive and antinutritive actions. Implications for nutrition and health. Phytochem Rev 11(2–3):227–244

Nishida C, Uauy R, Kumanyika S, Shetty P (2004) The joint WHO/FAO expert consultation on diet, nutrition and the prevention of chronic diseases: process, product and policy implications. Public Health Nutr 7(1a):245–250

PDA (U.S. Food and Drug Administration) (1999) Final rule. Federal Register 64:57699

Pedrosa MM, Cuadrado C, Burbano C, Allaf K, Haddad J, Gelencsér E, Takács K, Guillamón E, Muzquiz M (2012) Effect of instant controlled pressure drop on the oligosaccharides, inositol phosphates, trypsin inhibitors and lectins contents of different legumes. Food Chem 131(3):862–868

Pekşen E, Artık C (2005) Antibesinsel Maddeler ve Yemeklik Tane Baklagillerin Besleyici Değerleri. J Fac Agric OMU 20(2):110–120

Price KR, Eagles J, Fenwick GR (1988) Saponin composition of 13 varieties of legume seed using fast atom bombardment mass spectrometry. J Sci Food Agric 42(2):183–193

Pusztai A, Bardocz S, Stanley WBE (2008) Uses of plant lectins in bioscience and biomedicine. Frontiers in Bioscience 13:1130–1140

Quemener B, Brillouer JM (1983) Ciceritol, a pinitol disaccharide from seeds of chickpea, lentil and white lupin. Phytochem 22:1745–1751

Rao PU, Belavady B (1978) Oligosaccharides in pulses: varietal differences and effects of cooking and germination. J Agric Food Chem 26(2):316–319

Roberts MF, Wink M (1998) Introduction. In: Roberts MF, Wink M (eds) Alkaloids. Biochemistry, ecology, and medicinal applications. Plenum Press, New York, pp 1–7

Scarafoni ALESSIO, Kumar J, Magni CHIARA, Sironi ELENA, Duranti MARCELLO (2005) Biologically active molecules and nutraceutical properties of legume seeds. IFLRC IV, pp 18–22

Şehirali S (1988) Yemeklik Dane Baklagiller, A.U. Ziraat Fakultesi Yayınları, 1089, Ders Kitabı 314

Sen Gupta D, Thavarajah D, Knutson P, Thavarajah P, McGee RJ, Coyne CJ, Kumar S (2013) Lentils (Lens culinaris L.), a rich source of folates. J Agric Food Chem 61(32):7794–7799

Sen Gupta D, Thavarajah D, McGee RJ, Coyne CJ, Kumar S, Thavarajah P (2016) Genetic diversity among cultivated and wild lentils for iron, zinc, copper, calcium and magnesium concentrations. Aust J Crop Sci 10(10):1381

Sharma B (1988) İnsan beslenmesinde mercimek ve nohut: Bugünkü durumu ve beklentiler (Lentils and chickpeas in human nutrition conditions: present state and prospects). In Herkes İçin Mercimek Sempozyumu (lentils for everyone symposium)(29–30 Eylül 1988), Marmaris/ Muğla, pp. 157–171)

Shewry PR, Tatham AS (1999) The characteristics, structures and evolutionary relationships of prolamins. In: Seed proteins. Springer, Dordrecht, pp 11–33

Shi J, Arunasalam K, Yeung D, Kakuda Y, Mittal G, Jiang Y (2004) Saponins from edible legumes: chemistry, processing, and health benefits. J Med Food 7(1):67–78

Singh U, Kherdekar MS, Jambunathan R (1982) Studies on Desi and Kabuli chickpea (*Cicer arietinum* L.) cultivars. The Levels of Amylase Inhibitors, Levels of Oligosaccharides and In Vitro Starch Digestibility. J Food Sci 47(2):1982.510–1982.512

Singh B, Singh JP, Shevkani K, Singh N, Kaur A (2017) Bioactive constituents in pulses and their health benefits. J Food Sci 54(4):858–870. https://doi.org/10.1111/j.1365-2621.1982.tb10113.x

Srinivasan A, Giri AP, Harsulkar AM, Gatehouse JA, Gupta VS (2005) A Kunitz trypsin inhibitor from chickpea (*Cicer arietinum* L.) that exerts anti-metabolic effect on podborer (*Helicoverpa armigera*) larvae. Plant Mol Biol 57(3):359–374

Tabatabai A, Li S (2000) Dietary fiber and type 2 diabetes. Clin Excellence Nurse Pract Int J NPACE 4(5):272–276

Tava A, Oleszek W, Jurzysta M, Berardo N, Odoardi M (1993) Alfalfa saponins and sapogenins: isolation and quantification in two different cultivars. Phytochem Anal 4(6):269–274

Tiwari BK, Singh N (2012) Pulse chemistry and technology. Royal Society of Chemistry, Cambridge

Toklu F, Ozkan H, Karakoy T, Coyne C (2015) Evaluation of advanced lentil lines for diversity in seed mineral concentration, grain yield and yield components. J Agric Sci 23:213–222

Trowell H, Burkitt D, Heaton K, (1985) Dietary fibre, fibre-depleted foods and disease. Academic Press

Upadhyaya HD, Thudi M, Dronavalli N, Gujaria N, Singh S, Sharma S, Varshney RK (2011) Genomic tools and germplasm diversity for chickpea improvement. Plant Genetic Resour 9(1):45–58

Venn BJ, Mann JI (2004) Cereal grains, legumes and diabetes. Eur J Clin Nutr 58(11):1443–1461

Wang TL, Domoney C, Hedley CL, Casey R, Grusak MA (2003) Can we improve the nutritional quality of legume seeds? Plant Physiol 131(3):886–891

White PJ, Broadley MR (2005) Biofortifying crops with essential mineral elements. Trends Plant Sci 10(12):586–593

White PJ, Broadley MR (2009) Biofortification of crops with seven mineral elements often lacking in human diets–iron, zinc, copper, calcium, magnesium, selenium and iodine. New Phytol 182(1):49–84

WHO (1999) Malnutrition worldwide. World Health Organization. http://www.who.int/nut/malnutrition_worldwide.htm, Geneva, pp 1–13

Yang XE, Chen WR, Feng Y (2007) Improving human micronutrient nutrition through biofortification in the soil–plant system: China as a case study. Environ Geochem Health 29(5):413–428

Improving Pigeonpea Quality: An Elevation Towards Nutritional Security

C. V. Sameer Kumar and H. B. Shruthi

Abstract Pigeonpea is an important grain legume usually consumed as split "dal" in India. Globally, it is cultivated for about 7 Mha area with a production and productivity of 6.8 MT and 969 kg/ha, respectively, in 2018. Hidden hunger and malnutrition are the serious challenges faced due to the consumption of less nutrient food. Though India is now called "Food Secured Nation," its nutritional security is yet to be retained. In this context, pigeon pea has evolved as a promising nutri-legume. Being an ideal crop for sustainable agriculture, it also accounts for the protein content of 20–25 g, 2.76 mg of zinc, and 5.23 mg of iron (per 100 g of seed), respectively. The recommended daily allowance (RDA) of 13 mg per day (children), 17 mg per day (adult) for iron and 7 mg per day (children), 12 mg per day (adult) for zinc is necessary. Consumption of pigeonpea as protein and micronutrients source only meets lower percent of RDA for these nutrients. Thus, a major intervention has to be made to enrich this crop in terms micronutrients and proteins ensuring the nutritional security in future days.

Keywords *Cajanus cajan* · iron · zinc · protein · raffinose family oligosaccharides · phytic acid · genetic biofortification

Introduction

Malnutrition is a big challenge faced by developing and underdeveloped countries. Though India is now tagged "Food Secured Nation," its nutritional security has yet to be retained. Since animal protein is beyond the reach of this group, their primary protein supply comes from the plant-based products. Among these, pigeonpea or

C. V. Sameer Kumar (✉) · H. B. Shruthi
International Crops Research Institute for the Semi-Arid Tropics, Patancheru, Telangana State, India

D. S. Gupta et al. (eds.), *Breeding for Enhanced Nutrition and Bio-Active Compounds in Food Legumes*, https://doi.org/10.1007/978-3-030-59215-8_11

red gram (*Cajanus cajan* (L.) *Millspaugh*) is an important food legume that can be grown under rainfed conditions with least inputs. In India, de-hulled split cotyledons of pigeonpea seeds are cooked to make *dal* (thick soup) for eating with bread and rice, while in Southern and Eastern Africa and South America, its whole dry seeds are used in a porridge-like recipe (Saxena et al. 2010). The fully grown seeds of pigeonpea when harvested green before losing their green color are used as fresh, frozen, or canned vegetable. It is an excellent source of starch, protein, calcium, manganese, crude fiber, fat, trace elements, minerals and well-balanced nutritionally. It contains fivefold higher levels of vitamins A and C (Faris et al. 1987). Being a protein reservoir fits well in traditional cereal, banana, or tuber-based food supplementing the limiting nutrients. Pharmaceutical supplementation, industrial fortification, dietary diversification, and biofortification (Meenakshi et al. 2007; Hanumanthappa et al. 2018) are the major possible intervention for combating malnutrition and hidden hunger. In this chapter, possibilities of seed-quality (protein and mineral) enrichment in pigeonpea are discussed to ensure the nutritional security in the nation.

Pigeonpea's Role in Balanced Diet

In pigeonpea, methionine, cystine, tryptophan, and threonine are the limiting essential amino acids, whereas in rice and wheat, lysine is the limiting amino acid. A food combining both cereals and pulses provides a balanced diet as they complement the amino acid profiles of each other (Saxena et al. 2010). The mutual quality compensation is closest to the ideal value when the ratio by weight of cereals to legume is roughly 70:30 (Hulse 1977). In Southern and Eastern Africa, this ratio is 90:10, reflecting shortage of protein in the diet. Daniel et al. (1970) studied supplementation of cereal diets with various proportions of pigeonpea and reported that supplementation of ration with pigeonpea significantly enhanced the nutritive value of diet. Supplementation of rice diet with 8.5% and 16.7% pigeonpea *dal,* markedly improved the quality of diet. Similarly, Kurien (1981) demonstrated that a supplement of pigeonpea in maize diet significantly improved the quality of food too.

Bidinger and Nag's (1981) investigation, unleashed the fact of pigeonpea being the most preferred pulse crop in Indian villages, and its consumption patterns differed widely by age group, farm size, and the village. The consumption rate was found linear with small farmers consuming the least amount and the large farmers the most. The National Institute of Nutrition in India recommends cereal/pulse ratio of 3:1 for very young children, 5:1 for women, and 6:1 for men. In most of the cases, rural diet standards could not be met. Bidinger and Nag (1981) further reported that 10% of protein and 5% of energy in the village diets came from pigeonpea alone. The maximum lysine provided from the diet was 21.7%. These values are low and reflects the lower consumption of legumes.

Quality Traits in Pigeonpea

Protein Content

The protein content of pigeonpea, in general, varies around 20–22%. Protein content is mainly controlled by additive genetic action (Saxena 2008). Extension of hybrid parent research in the direction of breeding high-protein A-lines can help in developing hybrids with 25–30% yield advantage and high (26–27%) protein content. Saxena and Sawargaonkar (2016) reported that newly bred pigeonpea lines have protein between 28 and 30% and yield good as cultivars. An estimate of protein yield showed that the cultivation of high-protein lines in one hectare will yield an additional 100,000 g protein for the farming families living under subsistence level.

Starch Content

Starch content is the principal constituent in pigeonpea seed after protein. Singh et al. (1984) showed that the starch content in pigeonpea cultivars ranged from 51.4 to 58.8%.

Anti-Nutritional Factors

Pigeonpea seeds contain anti-nutritional factors like oligosaccharides (raffinose and verbascose), polyphenols (phenols and tannins), phytolectins, and enzyme inhibitors such as trypsin, chymotrypsin, and amylase (Saxena et al. 2010). According to Kamath and Belavady (1980), pigeonpea seeds also have some amounts of unavailable carbohydrates which adversely affect the bioavailability of certain vital nutrients. Some of the anti-nutritional factors such as phytolectins are heat sensitive and are destroyed during cooking. Godbole et al. (1994) reported protease inhibitors in seven-day-old seeds, while Ambekar et al. (1996) found that such inhibitors are either not synthesized or inactive up to 28 days of the seed development. No other plant part except seed exhibited trypsin or chymotrypsin inhibitors (Mutimani and Paramjyothi 1995). The white-seeded pigeonpea cultivars contain relatively less amounts of polyphenols. Such cultivars are preferred in many countries where de-hulling facilities are not available and whole seeds are consumed. In comparison to the white-seeded cultivars, the red-seeded types contain three times greater quantity of polyphenols (Singh et al. 1984). Similarly, the enzyme inhibition activity was also greater in the colored seeds of pigeonpea. Since in India, almost entire pigeonpea production (3.2 m tones) is de-hulled and converted into *dal* for consumption, the tannins present in the colored seed coat pose no nutritional problems.

Cooking Quality

Pigeonpea seeds in the form of either dry, green, or split peas are invariably consumed after cooking. Therefore, besides various nutritional aspects, the cooking time and other related parameters assume at most importance. Consumers always prefer *dal* that cooks faster and produces more volume upon cooking with high consistency and flavor. Cooking time of *dal* is independent of taste and flavor (Manimekalai et al. 1979). Narasimha and Desikachar (1978) and Pal (1939) exported a positive association of cooking time of pigeonpea seeds with their calcium and magnesium contents. According to Salunkhe (1982), cooking of pigeonpea, improved the bioavailability of nutrients as well as destroyed some anti-nutritional factors. Heat treatment of pigeonpea seeds is also known to enhance their starch digestibility. The lines, which take long time to cook, generally face the danger of losing important vitamins from food. The fermentation of seeds helps in reducing inhibitory activity of digestive enzymes (Rajalakshmi and Vanaja 1967). Geervani (1981) reported that thiamine and riboflavine were destroyed by heat but niacin content was unaltered during boiling, pressure cooking and roasting of pigeonpea seeds. She further found that the availability of lysine and methionine decreased on roasting but the available methionine increased on boiling and pressure cooking.

Biofortification

Biofortification involves developing crop varieties with superior nutrient qualities. It includes both increasing nutrient levels in the edible parts of food crops and their bioavailability when consumed. Biofortification can be achieved by two approaches: conventional breeding and agronomic biofortification. In conventional breeding method, locally adopted high-yielding varieties are crossbred with the variety with naturally rich nutrients to produce high-yielding and nutritious plants. Agronomic biofortification involves application of micronutrients, iron and zinc, to the soil or through foliar feeding (Hanumanthappa et al. 2018). If there is sufficient genetic variation for the density of micronutrients in edible parts of the crop, biofortification can be achieved through plant breeding (Mayer et al. 2008), wherein new crop varieties with elevated nutrient content can be developed for cultivation and production of nutrient-dense foods. Biofortification can be also be achieved through transgenic technologies. However, several countries across the globe have imposed restriction on deploying transgenic technology for food crops. Thus, currently, biofortification is being achieved through agronomic management practices and conventional breeding.

Traits for Biofortification

The pigeonpea crop accosts for protein content of 20–25 g, 2.76 mg of zinc and 5.23 mg of iron (per 100 g of seed) (Janila et al. 2016). The recommended daily allowance (RDA) of 13 mg per day (children) and 17 mg per day (adult) for iron and 7 mg per day (children) and 12 mg per day (adult) for zinc is necessary. But, in India, the consumption of 7gm/day/person of pigeonpea as food provides daily per capita iron content of 14.93 mg (adult), which is less than the RDA (Hanumanthappa et al. 2018). Thus, a major intervention has to be made to enrich the crop in terms Zn and Fe content.

Genetic Variability for Fe and Zn Content in Pigeonpea

Acceptable range of variability is found in pigeonpea for zinc and Iron. As reported by Upadhyaya et al. (2013), evaluation of pigeonpea mini-core collection resulted in the identification of high zinc and iron content accessions. ICP 4029, ICP 6929, ICP 6992, ICP 7076, ICP 10397, ICP 11690, ICP 12298, and ICP 12515 are the identified high zinc content (>40 ppm) pigeonpea accessions. Likewise, ICP 2698, ICP 11267, ICP 14444, and ICP 14976 are identified pigeonpea accessions for high iron content (>40 ppm). Utilization of these accessions in the future breeding program undoubtedly yields promising high seed iron and zinc pigeonpea varieties.

Analytical Methods

Biofortification requires developing or adapting cost-effective and rapid high-throughput analytical techniques for micronutrients, as thousands of samples need to be tested for mineral or vitamin content each season. These trait diagnostics include near-infrared spectroscopy (NIRS) and colorimetric methods for carotenoid analysis. For mineral analysis, X-ray fluorescence spectroscopy (XRF) emerged as the method of choice, as it requires minimal pre-analysis preparation and allows for nondestructive analysis (Paltridge et al. 2012a, 2012b; Howarth and Saltzman 2017).

Overview of Pigeonpea Biofortification

(a) **Agronomical Approach**

Hanumanthappa et al. (2018) have taken up an experiment to study the effect of micronutrient fertilization on the increase of iron and zinc content of pigeonpea genotypes. Results revealed that foliar application of iron at 0.5% and zinc at 0.5%

effectively improved the iron and zinc content in pigeonpea genotypes. However, among the genotypes, ICPL 96061 for iron and GRG-160 for Zn showed good response to foliar application. Thus, they concluded that foliar application of iron and zinc offers a practical and useful option to improve Fe and Zn content in pigeonpea genotypes.

(b) **Microbial Approach**

Gopalakrishnan et al. (2016) used seven strains of bacteria (*Pseudomonas plecoglossicida* SRI-156, *Brevibacterium antiquum* SRI-158, *Bacillus altitudinis* SRI-178, *Enterobacter ludwigii* SRI-211, *E. ludwigii* SRI-229, *Acinetobacter tandoii* SRI-305, and *Pseudomonas monteilii* SRI-360, demonstrated previously for control of charcoal rot disease in sorghum and plant growth promotion (PGP) in rice) to evaluate their plant growth promotion and biofortification traits in chickpea and pigeonpea under field conditions. When the harvested grains were evaluated for their mineral contents, iron (up to 18 and 12%), zinc (up to 23 and 5%), copper (up to 19 and 8%), manganese (up to 2 and 39%), and calcium (up to 22 and 11%) contents in chickpea and pigeonpea, respectively, were found enhanced in test bacteria inoculated plots over the uninoculated control plots. Thus, the study concluded that selected bacterial isolates not only have the potential for PGP in cereals and legumes but also for biofortification of mineral nutrients.

Trait-Specific Breeding Approach

Conventional breeding forms the best approach for biofortification, provided clarity on potential correlation and linkage drags. Mishra and Acharya (2017) conducted an experiment to study the role of target traits on the enhancement of seed iron and zinc concentration in pigeonpea. They inferred that breeding for high seed iron and zinc concentrations advocates to evaluate their association with various nutritional and anti-nutritional traits, which will assist in the selection process. The results indicated that bolder seeds with high zinc concentration and lipid content and lesser contents of phytic acid should be aimed for enhancement of seed iron concentration, while seed protein content can be used as a secondary variable for selection of this trait. On the other hand, small seeds with high lipid and iron concentrations as well as lesser contents of phytic acids and proteins should be targeted for improvement of seed zinc concentration in pigeonpea.

Transgenic Approach

Transgenic approach is the advanced approach in biofortification. Bhatnagar et al. (2011) in their paper entitled "Crop Biofortification Through Genetic Engineering: Present Status and Future Directions" have elaborated on the transgenic approach of biofortification in pigeonpea. At ICRISAT, initially, a single *psy1* gene from maize was used to develop transgenic pigeonpea for enhanced level of β-carotene using the *Zm psy1* gene driven by the oleosin promoter through *Agrobacterium*-mediated genetic transformation. Over 140 putative transgenic pigeonpea events with maize *psy1* were developed and characterized at the molecular level for the integration and expression of the transgenes. Total carotenoid content in seeds from the primary to putative transgenic pigeonpea plants was estimated spectrophotometrically and twofold increases in total carotenoid content were observed in several transgenic events over the non-transgenic (control) pigeonpea plant. These 11 events showed two- to threefold increases in β-carotene levels (6–11 μg/g in transgenic events, in contrast to 2 μg/g in the untransformed control) evidenced using HPLC analysis. Studies also indicated that the transgenic pigeonpea events had much higher lutein content over the controls among the individual carotenoids. Besides, β-*lycopene cyclase* gene was cloned from tomato and used in combination with the *psy1* gene under the control of CaMV35S promoter to further improve β-carotene levels in transgenic groundnut. Efforts are underway to develop marker-free pigeonpea transgenic plants carrying both maize *psy1* and tomato β-*lyc* genes to meet the target levels of β-carotene in this important pulse crop (Bhatnagar et al. 2011).

Conclusion

The current world population stands at an estimated 7.3 billion (United Nations 2015) and is projected to increase by 2 billion over the next four decades. Concomitant to this growth is the challenge of providing sustenance amidst dwindling resources. Currently, food production is adequate at approximately four billion metric tons per annum, yet about 870 million people still suffer from chronic malnutrition due to factors like unequal distribution of food, wastage and poor diets (FAO 2012; IMECHE 2013; Grace et al. 2017). Pigeonpea being a protein-rich nutri-legume forms a complementary cereal supplement in developing and underdeveloped countries. The latter's seed quality enrichment through micronutrient biofortification would serve as a weapon in combating malnutrition as well as hidden hunger. Therefore, cost-effective biofortification, an agri-based method of addressing micronutrient deficiency through crop improvement approaches, has to be aggrandized.

References

Ambekar SS, Patil SC, Giri AP, Kachole MS (1996) Trypsin and amylase inhibitors in pigeonpea seeds. Int Chickpea Pigeonpea Newslett 3:106–107

Bhatnagar M, Bhatnagar-Mathur P, Reddy DS, Anjaiah V, Sharma KK (2011) Crop biofortification through genetic engineering: present status and future directions. Genomics Crop Improv Relevance Reserv:392–407

Bidinger PD, Nag B (1981) The role of pigeonpeas in village diets. Proceeding of International Workshop on Pigeonpeas, Patancheru, 15–19 December 1980, 1, 357–364
Daniel VA, Narayanaswamy D, Desai BLM, Kurien S, Swaminathan M, Parpia HAB (1970) Supplementary value of varying levels of red gram (Cajanus cajan) to poor diets based on rice and ragi. Indian J Nutr Dietetics 7:358–362
FAO (2012) The state of food insecurity in the world. FAO Economic and Social Development Department, Rome
FAOSTAT (2018) Online agriculture statistics. http://www.faostat.org
Faris DG, Saxena KB, Mazumdar S, Singh U (1987) Vegetable Pigeonpea: a promising crop for India. ICRISAT, Patancheru
Geervani P (1981) Nutritional evaluation of Pigeonpea (Variety Hyderabad 3A) processed by traditional methods. Proceedings of International Workshop on Pigeonpeas. Patancheru, 15–19 December, 1980, 2, 427–434
Godbole SA, Krishna TG, Bhatia CR (1994) Changes in protease inhibitory activity from pigeonpea (Cajanus cajan (L.) Millspaugh) during seed development and germination. J Sci Food Agric 66:497–501
Gopalakrishnan S, Vadlamudi S, Samineni S, Sameer Kumar CV (2016) Plant growth-promotion biofortification of chickpea pigeonpea through inoculation of biocontrol potential bacteria, isolated from organic soils. Springer Plus 5(1882):01–11. ISSN 2193-1801
Grace ZH, Sudipta S, Das, B, Hoang ML, Karbaschi MR, Alexander AT, Williams B, Sagadevan G, Mundree SG (2017) Finger on pulse: Iron pumping into chickpea. Front Plant Sci, pp 1–17. https://doi.org/10.3389/fpls.2017.01755
Hanumanthappa D, Vasudevan, Maruthi SN, Shakuntala JB, Muniswamy NM, Sangeeta IM (2018) Enrichment of Iron and zinc content in pigeonpea genotypes through agronomic biofortification to mitigate malnutrition. Int J Curr Microbiol Appl Sci 7:4334–4342. ISSN: 2319-7706
Howarth EB, Saltzman A (2017) Improving nutrition through biofortification: a review of evidence from harvest plus, 2003 through 2016. Glob Food Sec 12:49–58
Hulse JH (1977) Problems of nutritional quality of pigeonpea and chickpea and prospects of research. In: Hulse JH, Rachie KO, Billingsley LW (eds) Nutritional standards and methods of evaluation for food legume breeders. International Development Research Center, Ottawa, pp 88–100
IMECHE (2013) Global food – waste not, want not. Wales
Janila P, Rupavatharam S, Sameer Kumar CV, Samineni S, Gaur PM, Varshney RK (2016) Technologies for intensification of production uses of grain legumes for nutrition security. Proc Indian Natl Sci Acad 82(5):1541–1553. ISSN 0370-0046
Kamath MV, Belavady B (1980) Unavailable carbohydrates of commonly consumed Indian foods. J Sci Food Agric 31:194–202
Kurien PP (1981) Advances in milling technology of pigeonpea. Proceedings of International Workshop on Pigeonpeas, Patancheru, 15–19 December 1980, 1, 321–328
Manimekalai G, Neelakantan S, Annapan RS (1979) Chemical composition and cooking quality of some improved varieties of red gram dal. Madras Agric J 66:812–816
Mayer JE, Pfeiffer WH, Bayer P (2008) Biofortified crop to alleviate micronutrient malnutrition. Curr Opin Plant Biol 11:166–170
Meenakshi JV, Nancy J, Johnson M, Victor DG, Hugo J, Joy Y, David N, Firdousi G, James GC, Erika, M (2007) How cost-effective is biofortification in combating micronutrient malnutrition? An ex-ante assessment. Harvest Plus Working Paper 2.pp 2–27
Mishra S, Acharya S (2017) Targeted traits for enhancement of seed Iron and zinc concentrations in Pigeonpea. Springer. https://doi.org/10.1007/s40011-017-0857-9
Mutimani VH, Paramjyothi S (1995) Protease inhibitors in some pigeonpea lines. Int Chickpea Pigeonpea Newslett 2:79–81
Narasimha HV, Desikachar HSR (1978) Objective methods for studying cooking ability of tur pulse (Cajanus cajan) and factors affecting varietal differences in cooking. J Food Sci Technol 15:47–50

Pal RK (1939) A review of the literature on the nutritive value of pulses. Indian J Agri Sci 9:133–137
Paltridge NG, Milham PJ, Ortiz-Monasterio JI, Velu G et al (2012a) Energy dispersive X-ray fluorescence spectrometry as a tool for zinc, iron and selenium analysis in whole grain wheat. Plant Soil 361:261–269
Paltridge NG, Palmer LJ, Milham PJ, Guild GE et al (2012b) Energy-dispersive X-ray fluorescence analysis of zinc and iron concentration in rice and pearl millet grain. Plant Soil 361:251–260
Rajalakshmi R, Vanaja K (1967) Chemical and biological evaluation of the effects of fermentation on the nutritive value of foods prepared from rice and legumes. Br J Nutr 21:467–473
Salunkhe DK (1982) Legumes in human nutrition: current status and future research needs. Curr Sci 51:387–394
Saxena KB (2008) Genetic improvement of pigeonpea—a review. Trop Plant Biol 1:159–178
Saxena KB, Sawargaonkar SL (2016) Genetic enhancement of seed proteins in pigeonpea – methodologies, accomplishments, and opportunities. Int J Sci Res 4:3–7. ISSN No. 2277-8179
Saxena KB, Vijay Kumar R, Sultana R (2010) Quality nutrition through pigeonpea: a review. Health 2(11):1335–1344. https://doi.org/10.4236/health.2010.211199
Singh U (1984) The inhibition of digestive enzymes by polyphenols of chickpea (Cicer arietinum L.) and pigeonpea [Cajanus cajan (L). Millspaugh]. Nutr Rep Int 29:745–753
Singh U, Jain KC, Jambunathan R, Faris DG (1984) Nutritional quality of vegetable pigeonpeas [Cajanus cajan (L.) Millspaugh]: dry matter accumulation, carbohydrates and proteins. J Food Sci 49:799–802
United Nations (2015) World Population Prospects, The 2015 Revision. United Nations Department of Economic and Social Affairs
Upadhyaya HD, Sharma S, Reddy KN, Saxena R, Varshney RK, Gowda CLL (2013) Pigeonpea. In: Singh M, Upadhyaya HD, Bisht IS (eds) Genetic and genomic resources of grain legume improvement. Elsevier, London, pp 181–202

Index

D. S. Gupta et al. (eds.), *Breeding for Enhanced Nutrition and Bio-Active Compounds in Food Legumes*, https://doi.org/10.1007/978-3-030-59215-8

The manufacturer's authorised representative in the EU is Springer Nature Customer Service Centre GmbH, Europaplatz 3, 69115 Heidelberg, Germany. If you have any concerns regarding our products, please contact ProductSafety@springernature.com

Printed and bound by CPI Group (UK) Ltd, Croydon, CR0 4YY
28/09/2026
02224090-0013